PLANT-BASED FUNCTIONAL FOODS AND PHYTOCHEMICALS

From Traditional Knowledge to Present Innovation

Innovations in Plant Science for Better Health: From Soil to Fork

PLANT-BASED FUNCTIONAL FOODS AND PHYTOCHEMICALS

From Traditional Knowledge to Present Innovation

Edited by

Megh R. Goyal, PhD, PE

Arijit Nath, PhD

Hafiz Ansar Rasul Suleria, PhD

First edition published 2021

Apple Academic Press Inc.
1265 Goldenrod Circle, NE,
Palm Bay, FL 32905 USA

4164 Lakeshore Road, Burlington,
ON, L7L 1A4 Canada

CRC Press
6000 Broken Sound Parkway NW,
Suite 300, Boca Raton, FL 33487-2742 USA

2 Park Square, Milton Park,
Abingdon, Oxon, OX14 4RN UK

First issued in paperback 2021

Library and Archives Canada Cataloguing in Publication

Title: Plant-based functional foods and phytochemicals : from traditional knowledge to present innovation / edited by Megh R. Goyal, PhD, Arijit Nath, PhD, Hafiz Ansar Rasul Suleria, PhD.

Names: Goyal, Megh Raj, editor. | Nath, Arijit, editor. | Suleria, Hafiz, editor.

Series: Innovations in plant science for better health.

Description: Series statement: Innovations in plant science for better health : from soil to fork | Includes bibliographical references and index.

Identifiers: Canadiana (print) 20200335154 | Canadiana (ebook) 20200335324 | ISBN 9781771889292 (hardcover) | ISBN 9781003055419 (ebook)

Subjects: LCSH: Phytochemicals. | LCSH: Functional foods. | LCSH: Plants, Edible. | LCSH: Traditional medicine.

Classification: LCC QK861 .P63 2021 | DDC 572/.2—dc23

Library of Congress Cataloging-in-Publication Data

Names: Goyal, Megh R., editor. | Nath, Arijit, editor. | Suleria, Hafiz, editor.

Title: Plant-based functional foods and phytochemicals : from traditional knowledge to present innovation / edited by Megh R. Goyal, PhD, Arijit Nath, PhD, Hafiz Ansar Rasul Suleria, PhD.

Description: First edition. | Palm Bay, FL, USA ; Burlington, ON, Canada : Apple Academic Press ; Boca Raton, FL, USA ; Abingdon, Oxon, UK : CRC Press, 2021. | Series: Innovations in plant science for better health: from soil to fork | Includes bibliographical references and index. | Summary: "Plant-Based Functional Foods and Phytochemicals: From Traditional Knowledge to Present Innovation covers the importance of the therapeutic health benefits of phytochemicals derived from plants. It discusses the isolation of potential bioactive molecules from plant sources along with their value to human health. It focuses on physical characteristics, uniqueness, uses, distribution, traditional and nutritional importance, bioactivities, and future trends of different plant-based foods and food products. Functional foods, beyond providing basic nutrition, may offer a potentially positive effect on health and cures for various disease conditions, such as metabolic disorders (including diabetes), cancer, and chronic inflammatory reactions. The volume looks at these natural products and their bioactive compounds that are increasingly utilized in preventive and therapeutic medications and in the production of pharmaceutical supplements and as food additives to increase functionality. It also describes the concept of extraction of bioactive molecules from plant sources, both conventional and modern extraction techniques, available sources, biochemistry, structural composition, and potential biological activities. Advanced extraction techniques, such as enzyme-assisted, microwave-assisted, ultrasound-assisted, pressurized liquid extraction, and supercritical extraction techniques, are described in this book. With contributions from many experienced researchers, this volume sheds new light on the potential of plant-based natural products for human health. Key features: Discusses the phytochemical and functional health benefits from plant-based foods Presents methods of isolation and extracion of potential bioactive molecules from plant sources Looks at the preservation of indigenous knowledge on functional foods Identifies novel plants and their bioactive compounds for treatment of different diseases Elucidates on the mechanisms of actions of the plants"-- Provided by publisher.

Identifiers: LCCN 2020042153 (print) | LCCN 2020042154 (ebook) | ISBN 9781771889292 (hardcover) | ISBN 9781003055419 (ebook)

Subjects: LCSH: Phytochemicals. | Functional foods. | Plants, Edible. | Traditional medicine.

Classification: LCC QK861 .P513 2021 (print) | LCC QK861 (ebook) | DDC 572/.2--dc23

LC record available at https://lccn.loc.gov/2020042153

LC ebook record available at https://lccn.loc.gov/2020042154

ISBN: 978-1-77188-929-2 (hbk)
ISBN: 978-1-77463-778-4 (pbk)
ISBN: 978-1-00305-541-9 (ebk)

OTHER BOOKS ON PLANT SCIENCE FOR BETTER HEALTH BY APPLE ACADEMIC PRESS, INC.

Book Series: *Innovations in Plant Science for Better Health: From Soil to Fork*
Editor-in-Chief: Hafiz Ansar Rasul Suleria, PhD

Assessment of Medicinal Plants for Human Health: Phytochemistry, Disease Management, and Novel Applications
Editors: Megh R. Goyal, PhD, and Durgesh Nandini Chauhan, MPharm

Bioactive Compounds of Medicinal Plants: Properties and Potential for Human Health
Editors: Megh R. Goyal, PhD, and Ademola O. Ayeleso

Bioactive Compounds from Plant Origin: Extraction, Applications, and Potential Health Claims
Editors: Hafiz Ansar Rasul Suleria, PhD, and Colin Barrow, PhD

Cereals and Cereal-Based Foods: Functional Benefits and Technological Advances for Nutrition and Healthcare
Editors: Megh Goyal, PhD, Kamaljit Kaur, PhD, and Jaspreet Kaur, PhD

Health Benefits of Secondary Phytocompounds from Plant and Marine Sources
Editors: Hafiz Ansar Rasul Suleria, PhD, and Megh Goyal, PhD

Human Health Benefits of Plant Bioactive Compounds: Potentials and Prospects
Editors: Megh R. Goyal, PhD, and Hafiz Ansar Rasul Suleria, PhD

Phytochemicals and Medicinal Plants in Food Design: Strategies and Technologies for Improved Healthcare
Editors: Megh R. Goyal, PhD, Preeti Birwal, PhD, and Santosh K. Mishra, Ph

Phytochemicals from Medicinal Plants: Scope, Applications, and Potential Health Claims
Editors: Hafiz Ansar Rasul Suleria, PhD, Megh R. Goyal, PhD, and Masood Sadiq Butt, PhD

Plant- and Marine-Based Phytochemicals for Human Health: Attributes, Potential, and Use
Editors: Megh R. Goyal, PhD, and Durgesh Nandini Chauhan, MPharm

Plant-Based Functional Foods and Phytochemicals: From Traditional Knowledge to Present Innovation
Editors: Megh R. Goyal, PhD, Arijit Nath, PhD, and Hafiz Ansar Rasul Suleria, PhD

Plant Secondary Metabolites for Human Health: Extraction of Bioactive Compounds
Editors: Megh R. Goyal, PhD, P. P. Joy, PhD, and Hafiz Ansar Rasul Suleria, PhD

The Role of Phytoconstitutents in Health Care: Biocompounds in Medicinal Plants
Editors: Megh R. Goyal, PhD, Hafiz Ansar Rasul Suleria, PhD, and Ramasamy Harikrishnan, PhD

The Therapeutic Properties of Medicinal Plants: Health-Rejuvenating Bioactive Compounds of Native Flora
Editors: Megh R. Goyal, PhD, PE, Hafiz Ansar Rasul Suleria, PhD, Ademola Olabode Ayeleso, PhD, T. Jesse Joel, and Sujogya Kumar Panda

ABOUT THE SENIOR EDITOR-IN-CHIEF

Megh R. Goyal, PhD, PE

Retired Professor in Agricultural and Biomedical Engineering, University of Puerto Rico, Mayaguez Campus; Senior Acquisitions Editor, Biomedical Engineering and Agricultural Science, Apple Academic Press, Inc.

Megh R. Goyal, PhD, PE, is a Retired Professor in Agricultural and Biomedical Engineering from the General Engineering Department in the College of Engineering at the University of Puerto Rico–Mayaguez Campus; and Senior Acquisitions Editor and Senior Technical Editor-in-Chief in Agriculture and Biomedical Engineering for Apple Academic Press, Inc. He has worked as a Soil Conservation Inspector and as a Research Assistant at Haryana Agricultural University and Ohio State University.

During his professional career of 49 years, Dr. Goyal has received many prestigious awards and honors. He was the first agricultural engineer to receive the professional license in Agricultural Engineering in 1986 from the College of Engineers and Surveyors of Puerto Rico. In 2005, he was proclaimed as "Father of Irrigation Engineering in Puerto Rico for the Twentieth Century" by the American Society of Agricultural and Biological Engineers (ASABE), Puerto Rico Section, for his pioneering work on micro irrigation, evapotranspiration, agroclimatology, and soil and water engineering. The Water Technology Centre of Tamil Nadu Agricultural University in Coimbatore, India, recognized Dr. Goyal as one of the experts "who rendered meritorious service for the development of micro irrigation sector in India" by bestowing the Award of Outstanding Contribution in Micro Irrigation. This award was presented to Dr. Goyal during the inaugural session of the National Congress on "New Challenges and Advances in Sustainable Micro Irrigation" held at Tamil Nadu Agricultural University. Dr. Goyal received the Netafim Award for Advancements in Microirrigation: 2018 from the American Society of Agricultural Engineers at the ASABE International Meeting in August 2018.

A prolific author and editor, he has written more than 200 journal articles and textbooks and has edited over 80 books. He is the editor of three book series published by Apple Academic Press: Innovations in Agricultural & Biological Engineering, Innovations and Challenges in Micro Irrigation, and Research Advances in Sustainable Micro Irrigation. He is also instrumental in the development of the new book series Innovations in Plant Science for Better Health: From Soil to Fork.

Dr. Goyal received his BSc degree in engineering from Punjab Agricultural University, Ludhiana, India; his MSc and PhD degrees from Ohio State University, Columbus; and his Master of Divinity degree from Puerto Rico Evangelical Seminary, Hato Rey, Puerto Rico, USA.

ABOUT CO-EDITORS

Arijit Nath, PhD

Visiting Research and Teaching Assistant,
Department of Food Engineering,
Faculty of Food Science, Szent István University,
Ménesist 44, HU–1118 Budapest, Hungary,
E-mails: arijit0410@gmail.com;
nath.arijit@etk.szie.hu

Arijit Nath, PhD, is a Visiting Research and Teaching Assistant in the Department of Food Engineering, Faculty of Food Science, Szent István University, Budapest, Hungary. Dr. Nath has held a variety of positions, including postdoctoral research fellow in the Department of Food Engineering at Corvinus University of Budapest, Hungary; a postdoctoral research fellow in the Department of Biochemical and Chemical Engineering at the Technical University of Dortmund, Germany; a research fellow in the Department of Refrigeration and Livestock Products Technology at Szent István University, Hungary; and chemical engineer in Soós Ernő Water Technology Research and Development Centre at University of Pannonia, Hungary. He has participated in several professional training and workshops in the field of biotechnology. He has published 20 research and review articles in international journals and10 book chapters. Dr. Nath has also presented the research outcomes in more than 50 international conferences and has received several best presenter awards at international conferences. He has also delivered lectures in several scientific communities. Due to his strong proficiency in the field of biochemical and food engineering, he is serving as a reviewer in various scientific SCI journals. Dr. Nath is a member of the Asia-Pacific Chemical, Biological & Environmental Engineering Society; Indian Institute of Engineers; Indian Institute of Chemical Engineers; Biotech Research Society of India; European Membrane Society; Indian Membrane Society and European Federation of Biotechnology. Dr. Nath is employed to teach several theory subjects and is co-supervisor for undergraduate, postgraduate and PhD students in the

Department of Food Engineering, Szent Istvan University, Hungary. His major research focus is on fermentation technology, enzymatic reaction, bioreactor design, probiotics and prebiotics, allergy-free food products, food waste valorization, membrane separation process, membrane integrated bioreactor, environmental benign process, and mathematical modeling of the bioprocess. For more details, contact him at: arijit0410@gmail.com.

Hafiz Ansar Rasul Suleria, PhD

Alfred Deakin Research Fellow, Deakin University, Melbourne, Australia; Honorary Fellow, Diamantina Institute Faculty of Medicine, The University of Queensland, Australia

Hafiz Anasr Rasul Suleria, PhD, is currently working as the Alfred Deakin Research Fellow at Deakin University, Melbourne, Australia. He is also an Honorary Fellow at the Diamantina Institute, Faculty of Medicine, The University of Queensland, Australia.

Recently he worked as a postdoc research fellow in the Department of Food, Nutrition, Dietetic and Health at Kansas State University, USA.

Previously, he has been awarded an International Postgraduate Research Scholarship (IPRS) and an Australian Postgraduate Award (APA) for his PhD research at the University of Queens School of Medicine, the Translational Research Institute (TRI), in collaboration with the Commonwealth and Scientific and Industrial Research Organization (CSIRO, Australia).

Before joining the University of Queens, he worked as a lecturer in the Department of Food Sciences, Government College University Faisalabad, Pakistan. He also worked as a research associate in the PAK-US Joint Project funded by the Higher Education Commission, Pakistan, and the Department of State, USA, with the collaboration of the University of Massachusetts, USA, and National Institute of Food Science and Technology, University of Agriculture Faisalabad, Pakistan.

He has a significant research focus on food nutrition, particularly in the screening of bioactive molecules—isolation, purification, and characterization using various cutting-edge techniques from different plant, marine, and animal sources; and *in vitro, in vivo* bioactivities; cell culture; and animal modeling. He has also done a reasonable amount of work on functional foods and nutraceuticals, food and function, and alternative medicine.

Dr. Suleria has published more than 50 peer-reviewed scientific papers in different reputed/impacted journals. He is also in collaboration with more than ten universities where he is working as a co-supervisor/special member for PhD and postgraduate students and is also involved in joint publications, projects, and grants. He is Editor-in-Chief for the book series Innovations in Plant Science for Better Health: From Soil to Fork, published by AAP.

Readers may contact him at: hafiz.suleria@uqconnect.edu.au.

CONTENTS

CONTRIBUTORS

Szilvia Bánvölgyi
Assistant Professor, Department of Food Engineering, Faculty of Food Science, Szent István University, Ménesist 44, HU-1118 Budapest, Hungary, E-mail: banvolgyi.szilvia@etk.szie.hu

Tesfaye F. Bedane
Postdoctoral Researcher, School of Agriculture and Food Science, University College Dublin, Dublin, *Ireland, E-mail:* tesfaye.bedane@ucd.ie

Csilla Benedek
Associate Professor, Department of Dietetics and Nutrition, Faculty of Health Sciences, Semmelweis University, Vas Street 17, Budapest, Hungary, E-mail: benedek.csilla@se-etk.hu

Renáta Gerencsérné Berta
Research Fellow, SoósErnő Water Technology Research Center, University of Pannonia, Zrínyi M. u. 18, Nagykanizsa, Hungary, E-mail: berta.r@sooswrc.hu

Prashant Kumar Biswas
Associate Professor, Department of Biochemical and Food Technology, Jadavpur University, Kolkata–7000032, India, E-mail: prasantakumar.biswas@jadavpuruniversity.in

Zsanett Bodor
PhD Research Scholar, Department of Physics and Control, Faculty of Food Science, Szent István University, Somlói Street 14–16, H-1118 Budapest, Hungary, E-mail: arscube@gmail.com

Vedant Vikrom Borah
Assistant Professor, Department of Life Science, Assam Don Bosco University, Tapesia Gardens, Kamarkuchi, Sonapur–782402, Assam, India, E-mails: monshbrh@gmail.com; vedant.borah@dbuniversity.ac.in

Dipankar Chakraborti
Associate Professor, Department of Genetics, University of Calcutta, 35 Ballygunge Circular Road, Kolkata–700019, West Bengal, India, Mob.: +91-9051322083, E-mail: dipankar1212@gmail.com

Rituparna Kundu Chaudhuri
Assistant Professor, Department of Botany, Krishnagar Government College, Krishnagar–741101, West Bengal, India, E-mail: rkc2821@gmail.com

Mahua Gupta Choudhury
Assistant Professor, Department of Life Science, Assam Don Bosco University, Tapesia Gardens, Kamarkuchi, Sonapur–782402, Assam, India, E-mail: mahua.guptachoudhury@gmail.com

Arpita Das
Assistant Professor, Department of Biotechnology, School of Life Science and Biotechnology, Adamas University, Barasat-Barrackpore Road, Kolkata–700126, India, E-mail: arpita_84das@yahoo.co.in

Nilanjan Das
MSc Student, Department of Biotechnology, St. Xavier's College (Autonomous), 30 Mother Teresa Sarani, Kolkata–700016, India, E-mail: nilanjandas.john@gmail.com

Rahel Suchintita Das
Assistant Professor, Department of Food Technology, Haldia Institute of Technology, HIT Campus, Purba Medinipur, Haldia–721657, West Bengal, India, E-mail: rahel.rsd@gmail.com

Francesco Donsì
Associate Professor, Department of Industrial Engineering, University of Salerno, via Giovanni Paolo II, 132 84084 Fisciano (SA), Italy, E-mail: fdonsi@unisa.it

Ildikó Galambos
Associate Professor, SoósErnő Water Technology Research Center, University of Pannonia, Zrínyi M. u. 18, Nagykanizsa, Hungary, E-mail: ildiko.galambos@gmail.com

Titas Ghosh
Project Fellow, Department of Biotechnology, School of Life Science and Biotechnology, Adamas University, Barasat-Barrackpore Road, Kolkata–700126, India, E-mail: titasghosh1993@gmail.com

Megh R. Goyal
Retired Faculty in Agricultural and Biomedical Engineering from College of Engineering at University of Puerto Rico-Mayaguez Campus; and Senior Technical Editor-in-Chief in Agricultural and Biomedical Engineering for Apple Academic Press Inc., PO Box 86, Rincon-PR–006770086, USA, E-mail: goyalmegh@gmail.com

Manisha Guha
MSc Student, Department of Biotechnology, St. Xavier's College (Autonomous), 30 Mother Teresa Sarani, Kolkata–700016, India, E-mail: guhamanisha@yahoo.in

Klára Pásztorné Huszár
Associate Professor, Department of Refrigeration and Livestock Product Technology, Faculty of Food Science, Szent István University, Ménesist 43–45, HU-1118 Budapest, Hungary, E-mail: Pasztorne.Huszar.Klara@etk.szie.hu

Andras Koris
Associate Professor, Department of Food Engineering, Faculty of Food Science, Szent István University, Ménesist 44, HU-1118 Budapest, Hungary, E-mail: Koris.Andras@etk.szie.hu

Zoltan Kovacs
Associate Professor, Department of Physics and Control, Faculty of Food Science, Szent István University, Somlói Street 14–16, H-1118 Budapest, Hungary, E-mail: kovacs.zoltan3@etk.szie.hu

Edit Márki
Associate Professor, Department of Food Engineering, Faculty of Food Science, Szent István University, Ménesist 44, HU-1118 Budapest, Hungary, E-mail: Marki.Edith@etk.szie.hu

Tsega Y. Melesse
Lecturer, Faculty of Chemical and Food Engineering, Bahir Dar Institute of Technology, Bahir Dar, Ethiopia, E-mail: shuambo@gmail.com

Arnia Sari Mukaromah
Lecturer, Department of Biology, Faculty of Science and Technology, UIN Walisongo, Semarang–50185, Indonesia, E-mail: arnia_sm@walisongo.ac.id

Arijit Nath
Visiting Research and Teaching Assistant, Department of Food Engineering,
Faculty of Food Science, Szent István University, Ménesist 44, HU-1118 Budapest, Hungary,
E-mails: arijit0410@gmail.com; nath.arijit@etk.szie.hu

Probin Phanjom
Assistant Professor, Department of Life Science, Assam Don Bosco University,
Tapesia Gardens, Kamarkuchi, Sonapur–782402, Assam, India,
E-mails: phanjom@gmail.com; probin.phanjom@dbuniversity.ac.in

Abinit Saha
Assistant Professor, Department of Biotechnology, School of Life Science and Biotechnology,
Adamas University, Barasat-Barrackpore Road, Kolkata–700126, India,
E-mail: abinitsaha@gmail.com

Xavier Savarimuthu
Vice-Principal and Head, Department of Environmental Science, St. Xavier's College (Autonomous),
30 Mother Teresa Sarani, Kolkata–700016, India; Visiting Faculty, Public Health Department,
Santa Clara University, 500 El Calmino Real, Santa Clara–95053, USA,
E-mail: sxavi2005@gmail.com

Mukesh Singh
Associate Professor, Department of Biotechnology, Department of Food Technology,
Haldia Institute of Technology, HIT Campus, Purba Medinipur, Haldia–721657, West Bengal, India,
E-mails: singhmukesh@hithaldia.in; msingh006@gmail.com

Hafiz Ansar Rasul Suleria
McKenzie Fellow, School of Agriculture and Food, Faculty of Veterinary and Agricultural Sciences,
The University of Melbourne, Parkville–3010, Victoria, Australia, Mob.: +61-470-439-670,
E-mail: h.suleria@hotmail.com

Fitria Susilowati
Lecturer, Department of Nutrition, Faculty of Psychology and Health Sciences, UIN Walisongo,
Semarang–50185, Indonesia, E-mail: fitria_susilowati@walisongo.ac.id

Ranjay Kumar Thakur
PhD Research Scholar, Assistant Professor, Department of Food Technology,
Haldia Institute of Technology, HIT Campus, Purba Medinipur, Haldia–721657, West Bengal, India,
E-mail: ktranjaybt2007@gmail.com

Gyula Vatai
Professor and Former-Dean, Department of Food Engineering, Faculty of Food Science,
Szent István University, Ménesist 44, HU-1118 Budapest, Hungary,
E-mail: Vatai.Gyula@etk.szie.hu

John-Lewis Zinia Zaukuu
PhD Research Scholar, Department of Physics and Control, Faculty of Food Science,
Szent István University, Somlói Street 14–16, H-1118 Budapest, Hungary,
E-mail: izaukuu@yahoo.com

ABBREVIATIONS

AAPH	2,2'-azobis (2-methylpropionamidine)
ABC transporters	ATP-binding cassette transporters
ABTS	2,2'-azino-bis (3-ethylbenzothiazoline-6-sulphonic acid)
ACE	angiotensin converting enzyme
AFE	accelerated fluid extraction
AGEs	advanced glycation end-products
ALAD	δ-aminolevulinic acid dehydratase
ALP	alkaline phosphatase
ALT	alanine amino transferase
AMP	adenosine mono phosphate
AMPK	adenosine monophosphate-activated protein kinase
AOAC	Association of Official Agricultural Chemists
AOC	antioxidant capacity
APCI	atmospheric pressure chemical ionization
apoA-I	anti-apolipoprotein A-I
Ar	argon
As	arsenic
As_2O_3	arsenic trioxide
ASE	accelerated solvent extraction
AST	aspartate aminotransferase
ATP	adenosine triphosphate
BAL	British anti-lewisite
BBB	blood-brain barrier
BCRP	breast cancer resistance protein
BFCs	bagasse fiber concentrates
BHT	butylated hydroxytoluene
BMI	body mass index
BP	blood pressure
BPE	background parenchymal enhancement
BRH	Biligiri Rangana hills
CBG	cytosolic β-glucosidase
CD4	cluster of differentiation 4
CE	capillary electrophoresis
CE	collision energy

CEC	capillary electrochromatography
CETP	cholesteryl ester transfer protein
CID	collision-induced dissociation
CL	chemi-luminescence
COX	cyclooxygenase
CRP	c-reactive protein
CSPs	cumin-seed peptides
Cu	copper
CUPRAC	cupric ion reducing antioxidant capacity
CVDs	cardiovascular diseases
CZE	capillary-zone electrophoresis
DAD	diode-array detectors
DLLME	dispersive liquid-liquid microextraction
DM	diabetes mellitus
DMPS	2,3-dimercaptopropane-1-sulfonate
DMSA	2,3-dimercaptosuccinic acid
DNA	deoxiribose nucleic acid
DOPA	L-3,4-dihydroxyphenylalanine
DPPH	1,1-diphenyl-2-picrylhydrazyl
EA	enzyme-assisted
EAAE	enzyme-assisted aqueous extraction
EACP	enzyme-assisted cold pressing
EAE	enzyme assisted extraction
EIC	extracted ion chromatogram
eNOS	endothelial NOS
ER+	estrogen-receptor-positive
ES	electron spray
ESE	enhanced solvent extraction
ESI	electro spray ionization
ET-1	endothelin-1
EU	European Union
FAB	fast atom bombardment
FAO	Food and Agriculture Organization
FBG	fasting blood glucose
FBS	fasting blood sugar
FCs	fiber concentrates
FDA	Food and Drug Administration
Fe	iron
Fe^{2+}	ferrous
Fe^{3+}	ferric ion

FFA	free fatty acid
FOSHU	foods for specified health used
FOXO	fork head box O
FPG	fasting plasma glucose
FRAP	ferric reducing antioxidant power
FTIR	Fourier transform infrared
G-CSF	granulocyte colony-stimulating factor
GGT	gangliosides, γ-glutamyl transpeptidase
GI	glycemic index
GLUT	glucose transporters
GLUT-4 mRNA	glucose transporter-4 messenger ribonucleic acid
GLUT4	glucose transporter type-4
GPx	glucose peroxidase
GSH	reduced glutathione
GSSG	oxidized glutathione
HAART	highly active anti-retroviral therapy
HAT	histone acetyl transferases
HAT	hydrogen atom transfer
HbA1c	hemoglobin A1c
HDL	high density lipoprotein
HHP	high hydrostatic pressure
HIF-1	hypoxia-inducible factor 1
HIV	human immunodeficiency virus
HO1	heme oxygenase 1
HOCl	hypochlorous acid
HPA	high pressure-assisted
HPH	high-pressure homogenization
HPLC	high performance liquid chromatography
HPSE	high-pressure solvent extraction
HPTLC	high-performance thin-layer chromatography
HSV	herpes simplex virus
HVEDA	high voltage electrical discharge assisted
IARC	International Agency for Research on Cancer
IDDM	insulin-dependent diabetes mellitus
IDFA	International Diabetes Federation Atlas
IFG	impaired fasting glucose
IGT	impaired glucose tolerance
IL-1, IL-6	interleukin-1 and interleukin-6
IL-1β	interleukin 1β
IMT	intima media thickness

Inos	inducible NOS
iNOS-NO	inducible nitric oxide synthase-nitric oxide
INSR	insulin receptor
IRAC	International Agency for Research on Cancer
ITREOH	International Training and Research Program in Environmental and Occupational Health
K	potassium
LAB	lactic acid bacteria
LC	liquid chromatography
LC-MS	liquid chromatography-mass spectrometry
LDA	linear discriminant analysis
LDL	low-density lipoprotein
LLE	liquid-liquid extraction
LPH	lactase phlorizin hydrolase
MA	microwave-assisted
MAE	microwave assisted extraction
MALDI-TOF	matrix-assisted-laser-desorption-ionization-time-of-flight
MAPK	mitogen-activated protein kinase
MAV	microwave-assisted extraction
MCF-7	Michigan cancer foundation-7
MCP	monocyte chemoattractant protein
MDA	malondialdehyde
MDR	multi-drug resistance
MEFA	moderate electric field-assisted
MEKC	micellar electro-kinetic chromatography
MEP	methylerythritol phosphate
MF	microfiltration
MiADMSA	monoisoamyl meso 2,3-dimercaptosuccinic acid
MIC	minimum inhibitory concentration
Mn	manganese
m-RNA	messenger ribonucleic acid
MRP	multidrug resistance protein
MS	mass spectrometry
MTCC	microbial type culture collection
MTHFR	methylenetetrahydrofolate reductase
MTT	3- (4,5-Dimethylthiazol-2-yl)-2,5-diphenyl tetrazolium bromide
MVA	microwave-assisted
MWCNTs	multi-walled carbon nanotubes
Na	sodium

$NaAsO_2$	sodium arsenite
NADPH	nicotinamide adenine dinucleotide phosphate
NCIM	national collection of industrial microorganisms
NF	nanofiltration
NFκB	nuclear factor kappa-B
NIDDM	non-insulin-dependent diabetes mellitus
NMR	nuclear magnetic resonance
nNOS	neuronal NOS
NO	nitric oxide
NOS	nitric oxide synthase
NPCA	negative pressure cavitation-assisted
NQO1	nadph quinone reductase
Nrf2	nuclear factor erythroid 2-related factor 2
OGTT	oral glucose tolerance test
OIV	International Organization of Vine and Wine
OPE	onion peel extract
ORAC	oxygen radical absorbance capacity
ox-LDL	oxidized low-density lipoprotein
PAH	phenylalanine hydroxylase
PBEF	pre-B cell colony-enhancing factor
PCA	principal component analysis
PCL	photochemi-luminescence
PCR	polymerized chain reaction
PEF	pulsed electric field
PEFA	pulsed electric field-assisted
PEFE	pulsed electric field extraction
PGI2	prostacyclin
PI3K	phosphatidylinositol-3'-kinase
PKC	protein kinase C
PLE	pressurized liquid extraction
PLS-DA	partial least squares discriminant analysis
PON-1	paraoxonase-*1*
PPAR-γ	peroxisome proliferator-activated receptor gamma
PSA	prostate specific antigen
PUFAs	polyunsaturated fatty acids
PVPP	polyvinylpolypyrrolidone
QOL	quality of life
Q-TOF-MS	quadrupole time-of-flight mass spectrometer
RARs	retinoic acid receptors
RBC	red blood cells

RBP	retinol binding protein
RNS	reactive nitrogen species
RO	reverse osmosis
ROS	reactive oxygen species
SAM	S-adenosyl-L-methionine
SC-CO_2	supercritical carbon dioxide
SDH	sorbitol dehydrogenase
SDS-PAGE	sodium dodecyl sulfate-polyacrylamide gel electrophoresis
SET	single electron transfer
SFE	supercritical fluid extraction
SGLT-2	sodium-glucose co-transporter-2
SNPs	single nucleotide polymorphisms
SOD	superoxide dismutase
SPE	solid phase extraction
SREBP-1c	sterol regulatory element binding protein
SSBs	single-strand breaks
SSF	solid state fermentation
STZ	streptozotocin
T2DM	type 2 diabetes mellitus
TAC	total antioxidant capacity
TAS	total antioxidant status
TBARS	thiobarbituric acid reactive substances
TCA	tricarboxylic acid
TCM	traditional Chinese medical
TDS	total dissolved solids
TEAC	trolox equivalent antioxidant capacity
TFC	total flavonoid content
TGX	tamarind xyloglucan
TNF-α	tumor necrotic factor-α
TOF	time-of-flight
TOS	total oxidant status
TOSC	total oxyradical scavenging capacity
TP	total polyphenol
TPC	total polyphenol content
TPTZ	2,4,6-tris (2-tripyridyl)-s-triazine
TRAP	total-radical trapping antioxidant parameter
TRP	tyrosinase-related protein
TSP	thermo-spray

TYR	tyrosinasc
UA	ultrasound-assisted
UAE	ultrasound assisted extraction
UF	ultra filtration
UHPLC	ultra-high-performance liquid chromatography
UNICEF	United Nations International Children's Emergency Fund
UROtsa	urothelial cell line
USAID	United States Agency for International Development
UTIs	urinary tract infections
UV	ultra violet
VEGF	vascular endothelial growth factor
WBC	white blood cells
WHO	World Health Organization
XBP	x-box binding protein
Zn	zinc

PREFACE 1

We introduce this new book volume under book series *Innovations in Plant Science for Better Health: From Soil to Fork*. This book mainly covers the current scenario of research and case studies; the importance of phytochemicals from plant-based on foods in therapeutics, under four main parts: Part I: Plant-Based Functional Foods; Part II: Role of Phytochemicals in Traditional Ethnomedicines; Part III: Biological Activities of Plant-Based Phytochemicals; and Part IV: Plant-Based Phytochemicals: Extraction, Isolation, and Healthcare.

This book mainly covers the isolation of potentially bioactive molecules from plant sources for their importance and health perspectives. The incorporation of functional foods, nutraceuticals, and bioactives in the daily diet is a beneficial endeavor to prevent the progression of chronic disorders. This book focuses on physical characteristics, uniqueness, uses, distribution, traditional importance, nutritional importance, bioactivities, and future trends of different plant-based foods and food products. Functional foods, beyond providing basic nutrition, may offer a potentially positive effect on health and cure various disease conditions such as metabolic disorders, cancer, and chronic inflammatory reactions. Natural products and their bioactive compounds are increasingly utilized in preventive and therapeutic medication. Bioactive compounds have been utilized for the production of pharmaceutical supplements and more recently as food additives to increase the functionality of foods. The book also describes the extraction of bioactive molecules from plant sources, both conventional and modern extraction techniques, available sources, biochemistry, structural composition, and potential biological activities. Advanced extraction techniques such as enzyme-assisted, microwave-assisted, ultrasound-assisted, pressurized liquid extraction, and supercritical extraction techniques are described in this book.

This book volume sheds light on the potential of both plant-based natural products for human health for different technological aspects, and it contributes to the ocean of knowledge on food science and nutrition. We hope that this compendium will be useful for students and researchers as well as for persons working in the food, nutraceuticals, and herbal industries.

The contributions of the cooperating authors to this book volume have been most valuable in the compilation. Their names are mentioned in each chapter and in the list of contributors. We appreciate you all for having patience with our editorial skills. This book would not have been written without the valuable cooperation of these investigators, many of whom are renowned scientists who have worked in the field of food science, biochemistry, and nutrition throughout their professional careers. We proudly welcome coeditor Arijit Nath to the editorial community on plant science for better health, and he brings his international experience.

The goal of this book volume is to guide the world science community on how bioactive compounds can alleviate us from various conditions and diseases.

We will like to thanks to editorial and production staff, and Ashish Kumar, Publisher, and President at Apple Academic Press, Inc., for making every effort to publish this book when all are concerned with health issues.

We express our admiration to our families and colleagues for their understanding and collaboration during the preparation of this book volume.

—Megh R. Goyal, PhD
—Hafiz Ansar Rasul Suleria, PhD
Editors

PREFACE 2

Plants and their parts, among all habitats around the world, are able to satisfy the demands of all types of gastronomy. In addition to the use of plant-based foods in low-income communities (unaware about healthy diet and health benefits), these foods have also received a significant position in the low-calorific diet chart of vegan people in high-income communities. Intentionally, the community prefers to turn towards plant-based functional foods, because they are nutritious, low-calorific, and have a potential for detoxification.

When there is an argument related with a genetically modified crop, traditional plant-based foods do not lose their position in main and side dishes. Such foods are considered as heritage items of culture and society. The contribution of plants or their parts (especially stem, pseudo-stem, seed, bark, leaves, fruits) as a source of natural remedies is noteworthy. For example, spices offer unique medicinal values for reducing risks and severities of several diseases. In addition to the preparation of functional foods, plants are also used to prepare several refreshment drinks and alcoholic beverages. For example, grapes, cereals, citrus, lavender, mint, tea, etc., have been used to prepare beverages.

The presence of phytochemicals in plants with exclusive chemical structure and phytochemical activities is responsible for many health benefits. Phytochemicals in the crude matrix have not only been implemented in primary health care, but their isolation and formulation for specific targets have boosted advanced pharmaceutical formulations. Research studies on pharmacology and bioinformatics have corroborated that plant-based bioactive compounds and herbal drugs are 'safe' and do not offer significant adverse side-effects. All beneficial outcomes enhance the utilization and consumption of plant-based foods, beverages, and medicines from their primitive status.

Production, export, and import of plant-based foods, beverages, and medicines provide developing perspectives in the economic sector. In a similar platform, the emergence of biochemical engineering is a boon for the development of new processes and equipment that offer revolutionary upgradation for the production of several plant-based foods, therapeutics, and biomolecules. As time progresses, shrewd revolt with new ideas and confidence offers a new arena in the process intensification and provides a rebellion for the manufacturing process of plant-based foods and extraction of phytochemicals.

As plant-based foods and phytochemicals have drawn attention from the field to "ready to eat food" items, there is therefore an urgent need to provide in-depth knowledge related to processing of different phytochemicals, plant based foods, and beverages along with their health benefits to a wide range of research communities, industrial sectors, and medical practitioners. Our expectation of this book volume is to interest researchers with pharmacology, food, chemical, and clinical backgrounds and promote interdisciplinary research. Furthermore, this book volume provides an opportunity to meet the present scenario of business related to herbal medicines and phytochemicals.

I express thanks to Prof. Megh R. Goyal, Senior Editor-in-Chief, for providing me the opportunity to join his team. In my budding stage of professional life, his constant encouragement and availability to complete this task has helped me to meet various aspects of academic, professional, and management skills, and help to bring about an exceptional book in plant science and human health. This book preparation offers a new shape in my academic and research career.

My special thanks to my parents and beloved sister for their inspiration and motivation. I am thankful to Almighty Supreme God, who always put me in His palm. I cannot forget to acknowledge Dr. Lawrence Abello, S. J. (renowned professor, inventor, and a devout companion of St. Mother Teresa) for his blessings and encouragement. If this book fulfills the expectations of cooperating authors and readers, then I can consider it a great tribute to Almighty Supreme God, Prof. Megh R. Goyal, my parents, Dr. Lawrence Abello, S. J.

—**Arijit Nath, PhD**
Co-Editor

PART I

Plant-Based Functional Foods

CHAPTER 1

SOYBEAN-BASED FUNCTIONAL FOODS THROUGH MICROBIAL FERMENTATION: PROCESSING AND BIOLOGICAL ACTIVITIES

ARIJIT NATH, TITAS GHOSH, ABINIT SAHA,
KLÁRA PÁSZTORNÉ HUSZÁR, SZILVIA BÁNVÖLGYI,
RENÁTA GERENCSÉRNÉ BERTA, ILDIKÓ GALAMBOS, EDIT MÁRKI,
GYULA VATAI, ANDRAS KORIS, and ARPITA DAS

ABSTRACT

Leguminous soybean (*Glycine max*) has a high concentration of edible protein and their bioavailability, in addition to inorganic minerals, vitamin C, vitamin K, and isoflavones. Soybean is popular for its functional ingredients and meat-like texture and flavor, despite a much lower energy density than the meat. Traditional non-fermented food products made from soybean are: soymilk, tofu, okara, soy flour, and yuba, whereas the common soybean-based fermented foods include tempeh, natto, soy sauce, miso, douchi, kinema, cheonggukjang, doenjang, kanjang, gochujang, and soy yogurt. Various bacteria, fungus, yeast, and mold are used to prepare soybean-based fermented products. Numerous research studies have confirmed that soybean-based fermented foods and peptides offer anti-obesity, anti-diabetic, anti-angiotensin converting enzyme, anti-oxidant, anti-microbial, and anti-cancer activities. This chapter focuses on technologies for preparation of soybean-based foods through microbial fermentation process. Biological activities of soybean-based foods (produced by microbial fermentation) have also been described.

1.1 INTRODUCTION

Leguminous soybean (*Glycine max*) is cultivated around the world due to its unique nutritional and therapeutic values and economic importance [24, 25]. According to Food and Agricultural Organization of the United Nations [108], remarkable producers of soybean are The United States, Canada, Brazil, Argentina, Paraguay, Uruguay, China, and India [108].

Soybean seeds are huge source of nutrients [91]. It consists of carbohydrates (~30% w/w), fat (~19% w/w), protein (~36% w/w), and wide range of essential minerals (iron, calcium, sodium, magnesium, zinc, manganese, selenium). Furthermore, vitamin C (~6 mg in 100 g) and vitamin K or phylloquinone (~47 μg in 100 g) are abundant in soybean [91, 103]. Four different types of soybean proteins with biochemical activities are: glycinin, β-conglycinin, lunasin, and lectin [22, 49]. Because of high amount of proteins, soybean is accepted as source of edible protein (32–42%) [118]. Genetic modification has improved the productivity of soybean and protein content and oil [22, 49, 103]. Soybean contains three groups of isoflavones (such as: daidzeins, genisteins, and glyciteins) and their respective β-glycosides (daidzin, genistin, and glycitin. These are referred to phytoestrogen with estrogenic properties. After consumption of soybean, intestinal microflora and glucosidases hydrolyze isoflavones and produce aglycones, daidzein, genistein, and glycitein. Later, these compounds are absorbed in the gut-wall or metabolized to other metabolites, such as, equol, and p-ethylphenol. The bioavailability of isoflavones can be increased by processing techniques, such as, steaming, roasting, cooking, and fermentation with microbial consortia [121].

Because of nutritional importance and taste, several non-fermented soybean-based food products (such as: soymilk, tofu, and yuba) and common soybean-based fermented foods (such as: tempeh, natto, soy sauce, miso, douchi, kinema, cheonggukjang, doenjang, kanjang, gochujang, and soy yogurt) are popular or several health benefits [24]. Peptides with lower molecular weight and modified proteins with large molecular weight through microbial fermentation process provide exclusive biological activities of soybean-based fermented foods. They offer anti-obesity, anti-angiotensin converting enzyme (ACE) activity, anti-oxidant, anti-microbial, and anticancer activities [13, 20, 28, 44, 97, 113]. However, bioactive peptides can also be obtained through enzymatic hydrolysis of proteins from human milk [1], fish, egg, oyster, cereals, legumes, and oilseeds [20, 28, 44, 56, 118, 120] due to presence of all essential amino acids in soybean-derived peptides [24, 100].

In this chapter, technologies have been discussed for preparation of foods from soybean through microbial fermentation process. The biological activities and nutritional aspects of soybean-based foods (produced through microbial fermentation process) are also discussed.

1.2 ROLE OF MICROBES IN SOYBEAN PROTEIN FERMENTATION

For production of soybean-based fermented foods (Figure 1.1), variety of microbes have been used, such as: Homo-fermentative lactic acid bacteria (LAB) (*Lactobacillus bulgaricus*, *Lactobacillus acidophilus*, *Streptococcus thermophilus*) and hetero-fermentative (*Lactobacillus plantarum*, *Lactobacillus casei*), *Bacillus coagulans*, *Bacillus subtilis* [6, 90], yeast (*Zygosaccharomyces rouxii*) [90], and molds (*Candida etchellsii*, *Clavaria versatilis*, and *Aspergillus oryzae*) [20, 22, 24, 103].

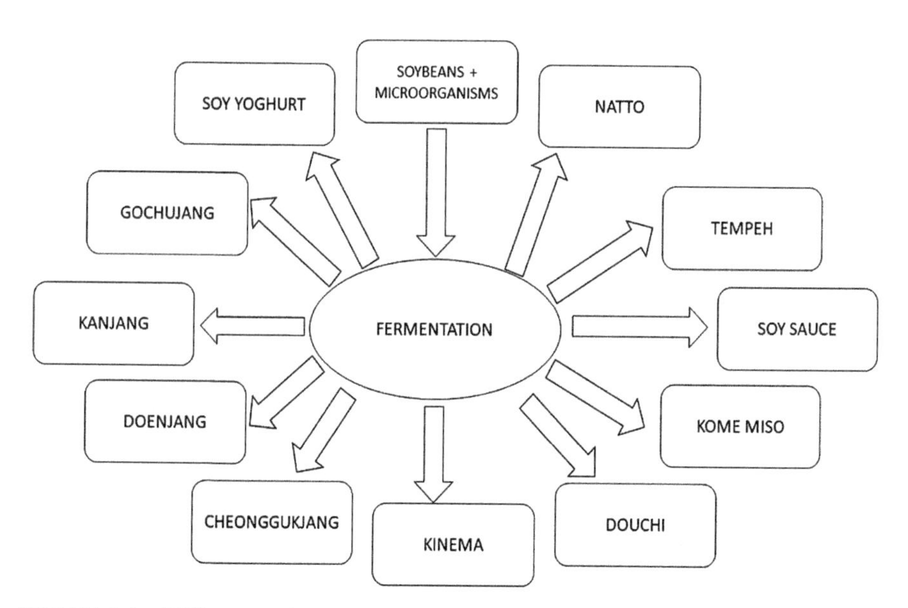

FIGURE 1.1 Different soybean-based foods, produced through the microbial fermentation process.

Source: Self-developed with concepts from Refs. [22, 24].

During fermentation, microorganisms break and convert the soybean-proteins. Most of these microorganisms are widely present in nature [103], even in human digestive system [6, 90]. They are recognized as safe and provide several health benefits to consumers. They are well known for their physiological significance and technological importance for the development of texture, test, and flavor of soybean-based fermented food products [10, 22, 24, 103].

In Figure 1.2, metabolic pathway in microbes for soybean protein fermentation is presented. Arrays of intracellular and extracellular biochemical reactions promote microbial fermentation process and synthesis of bioactive peptides. Both exopeptidase and endopeptidase are responsible for soybean protein hydrolysis. When exopeptidase breaks the soybean proteins or polypeptides in a random basis, then there is a production of more free amino acids. And endopeptidase cleaves the particular or specific amino acid in peptide chain and produces small peptides [13, 20, 28, 44, 118]. Smaller peptides can be uptaken by microbes and free amino acids are produced by intracellular hydrolysis. Amino acids are capable to enter from abiotic phase to biotic phase through membrane protein transporters [10, 74, 98].

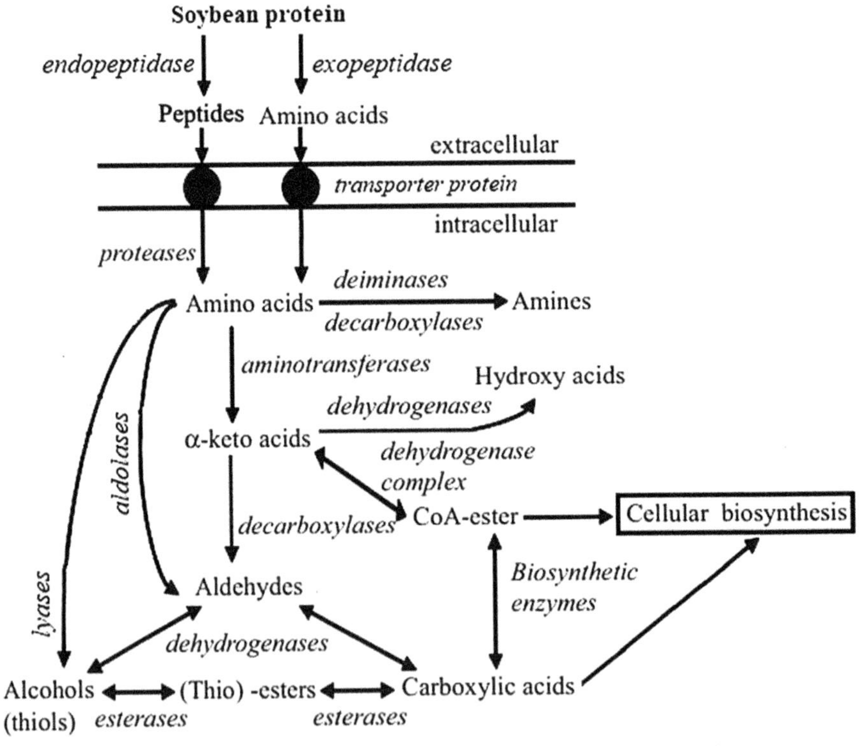

FIGURE 1.2 Metabolic pathway for soybean protein fermentation in microbes.

Source: Self-developed with concepts from Refs. [10, 98].

Before addition of microbes in fermentation containers, cleaned, and soaked soybean seeds are generally used [24]. Microbes, in their exponential growth phase, are applied as starter inoculum for the fermentation process. The yield of fermentation process depends on some factors, such as [22, 24, 103]: (a) duration of the fermentation, (b) fermentation condition, (c) type of microbial strain and the activity of their synthesized enzymes and (d) source of soybean proteins.

1.3 SOYBEAN PRODUCTS THROUGH FERMENTATION

Soybean products through fermentation have been used mostly in eastern Asian countries [24, 25]. Soybean-based fermented foods are enriched with biologically active peptides, amino acids, and biogenic amines [22, 24, 100, 103]. The traditional microbial fermented soybean-based foods are classified as:

- Salt-fermented soybean-based food products: kanjang, douchi, kome miso, gochujang, and soy sauce;
- Non-salted soybean-based fermented food products, such as: natto, tempeh, kinema, chungkkjang, soy yogurt and doejang.

1.3.1 NATTO

Natto is one of the traditional foods in the Japanese community [24, 31]. Primarily, soybean seeds are cleaned with water and soaked in water for 12–20 hours. Then, soybean seeds are boiled or steamed for 35–45 minutes. Subsequently, steamed soybean seeds are mixed with an inoculum of *Bacillus subtilis* spores (10^4–10^6 spores g^{-1}) and are kept at temperature 35–47°C for 12–20 hours for fermentation [31, 58]. The sample is allowed to cool and is kept in a refrigerator for the aging process. In this process, the smaller size of soybean seeds is preferred, because *Bacillus subtilis* has more opportunity to ferment the entire beans [24, 32, 58, 77]. In Figure 1.3, the methodology for preparation of natto from soybean is presented.

1.3.2 TEMPEH

Tempeh is one of the traditional Indonesian dishes with very high concentration of vitamin B_{12}. The four major steps for tempeh preparation are: (a) boiling, (b) soaking, (c) inoculation of microbes and (d) incubate at room temperature (~27°C) [24]. Primarily, soybean seeds are boiled for 5–10 minutes and subsequently they are soaked in cold water for 15–17 hours. After soaking, water is drained-out and is followed by dehulling of the boiled soybean seeds. Lastly, sample is inoculated with *Rhizopus* sp. and fermentation is performed for 35–37 hours under the punctured polymer cover at room temperature. Fermentation of soybean seeds with *Rhizopus* generates flavor, improves the texture and nutritional value of tempeh [89]. Wide range of filamentous fungal species was also isolated in tempeh, such as [102]: *Rhizopus oligosporus*, *Rhizopus stolonifer*, *Rhizopus arrhizus*, *Rhizopus oryzae*, *Rhizopus formosaensis*, and *Fusarium* sp. Some bacteria (such as: non-pathogenic *Klebsiella pneumoniae* and *Citrobacter freundii*) are key producers of vitamin B_{12} in tempeh [24, 58]. Cooked tempeh is eaten instead of its native form. In Figure 1.4, methodology for preparation of tempeh from soybean is presented.

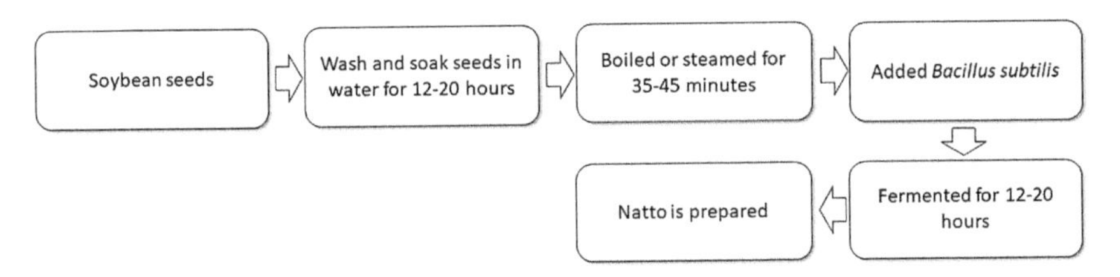

FIGURE 1.3 Methodology for preparation of natto from soybean.
Source: Self-developed with concepts from Refs. [24, 31, 32, 58, 77].

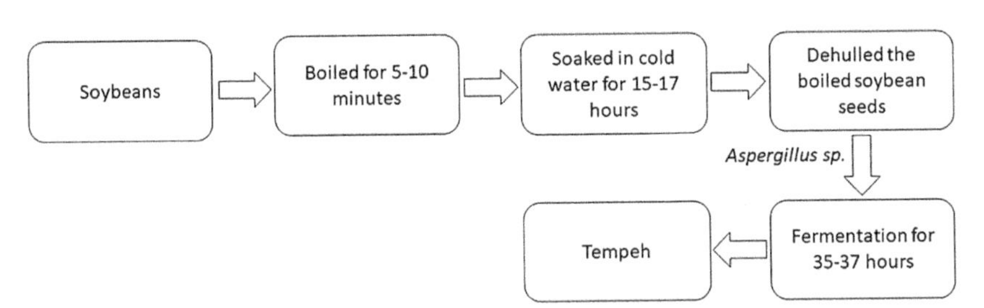

FIGURE 1.4 Methodology for preparation of tempeh from soybean.
Source: Self-developed with concepts from Refs. [24, 58, 89].

1.3.3 SOY SAUCE

Soy sauce originated in China prior to ~500 years B. C. [24]. In first step, steamed, and presoaked soybean seeds are mixed with roasted wheat flour.

For Chinese and Japanese variants, their mixing ratios are 4:1 and 1:1, respectively. Subsequently, the mixture is fermented with fungal consortia (such as: *Aspergillus oryzae* or *Aspergillus sojae*) to prepare 'Koji,' the first fermented product [26]. In the next step, sodium chloride solution at a concentration 16–18% w/v is mixed with kioji and mixture is allowed to ferment. In this fermentation step, kojiis converted to moromi [18, 24, 36]. During the moromi phase, the microbial community is changed from filamentous fungi to salt-tolerant LAB and acidophilic yeast. Lactic acid bacteria (such as: *Weissella* sp., *Lactobacillus* sp., *Streptococcus* sp. and *Tetragenococcus* sp.) have been generally detected in moromi. These bacteria reduce the pH of moromi, which supports the growth of several acidophilic yeasts, such as *Candida etchellsii, Zygosaccharomyces rouxii,* and *Candida versatilis* [24, 105]. In the fermentation process, yeast produces alcohol and volatile flavor compounds [18, 76, 110]. Soy sauce preparation from soybean is presented in Figure 1.5.

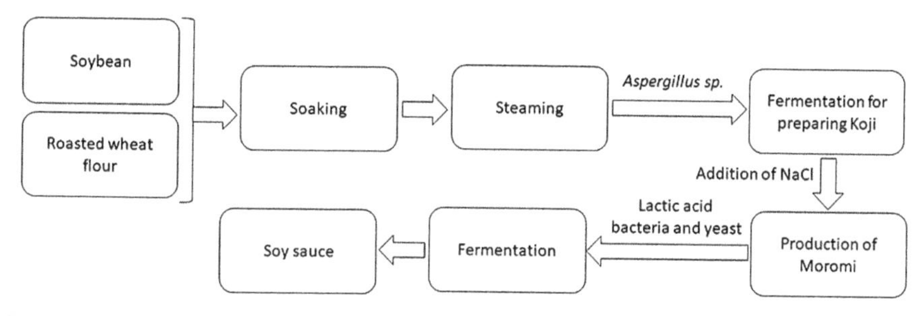

FIGURE 1.5 Methodology for preparation of soy sauce from soybean.
Source: Self-developed with concepts from Refs. [24, 36, 38, 105].

1.3.4 KOME MISO

Kome miso is a traditional Japanese food, which is a salt-fermented soybean-based food [22]. Primarily, koji is prepared by mixing boiled rice with filamentous fungus, such as, *Aspergillus oryzae.* It produces various proteolytic enzymes that break the existing soy protein and produce bioactive peptides. In a subsequent process, steamed soybean is added to koji for further fermentation. After sufficient fermentation by *Aspergillus oryzae*, fermentation product is considered for aging to prepare the kome miso [22, 24]. In Figure 1.6, the methodology for the preparation of kome miso is presented.

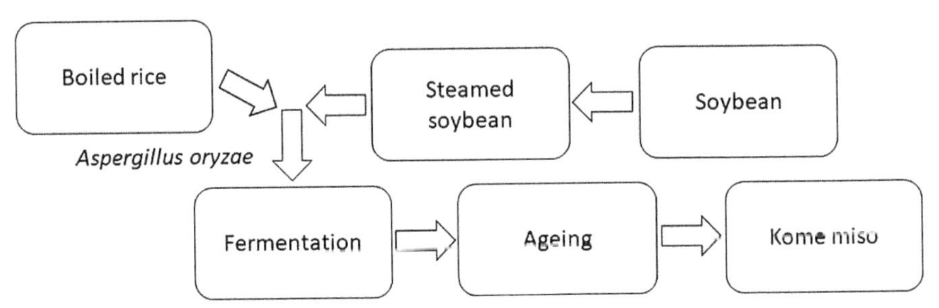

FIGURE 1.6 Methodology for preparation of Kome miso from soybean.
Source: Self-developed with concepts from Refs. [22, 24].

1.3.5 DOUCHI

Douchi is a traditional soybean-based fermented Chinese food [24]. There are three types of microorganisms are used for douchi preparation, such as: *Mucor racemosus*, bacteria *(Bacillus subtilis var. natto)*, and fungus *(Rhizopus oligosporus* and *Aspergillus oryzae)*. Douchi, fermented by *Aspergillus* sp. is the most ancient. This method is widely used in China. Douchiis used as a complementary and alternative medicine for the treatment of heart-related diseases, such as, heartburn, inflammation, and common cold.

To produce douchi, soybean seeds are soaked in water for 4 hours at temperature ~30°C. Then, soybeans are steamed at temperature ~121°C for 50 minutes. Subsequently, temperature is reduced and a starter culture, i.e., *Aspergillus* sp. is added to the soybeans. The mixture is left for fermentation for 48 hours at temperature ~30°C. Then the product is washed with 10% water and mixed with 16% kitchen salt [112]. Additionally, mixture of ginger, shallot, and garlic is added. The mixture is then sealed and kept for aging for 15 days at room temperature (~27°C). After 15 days, douche is produced. Addition of salt is generally done after the fermentation. It changes the taste of the product and control the microbial growth in douchi [24, 58, 112, 122]. In Figure 1.7, preparation of douchi from soybean is presented.

1.3.6 KINEMA

Kinema is a soybean-based fermented food among ethnic community in Nepal [103]. It has a sticky texture, grey color, slightly alkaline and ammoniacal flavor [22, 24, 77]. Several microorganisms (such as: *Bacillus licheniformis, Bacillus subtilis, Bacillus cereus, Bacillus thuringiensis, Bacillus circulans,*

Bacillus sphaericus, Enterococcus faecium, and *Candida sp.*) are used for preparation of kinema [104]. In Nepal, for traditional preparation of kinema, small-sized (~6 mm) yellow cultivar soybean seeds are used. Initially, soybean seeds are washed with water and soaked in water for overnight (~8 hours). Subsequently, soaked seeds are separated from water and are boiled with pure water for ~2.5 hours until the texture becomes soft. Then water is removed and soft soybean seeds are cracked lightly using a wooden mortar and a wooden pestle. This process may increase the surface area of substrate, which accelerates the fermentation process [24, 68]. Firewood ash (approximately 1%) is introduced to cracked soybeans to make it alkaline. Soybean gravels are placed in a bamboo basket, lined by leaves of fresh fern (*Glaphylopteriolopsis erubescens*). This basket is covered with a jute bag and natural indigenous fermentation is allowed on an earthen oven at temperature 25–40°C. In the summer season, fermentation process may be completed in 1–2 days because of high heat during day time; whereas in winter season, fermentation process may take 2–3 days. Kinema is generally eaten as a curry with steamed rice. To prepare kinema curry, fresh kinema is mixed with chopped vegetables and spices (onion, tomato, and turmeric), and is fried with vegetable oil [24, 103, 104]. Figure 1.8 shows the methodology for preparation of kinema.

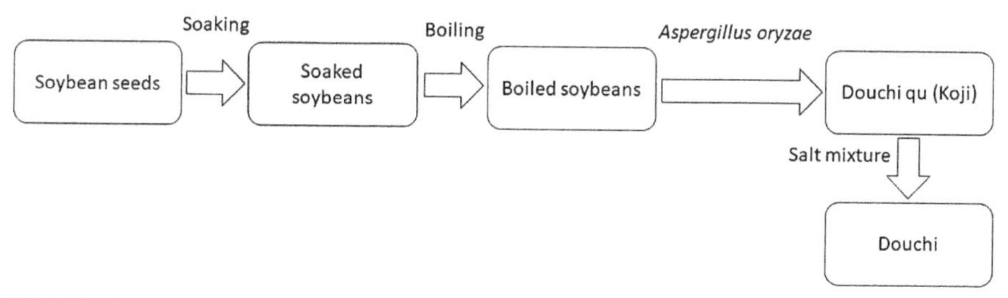

FIGURE 1.7 Methodology for preparation of douchi from soybean.
Source: Self-developed with concepts from Refs. [24, 58, 112].

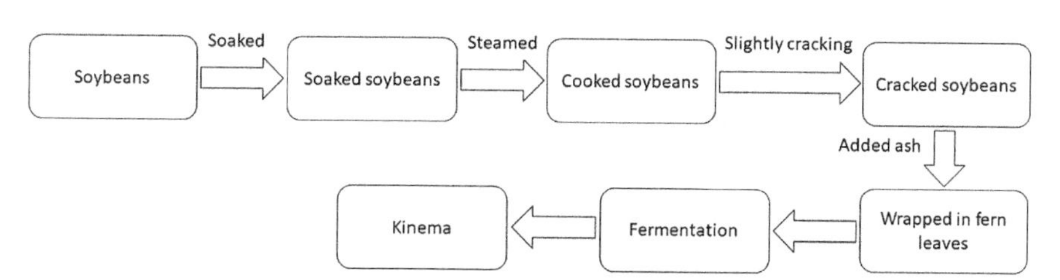

FIGURE 1.8 Methodology for preparation of kinema from soybean.
Source: Self-developed with concepts from Refs. [22, 24, 68, 103, 104].

1.3.7 CHUNGKUKJANG (SOYBEAN PASTE)

Chungkukjang is traditional Korean fermented soybean-based food [22, 24]. To prepare chungkukjang, soybean seeds are cleaned, and are soaked in pure water for 18 hours at 10°C. Then, soaked seeds are cooked for 30 min at ~121°C. After cooling, seeds are inoculated with 5% *Bacillus licheniformis* and incubated at ~40°C for 3 days. Traditionally, fermented chung kukjang with anti-diabetic activity contains *Bacillus subtilis*, *Bacillus amyloliquefaciens*, and *Bacillus licheniformis* [50, 54].

1.3.8 DOENJANG (KOREAN FERMENTED SOY PASTE)

Korean traditional soybean paste is known as doenjang [22, 24]. Soybean seeds are first fermented and then boiled in 5% (w/v) salted water at 100°C. The solid mass of soybean, which is obtained by draining the water, is known as meju. Doenjang is prepared by utilizing this meju. Boiling in brine causes the degradation of soy proteins and production of organic acids, amino acids and minerals [74, 86]. To continue the fermentation process, more cooked cereals (such as: rice, wheat) are added into meju and the mixture is allowed to maturate for 3–6 months. Fungi is proliferated in meju [55] and enzymes (synthesized by *Bacillus* spp.) enhance the maturation of doenjang [30]. Freshly prepared doenjang is usually light brown and is turned to darker with time [71].

1.3.9 KANJANG

Kanjang is a by-product of Korean fermented soybean [22, 24]. It is a sauce from soybean. Traditionally, doenjang, and kanjang are prepared through similar process. Meju is boiled in 5% of brine and is kept for incubation at ~27°C for 5–6 months; then the liquid part of the brine is called kanjang. Normally, fermentation, and maturation take about 3–6 months. Proteases from fungus, *Aspergillus* sp. And *Bacillus* sp. breakdown soybean proteins to amino acids. It creates unique savory-type taste [9]. Produced amino acids react with saccharides; and malanoidine (a brown-colored substance) is generated [71]. However, kanjang is produced from defatted soybean and wheat flour. Main components of kanjang are saccharides, peptides, amino acids and alcohol. Organic acids affect the organoleptic properties of kanjang [22, 24, 96].

1.3.10 GOCHUJANG

Gochujang is a traditional Korean soybean food similar to kanjang (fermented soy sauce) and doenjang (fermented soy paste). To prepare gochujang, glutinous rice (~20% w/w) is considered for soaking in water for 24 hours. Subsequently, water is removed and soaked rice is grounded. Malt is soaked in water (~5%w/w) for 6 hours and is boiled at ~100°C. Subsequently, water is filtrated from the boiled malt. A mixture of malt and glutinous rice is boiled with water for 30 minutes. Subsequently, its temperature is reduced from ~100°Cto room temperature (~27°C). In next step, this item is heated at 50°C to allow saccharification [7, 87, 95]. Powdered meju (5.5% w/w) [7], salt (12.8% w/w) [87] and red pepper (25.0%w/w) [116] are added to the mixture. This mixture is fermented for 3–6 months [46]. Meju is used as a starter culture for gochujang preparation. The important bacteria in meju for gochujang preparation are *Bacillus licheniformis, Bacillus pumilus,* and *Bacillus subtilis.* The microorganisms proliferate in meju and their synthesized enzymes hydrolyze protein in soybean and starch in wheat. The time for natural fermentation with wild microorganisms to prepare traditional gochujang meju usually takes 2–3 months. After fermentation, the remaining solid part is dried and the pulverized substances are considered for preparation of gochujang. During the fermentation, proteins, and starch are converted to several biomolecules, thus creating an umami (savory) and sweet flavor [7, 87, 42]. Methodologies for preparation of chungkukjang, doenjang, gochujang, and kanjang are presented in Figure 1.9.

1.3.11 SOY YOGURT

One of most common soybean-based food is soy yogurt. Soybean milk is produced by blanching of crude soybean milk, dehulling, blending, and subsequently filtration of milk to remove the suspended solids and solid parts. Subsequently, soy-milk along with sucrose are pasteurized at temperature 85°C for 15 minutes and subsequently it is cooled to 42°C.

Soy yogurt is prepared by fermentation of soybean milk with acid forming LAB (such as: *Lactobacillus bulgaricus* and *Streptococcus thermophiles). During fermentation, LAB produce organic acids, which help to reduce the pH of milk. At isoelectric point, soybean proteins are coagulated creating the gel* [15, 24]. To improve the texture of yogurt, starch is added during pre-fermentation or post-fermentation process. Fermentation process is performed for 6 hours with starter culture at 42°C. Subsequently, soft gel of yogurt is considered for maturation of texture at ~7°C for 12 hours [80]. In Figure 1.10, steps for preparation of soybean-based yogurt are presented.

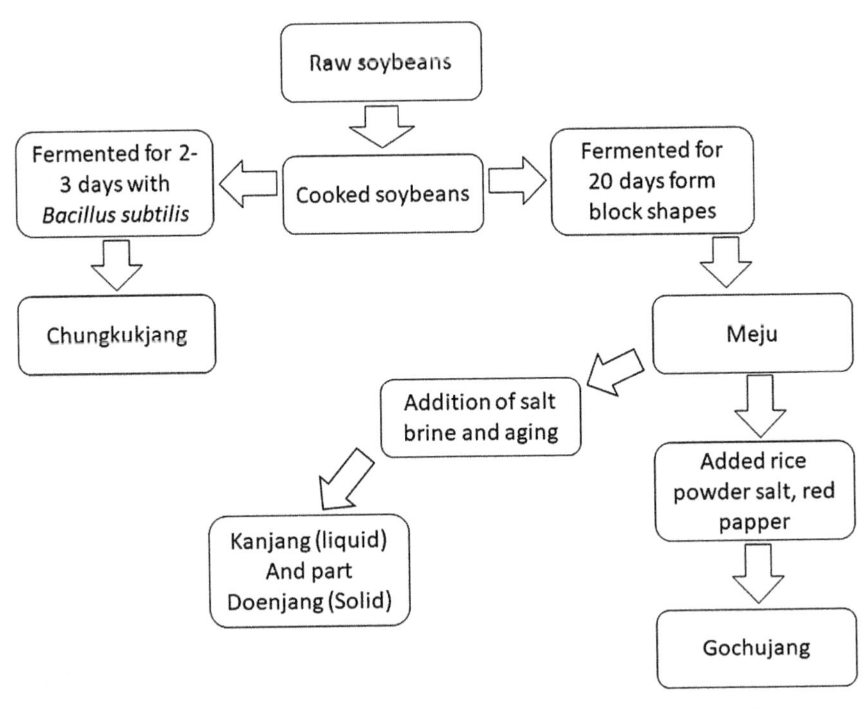

FIGURE 1.9 Process for preparation of chungkukjang, kanjang, doenjang, and gochujang.
Source: Self-developed with concepts from Refs. [9, 22, 24].

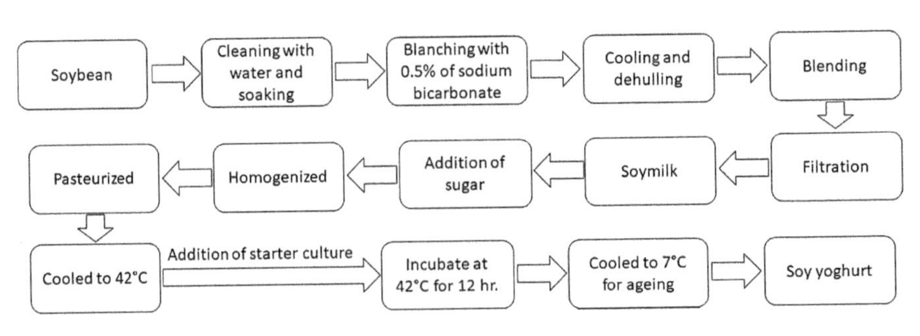

FIGURE 1.10 Preparation of soy yogurt.
Source: Self-developed with concepts from Refs. [22, 24, 80].

1.4 BIOLOGICAL ACTIVITIES OF SOYBEAN-BASED FERMENTED FOODS

Consumption of soybean-based fermented foods offers many physiological effects to consumers. Soybean-based food products are generally main or side dishes in China, Japan, and Korea [22, 24, 25]. Recently, soybean-based food products have been introduced in diet among western communities [78, 113]. The isoflavones and peptides in soybean-based fermented foods are responsible for offering therapeutic activities [13, 20, 28, 44, 97]. Consumption of soybean-based food products are recognized for the reduction of occurrence or severity of several diseases and offer modest health benefits [13, 20, 28, 44, 97, 113]. The therapeutic activities (such as: anti-obesity, anti-angiotensin converting enzyme activity, anti-oxidation activity, anti-microbial activity and anti-cancer activity) have been offered by soybean-based fermented foods and peptides are discussed in this section (Figure 1.11).

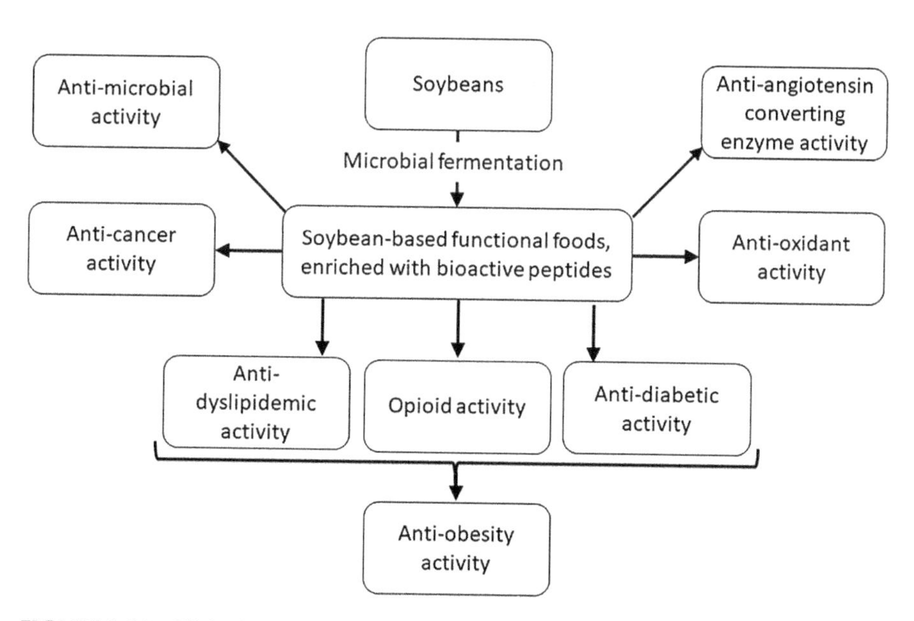

FIGURE 1.11 Biological activities, provided by different types of soybean-based fermented food products.

Source: Self-developed with concepts from Refs. [5, 13, 20, 28, 44, 97, 113].

1.4.1 ANTI-OBESITY ACTIVITY

Obesity or overweight is strongly correlated with the consumption of high caloric food, dyslipidemia (accumulations of triglyceride and cholesterol in systems and organs) and diabetes. It is also related to heredity [8, 48]. lack of physical movement, endocrine disorders and medications. If obesity-associated disorders are not considered as serious issue, it can lead to life-threatening situations [114, 119].

1.4.2 OPIOID ACTIVITY

Soybean-derived peptides have activity on nerves system. The β-subunit of soybean protein β-conglycinin contains the sequence Thr-Pro-Phe-Val (an opioid peptide with morphine-like activity). Three peptides (such as: Thr-Pro-Phe-Val-Val, Thr-Pro-Phe-Val-Val-Asn, and Thr-Pro-Phe-Val-Val-Asn-Ala) with anxiolytic activity have been identified. These peptides can suppress the urge of food intake and small intestinal transit. Furthermore, Thr-Pro-Phe-Val-Val has influence on oral food intake, blood glucose and triglyceride levels. Similarly, β-conglycinin-derived peptide (Val-Arg-Ile-Arg-Leu-Leu-Gln-Arg-Phe-Asn-Lys-Arg-Ser) can suppress the urge for food intake and gastric emptying [5, 92].

1.4.3 ANTI-DYSLIPIDEMIC ACTIVITY

The soybean-derived bioactive peptides can reduce the risk of dyslipidemia (accumulation of cholesterol and triglycerides. Their accumulations are recognized as one of the causes of obesity [3, 8, 45, 114, 117].

Soybean-derived peptides can reduce lipid synthesis and lipid accumulation through versatile mechanisms. Following peptides have been identified for biochemical activities [33, 51, 82, 92], such as:

- Reduce lipoprotein lipase activity;
- Lower activity of fatty acids synthase;
- Suppress the oxidative stress;
- Reduce the growth of preadipocytes;
- Reduce the activities of sterol regulatory element-binding proteins-1c, peroxisome proliferator-activated receptor gamma and fatty acid synthase several di-peptides (such as: Lys-Ala, Val-Lys, and Ser-Tyr [34], tri-peptide Pro-Gly-Pro [92], tetra-peptide Leu-Pro-Tyr-Pro

[51]) and long-chain peptide (such as: Leu-Pro-Tyr-Pro-Arg [92], Ile-Ala-Val-Pro-Gly-Glu-Val-Ala, Ile-Ala-Val-Pro-Thr-Gly-Val-Ala [51, 92]).

In Figure 1.12, biochemical mechanisms of anti-dyslipidemic activity due to soybean-derived peptides are presented.

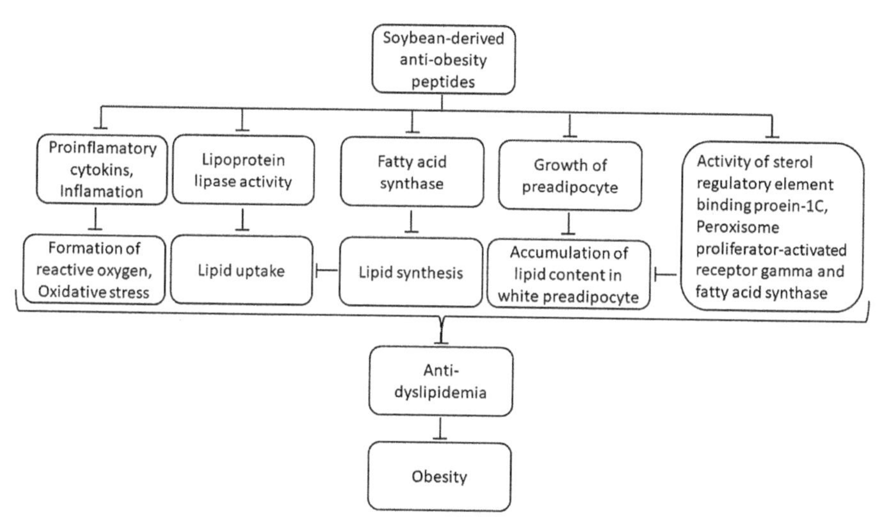

FIGURE 1.12 Biochemical mechanisms of anti-dyslipidemic activity offered by soybean-derived peptides.

Source: Self-developed with concepts from Refs. [5, 51, 82, 92].

1.4.4 ANTI-DIABETIC ACTIVITY

Several soybean-derived peptides have anti-diabetic activity in addition to their anti-obesity activity [48, 59, 75]. Soybean-derived peptides (such as: Leu-Pro-Tyr-Pro, Iso-Ala-Val-Pro-Gly-Glu-Val-ala, and Iso-Ala-Val-Pro-Thr-Gly-Val-Ala) help glucose metabolism by supporting glucose uptake in liver cells by activation of glucose transporters (GLUT 1 and GLUT 4) [5]. The peptide, Iso-Ala-Val-Pro-Thr-Gly-Val-Ala, inhibits activity of dipeptidyl peptidase IV that is responsible for hydrolysis of peptide hormone glucagon-like peptide-1 and glucose-dependent insulinotropic polypeptide [52]. Furthermore, soybean-derived peptides can suppress the activity of α-glucosidase that is located at brush border of the enterocytes of jejunum in small intestine. It is a key enzyme for producing monosaccharide and their absorption on the gut-wall.

Inhibition of α-glucosidase activity affects secretion and activity of glucagon-like protein-1. Therefore, it can suppress postprandial hyperglycemia [5]. Soybean-derived peptides control the diabetes by controlling appetite through regulating the hormones, such as: ghrelin, CCK, and peptide YY [59, 75, 92]. Also, soybean-derived peptides can protect pancreas and its activity from inflammation and oxidative stress [5]. It has been reported that soybean-based fermented food products (such as: natto [106], chungkookjang [38, 46, 47], and tauchi [21]) were able to reduce the risks of diabetes in humans and mouse models. In Figure 1.13, biochemical mechanisms of anti-diabetic activity due to soybean-derived peptides are presented.

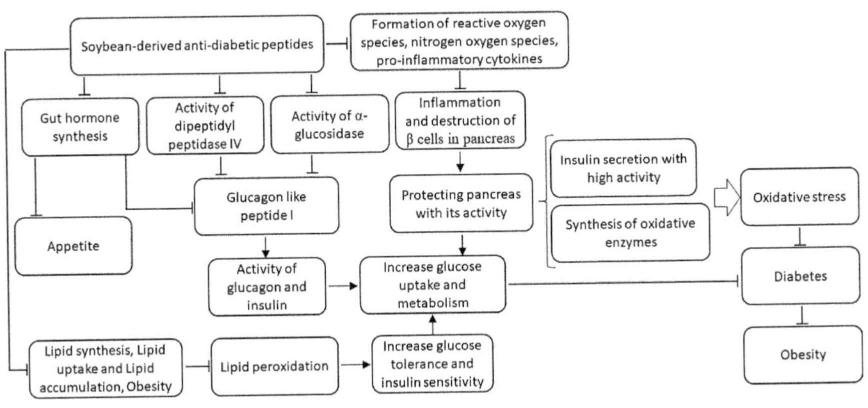

FIGURE 1.13 Biochemical mechanisms of anti-diabetic activity offered by soybean-derived peptides.

Source: Self-developed with concepts from Refs. [5, 59, 75, 92].

1.4.5 ANTI-ANGIOTENSIN CONVERTING ENZYME ACTIVITY

Due to presence of an angiotensin-converting enzyme (ACE) in soybean products, Angiotensin-I is converted to Angiotensin-II in the rennin-angiotensin pathway, which promotes vasoconstriction and high blood pressure (BP). Also, ACE supports the transformation of bradykinin into inactive metabolites [17].

Several di-peptides (such as: Ala-Trp, Gly-Trp, Ala-Tyr, Ser-Tyr, Gly-Tyr, Ala-Phe, Val-Pro, Ala-Ile, and Val-Gly [73]) and tri-peptides (such as: Val-Pro-Pro, Ile-Pro-Pro, Ala-Phe-His, Ile-Phe-Leu, Ile-Phe-Tyr, Leu-Phe-Tyr, His-His-Leu [5, 92]) with angiotensine converting inhibitory activity have been identified in several soybean-based fermented food products

[15,67,114]. Furthermore, the angiotensin-converting inhibitory activity of several long-chain peptides (e.g.: Ala-Asp-Phe-Val-Leu-Asp-Asn-Glu-Gly-Asn-Phe-Leu-Glu-AsnGly-Gly-Thr-Tyr-Tyr-Ile, Phe-Phe-Tyr-Tyr, Tyr-Val-Val-Phe-Lys, and Lle-Pro-Pro-Gly-Val-Pro-Try-Trp-Thr) in soybean-based fermented foods have been investigated [92]. In peptide chain, presence of Val, Ala, Ile at the N-terminal end and Pro at the C-terminal end have been recognized for potential of ACE inhibitory activity [5]. Figure 1.14 shows biochemical mechanisms of vasodilation and vasoconstriction that are offered by soybean-derived peptides.

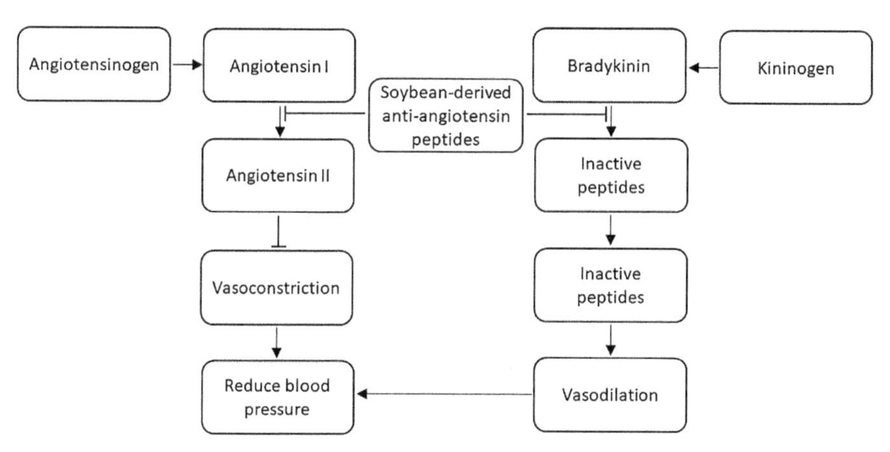

FIGURE 1.14 Biochemical mechanisms of vasodilation and vasoconstriction offered by soybean-derived peptides.

Source: Self-developed with concepts from Refs. [5, 43, 62, 93, 94].

1.4.6 ANTI-OXIDATION ACTIVITY

Antioxidants help to protect cells against oxidative stress. Free radicals (such as: reactive oxygen species-peroxides, superoxide, hydroxyl radical, singlet oxygen, and alpha-oxygen) and reactive nitrogen species (nitric oxide) as metabolic byproducts are produced due to unhealthy lifestyle, smoking, high alcohol consumption, over stress, inflammation, suffering with chronic metabolic disease. Their generation and accumulation are cause of oxidative damage of cell organelles, including DNA. Antioxidants operate by triggering and neutralizing free radicals in cell through (a) donating electron and (b) activation of transcription factor DAF-16 [5, 27].

Soybean-derived bioactive peptides playas anti-oxidants in damaged or unhealthy cells that are under oxidative stress. As antioxidants, isoflavones, and lunasin may suppress risk of cancer by inhibiting DNA damage [5]. Genistein among all of soybean isoflavones might support the synthesis and activity of antioxidant enzyme superoxide dismutase (SOD), which inhibit accumulations and activities of free radicals [92]. Due to enzymatic hydrolysis of soy protein, more R-groups in peptide chain are exposed, which offer free radical scavenging. The C- or N-terminal end amino acids in bioactive peptide sequence donate electrons to free radicals and neutralize the activity of reactive oxygen species and protect the cells and tissues from oxidative damage [23, 92]. Anti-oxidation activity in several soybean-based fermented food products (such as: soy milk [101], gochujang [116], kinema [68], soy sauce [69], dauchi [111], and kanjang [96]) have been validated. Figure 1.15 indicates biochemical mechanisms of anti-oxidation activity that are being offered by soybean-derived peptides.

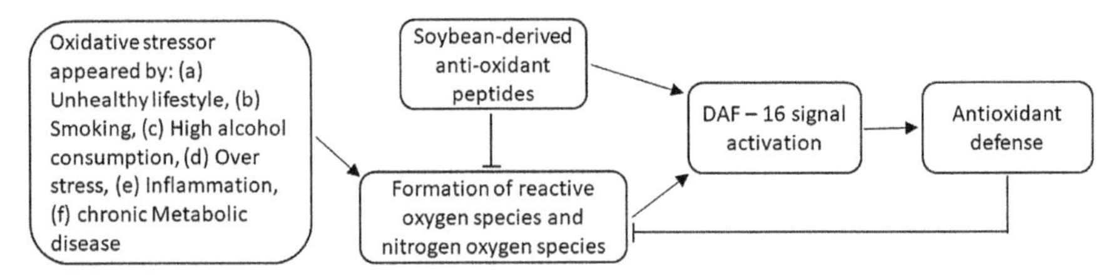

FIGURE 1.15 Biochemical mechanism of anti-oxidation activity of soybean-derived peptides.
Source: Self-developed with concepts from Refs. [5, 27, 92].

1.4.7 ANTI-MICROBIAL ACTIVITY

Multidrug-resistant mechanisms by chemically synthesized antimicrobial compounds are [35, 115] are:

- Genetic modifications, such as: (i) mutational resistance; and (ii) horizontal gene transfer;
- Mechanistic based modifications, such as: (i) partially modifications in the antibiotic molecule; (ii) decrease antibiotic penetration and efflux; (iii) changes in target sites; and (iv) global cell adaptations.

Bioactive peptides display anti-bacterial, anti-viral, anti-fungal, and/or anti-parasitic activities. They often play a major role in innate immunity [83]. Some soybean-based peptides offer anti-microbial activities. Similar to the anti-microbial mechanism of antibiotics, anti-bacterial peptides do not inhibit bacterial peptidoglycan synthesis or genetic transformation. Instead, they bind with membrane proteins and create complex as a primary step in the antimicrobial mechanism [115]. They make pores in cell layer/membrane from cellular items (cell organelles) and are leaked to abiotic phase [107, 109]. It has been reported that the anti-bacterial mechanism of soybean-derived peptide is detergent-like. Phospholipids in cellular membrane, teichoic acid and lipoteichoic acids in peptidoglycans of Gram-positive bacteria participate in electrostatic interaction with functional groups of peptides from soybean [63]. Interaction between anti-bacterial peptides with cytoplasmic membrane frequently lead to lipid segregation in cell membrane, which affects membrane permeability, inhibits cell division and leads to delocalization of essential membrane proteins [115]. It was reported that the level of binding of peptides with bacterial cell membrane depends on the sequence and concentration of antimicrobial peptides and nature of the bacterial cell membrane [88].

Anti-microbial activity of conglycinin- and glycinin-derived peptides has been proven against both Gram-negative and Gram-positive bacteria [109]. Dhayakaran et al. [14] found that soybean-based peptides (such as: pro-gly-thr-ala-val-phe-lys and iso-lys-ala-phe-lys-glu-ala-thr-lys-val-asp-lys-val-val-val-luc-try-thr-ala) have significant effects against *Listeria monocytogenes* and *Pseudomonas aeruginosa* [14]. Anti-bacterial activity of tempeh [70] and miso [79] have been proven. Figure 1.16 shows the biochemical mechanism of anti-microbial activity, offered by soybean-derived peptides.

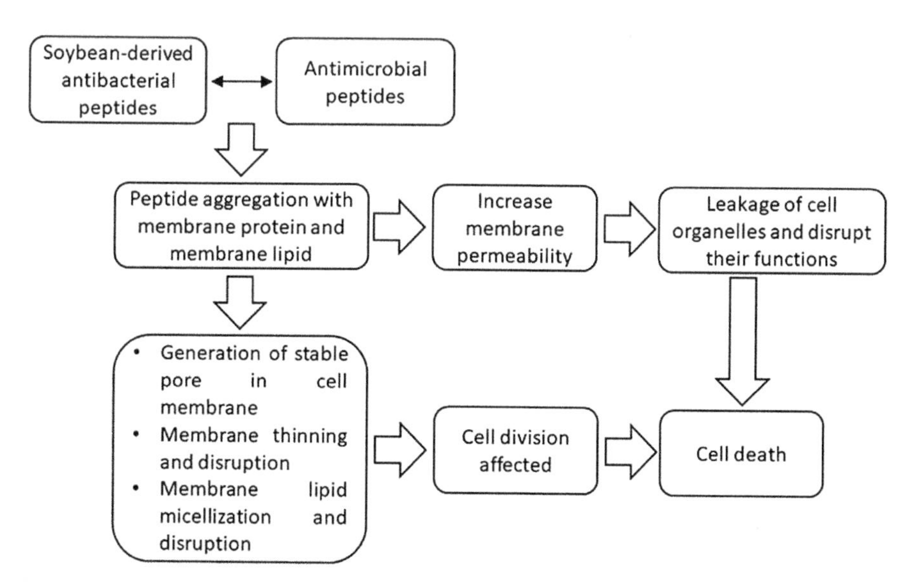

FIGURE 1.16 Biochemical mechanism of anti-microbial activity, offered by soybean-derived peptides.

Source: Self-developed with concepts from Refs. [88, 115].

1.4.8 ANTI-CANCER ACTIVITY

Cancer is partially related to activation of pro-inflammatory agents, pro-oxidant, and immunosuppressive mechanisms that are prone to growth of tissues in an abnormal way and often formation of tumor [4, 60, 65]. Several research studies have shown that some soybean-derived peptides offer anti-cancer activity [11, 39, 40, 53, 84]. Mainly two peptides (known as lunasin [12, 60, 65, 81, 84] and Bowman-Birk inhibitor [19, 64, 99]) have shown to have anti-cancer properties.

Bowman-Birk inhibitor suppresses reactive-oxygen-species induced mitochondrial damage after proteasomal inhibition and angiogenesis [19, 64, 99]. Amino acid sequence in lunasin peptide is almost similar to Bowman-Birk inhibitor. Structurally, lunasin is 43 amino acid having nine aspartic acid residues at the C-terminal end with a cell adhesion motif preceding to it. Tri-peptide RGD (Arg-Gly-Asp) from the sequence helps its binding with the non-acetylated H3 and H4 histones to prevent their acetylation and anti-carcinogenic activity [65, 84]. It has been reported that Bowman-Birk inhibitor has a role in the protection of lunasin from gastrointestinal degradation [12, 85].

Soybean-derived peptide has anti-cancer and anti-oxidation activities [12, 60, 64]. Presence of small amount of free radicals in tissues increases the risk of cell damage in DNA, which leads to increase the risks of cancer [27]. Soybean-based fermented food products (such as: soy sauce, natto, etc.), increase the activity of natural killer cells and cytokine regulation [65]. Soybean-derived peptides (such as: lunasin) and other similar peptides inhibit inflammation in lipopolysaccharide-induced macrophages. It suppresses the nuclear factor-kappa-B formation pathway. Furthermore, it has been proven that peptide (with molecular weight 5 kDa) may inhibit the formation of most potent pro-inflammatory markers, including formations and expressions of interleukin-6, nuclear factor-kappa B, nitric oxide, cyclooxygenase-2, nitric oxide synthase (NOS), nuclear translocation, and p50 nuclear translocation [60, 65]. The research study has been attempted to prove the potentiality of soybean isoflavones to the suppression of breast cancer [92], due to soybean-based foods [119]. Soy isoflavone supplementation reduces the increase of the concentration of serum prostate-specific antigen (PSA) along with growth of tumor in prostate and cancer in patients [121].

The biochemical activities of different soybean-based fermented food products are presented in Table 1.1.

TABLE 1.1 Biochemical Activities of Soybean-Based Fermented Food Products

Soybean-Based Food	Microorganism Used	Biological Activities	References
Natto	*B. natto* O9516	Anti-hypertensive	[31]
Tempeh	*R. microspores*	Anti-oxidant	[23]
	E. faecium LMG 19827, *E. faecium* LMG 19828	Anti-microbial	[70]
Soy sauce	—	Anti-tumor	[39, 53]
	A. sojae	Angiotensin-converting enzyme inhibitory	[73]
	—	Angiotensin-converting enzyme inhibitory	[41]
	—	Anti-bacterial	[61]
	—	Anti-cancer	[2, 37, 72]
Douchi	*B. subtilis natto*, *B. subtilis* B1	Angiotensin-converting enzyme inhibitory	[57]
	A. oryzae, *M. wutungkiao*	Angiotensin-converting enzyme inhibitory	[57]
	A. egypticus	Anti-hypertensive	[122]
	A. oryzae	Anti-oxidant	[111]
Miso	*Lactococcus sp.* GM005	Anti-bacterial	[79]
Kinema	*B. subtilis*	Anti-oxidant	[68]
Soy yogurt	*Lb. delbrueckii* LB 1466, *S. thermophilus* St 1342, *Lb. acidophilus* LAFTI L10, *B. lactis* LAFTI B94, *Lb. paracasei* LAFTI L26	Angiotensin-converting enzyme inhibitory	[15]
Cheonggukjang	*B. licheniformis* SCD 111067P	Anti-diabetic	[47]
	B. subtilis CSY 191	Anti-cancer	[54]
Meju	*A. oryzae*	Anti-microbial	[6]
	B. amyloliquefaciens FSE-68	Anti-microbial	[6]
Doenjang	*B. subtilis* CSY191	Anti-cancer	[54]

TABLE 1.1 *(Continued)*

Soybean-Based Food	Microorganism Used	Biological Activities	References
	B. subtilis SCB3	Angiotensin-converting enzyme inhibitory	[30]
	—	Anti-cancer	[11]
	A. oryzae BCRC 30222, *A. sojae* BCRC 30103, *A. awamori*, *R. azygospous* 31158, *Rhizopus sp.* No. 2	Anti-cancer	[29]
Kanjang	—	Anti-oxidant	[69, 96]
	—	Anti-tumor	[11, 87]
	—	Anti-cancer	[40]
Gochujang	*B. amyloliquefaciens* CJ 3–27 (KCCM 11317P), *B. amyloliquefaciens* CJ 14-6 (KCCM 11718P), *A. oryzae* CJ 1354 (KCCM 11300P)	Anti-obesity	[95]
	—	Anti-oxidant	[116]

1.5 SUMMARY

Soybean-based food products are good source of edible proteins due to presence of all types of essential amino acids for human nutrition. The fermented soybean foods are tempeh, natto, soy sauce, miso, douchi, kinema, cheonggukjang, doenjang, kanjang, gochujang, and soy yogurt. The bacteria, fungus, yeast, and mold are used to prepare these products. Also, such products offer anti-obesity, anti-angiotensin, anti-oxidant, anti-microbial, and anti-cancer activities.

ACKNOWLEDGMENTS

Authors of this chapter acknowledge Chancellor at Adams University, Kolkata, India for the infrastructure and facilities. They also thank Rajib Majumder of the Department of Biotechnology, Adams University for preparing computer graphics.

KEYWORDS

- **anti-cancer**
- **anti-obesity**
- **anti-oxidant activity**
- **bioactive peptides**
- **functional foods**
- **microbial fermentation**

REFERENCES

1. Aguilar-Toalá, J. E., Santiago-López, L., & Peres, C. M., (2017). Assessment of multifunctional activity of bioactive peptides derived from fermented milk by specific *Lactobacillus plantarum* strains. *Journal of Dairy Science, 100*, 65–75.
2. Benjamin, H., Storkson, J., Nagahara, A., & Pariza, M. W., (1991). Inhibition of benzo [a] pyrene-induced mouse fore stomach neoplasia by dietary soy sauce. *Cancer Research, 51*, 2940–2942.

3. Brij, P. S., Shilpa, V., & Subrota, H., (2014). Functional significance of bioactive peptides derived from soybean. *Peptides, 54,* 171–179.
4. Burris, R. L., Ng, H. P., & Nagarajan, S., (2014). Soy protein inhibits inflammation-induced VCAM-1 and inflammatory cytokine induction by inhibiting the NF-kappa B and AKT signaling pathway in apolipo protein E-deficient mice. *European Journal of Nutrition, 53,* 135–148.
5. Chatterjee, C., Gleddie, S., & Xiao, C. W., (2018). Soybean bioactive peptides and their functional properties. *Nutrients, 10,* 1211–1126.
6. Cho, S. J., Oh, S. H., Pridmore, R. D., Juillerat, M. A., & Lee, C. H., (2003). Purification and characterization of proteases from *Bacillus amyloliquefaciens* isolated from traditional soybean fermentation starter. *Journal of Agriculture and Food Chemistry, 51,* 7664–7670.
7. Choi, J. Y., Lee, T. S., & Noh, B. S., (2000). Quality characteristics of the gochujang prepared with mixture of meju and koji during fermentation. *Korean Journal of Food Science and Technology, 32,* 125–131.
8. Choi, J. H., Pichiah, P. B. T., Kim, M. J., & Cha, Y. S., (2016). Cheonggukjang: Soybean paste fermented with *B. Licheniformis*-67 prevents weight gain and improves glycemic control in high fat diet induced obese mice. *Journal of Clinical Biochemistry and Nutrition, 59,* 31–38.
9. Choi, K. K., Cui, C. C., Ham, S. S., & Lee, D. S., (2003). Isolation, identification, and growth characteristics of main strain related to meju fermentation. *Journal of the Korean Society of Food Science and Nutrition, 32,* 818–824.
10. Christensen, J. E., Dudley, E. G., Pederson, J. A., & Steele, J. L., (1999). Peptidases and amino acid catabolism in lactic acid bacteria (LAB). *Antonie van Leeuwenhoek, 76,* 217–246.
11. Chung, K. S., Yoon, K. D., Kwon, D. J., Hong, S. S., & Choi, S. Y., (1997). Cytotoxicity testing of fermented soybean products with various tumor cells using MTT assay. *Korean Journal of Applied Microbiology and Biotechnology, 25* (5), 477–482.
12. Cruz-Huerta, E., Fernandez-Tome, S., Arques, M. C., Amigo, L., Recio, I., Clemente, A., & Hernandez-Ledesma, B., (2015). The protective role of the Bowman-Birk protease inhibitor in soybean lunasin digestion: The effect of released peptides on colon cancer growth. *Food and Function, 6,* 2626–2635.
13. De-Mejia, E. G., & De-Lumen, B. O., (2006). Soybean bioactive peptides: New horizon in preventing chronic diseases. *Sexuality, Reproduction, and Menopause, 4,* 91–95.
14. Dhayakaran, R., Neethirajan, S., & Weng, X., (2016). Investigation of the antimicrobial activity of soy peptides by developing a high throughput drug-screening assay. *Biochemistry and Biophysics Reports, 6,* 149–157.
15. Donkor, O. N., Henriksson, A., Vasiljevic, T., & Shah, N. P., (2005). Probiotic strains as starter cultures improve angiotensin-converting enzyme inhibitory activity in soy yogurt. *Journal of Food Science, 70,* 375–381.
16. Dziuba, J., Iwaniak, A., & Minkiewicz, P., (2003). Computer-aided characteristics of proteinsas potential precursors of bioactive peptides. *Polimery, 48,* 50–53.
17. Erdmann, K., Cheung, B. W., & Schroder, H., (2008). The possible roles of food-derived bioactive peptides in reducing the risk of cardiovascular disease. *The Journal of Nutritional Biochemistry, 19,* 643–654.
18. Feng, Y., Cui, C., Zhao, H., Gao, X., Zhao, M., & Sun, W., (2013). Effect of koji fermentation on generation of volatile compounds in soy sauce production. *International Journal of Food Science and Technology, 48,* 609–619.

19. Fereidunian, A., Sadeghalvad, M., Oscoic, M. O., & Mostafaie, A., (2014). Soybean Bowman-Birk protease inhibitor (BBI): Identification of the mechanisms of BBI suppressive effect on growth of two adenocarcinoma cell lines, AGS, and HT29. *Archives of Medical Research, 45*, 455–461.
20. Fields, K., Falla, T. J., Rodan, K., & Bush, L., (2009). Bioactive peptides: Signaling the future. *Journal of Cosmetic Dermatology, 8,* 8–13.
21. Fujita, H., & Yamagami, T., (2001). Fermented soybean-derived touchi-extract with anti-diabetic effect via alpha-glucosidase inhibitory action in a long-term administration study with KKAyMice. *Life Sciences, 70*, 219–227.
22. Garcia, M. C., Torres, M., Marina, M. L., & Laborda, F., (1997). Composition and characterization of soybean and related products. *Critical Reviews in Food Science and Nutrition, 37*, 361–391.
23. Gibbs, B. F., Zoygman, A., Masse, R., & Mulligan, C., (2004). Production and characterization of bioactive 643 peptides from soy hydrolysate and soy-fermented food. *Food Research International, 37*, 123–127.
24. Golbitz, P., (1995). Traditional soyfoods: Processing and products. *The Journal of Nutrition, 125,* 570–572.
25. Gyoung-Ah, L., Gary, W., & Crawford, L. L., (2011). Archaeological soybean (*Glycine max*) in East Asia: Does size matter? *PLoS One, 6*, 26720–26732.
26. Gyu-Min, L., Dong-Ho, S., Sung, J., & Choong-Hwan, L., (2016). Metabolomics provides quality characterization of commercial gochujang (Fermented Pepper Paste). *Molecules, 21*, 921–935.
27. Halliwell, B., (1994). Free radicals, antioxidants, and human disease: Curiosity, cause or consequence. *Lancet, 344,* 721–724.
28. Hou, Y., Wu, Z., Dai, Z., & Wang, G., (2017). Protein hydrolysates in animal nutrition: Industrial production, bioactive peptides and functional significance. *Journal of Animal Science and Biotechnology, 8,* 24–37.
29. Hung, Y. H., Huang, H. Y., & Chou, C. C., (2007). Mutagenic and antimutagenic effects of methanol extracts of unfermented and fermented black soybean. *International Journal of Food Microbiology, 118,* 62–68.
30. Hwang, J., (1997). Angiotensin-I converting enzyme inhibitory effect of doenjang fermented by *B. subtilis* SCB-3 isolated from meju, Korean traditional food. *Journal of the Korean Society Food Science and Nutrition, 26*, 776–783.
31. Ibe, S., Yoshida, K., Kumada, K., & Tsurushiin, S., (2009). Antihypertensive effects of natto: Traditional Japanese fermented food in spontaneously hypertensive rats. *Food Science and Technology Research, 15*, 199–202.
32. Ikeda, H., & Tsuno, S., (1985). The component changes during the manufacturing process of natto, part II: Water-soluble protein. *Shokumotsu Gukkaishi (Kyoto Joshi Daigaku) (In Japanese), 40*, 27–37.
33. Inoue, N., Fujiwara, Y., Kato, M., & Funayama, A., (2015). Soybean beta-conglycinin improves carbohydrate and lipid metabolism in Wister rats. *Bioscience, Biotechnology, and Biochemistry, 79*, 1528–1534.
34. Inoue, N., Nagao, K., Sakata, K., Yamano, N., & Gunawardena, P. E., (2011). Screening of soy protein-derived hypotriglyceridemic di-peptides *in vitro* and *in vivo. Lipids in Health and Disease, 10*, 85–91.
35. Jenssen, H., Hamill, P., & Hancock, R. E. W., (2006). Peptide antimicrobial agents. *Clinical Microbiology Reviews, 19* (3), 491–511.

36. Kataoka, S., (2005). Functional effects of Japanese style fermented soy sauce (Shoyu) and its components. *Journal of Bioscience and Bioengineering, 100*, 227–234.
37. Kataoka, S., Liu, W., & Albright, J., (1997). Inhibition of benzo [a] pyrene-induced mouse forestomach neoplasia and reduction of H_2O_2 concentration in human polymorphonuclear leucocytes by flavor components of Japanese style fermented soy sauce. *Food Chemistry and Toxicology*, *35*, 449–457.
38. Kim, D. J., Jeong, Y. J., Kwon, J. H., & Moon, K. D., (2008). Beneficial effect of chungkukjang on regulating blood glucose and pancreatic beta-cell functions in C75BL/KsJ-db/db mice. *Journal of Medicinal Food, 11,* 215–223.
39. Kim, S. E., Pai, T., & Lee, H. J., (1998). Cytotoxic effects of the peptides derived from traditional Korean soy sauce on tumor cell lines. *Food Science Biotechnology*, 7, 75–79.
40. Kim, S. J., Jung, K. O., & Park, K. Y., (1999). Inhibitory effect of kochujang extracts on chemically induced mutagenesis. *Journal of Food Science and Nutrition, 4*, 38–42.
41. Kinoshita, E., Yamakoshi, J., & Ikuchi, M., (1993). Purification and identification of anangiotensin-I converting enzyme inhibitor from soy sauce. *Bioscience, Biotechnology, and Biochemistry*, *57*, 1107–1110.
42. Ko, B. K., (2006). *Preparation and Physicochemical Properties of Abalone Gochujang* (p. 231). PhD Thesis; Chonnam National University, Gwangju, Korea.
43. Kong, X. Z., Guo, M. M., Hua, Y., Dong, C., & Zhang, C., (2008). Enzymatic preparation of immunomodulating hydrolysates from soy proteins. *Bioresource Technology*, *99*, 8873–8879.
44. Korhonen, H., & Pihlanto, A., (2003). Food-derived bioactive peptides: Opportunities for designing future foods. *Current Pharmaceutical Design*, *9*, 1297–1308.
45. Kwak, C. S., Park, S. C., & Song, K. Y., (2012). Doenjang: Fermented soybean paste, decreased visceral fat accumulation and adipocyte size in rats fed with high fat diet more effectively than non-fermented soybean. *Journal of Medicinal Food*, *15*, 1–9.
46. Kwon, D. Y., Jang, J. S., Lee, J. E., Kim, Y. S., Shin, D. W., & Park, S., (2006). The isoflavonoid aglycone-rich fractions of chungkookjang, fermented unsalted soybeans, enhance insulin signaling and peroxisome proliferator-activated receptor-G activity *In Vitro. Biofactors, 26,* 245–258.
47. Kwon, D. Y., Jang, J. S., Hong, S. M., Lee, J. E., Sung, S. R., & Park, H. R., (2007). Long-term consumption of fermented soybean-derived chungkookjang enhances insulinotropic action unlike soybeans in 90% Pancreatectomized diabetic rats. *European Journal of Nutrition*, *46*, 44–52.
48. Kwon, D. Y., Daily, J. W., Kim, H. J., & Park, S., (2010). Antidiabetic effects of fermented soybean products on type-II diabetes. *Nutrition Research*, *30*, 1–13.
49. Kwon, D. Y., Oh, S. W., Lee, J. S., & Yang, H. J., (2002). Amino acid substitution of hypocholesterolemic peptide originated from glycinin hydrolyzate. *Food Science and Biotechnology*, *11*, 55–61.
50. Kwon, G. H., Lee, H. A., Park, J. Y., & Kim, J. S., (2009). Development of a RAPD-PCR method for identification of bacillus species isolated from cheonggukjang. *International Journal of Food Microbiology*, *129*, 282–287.
51. Lammi, C., Zanoni, C., & Arnoldi, A., (2015). Iavpgeva, iavptgva, and LPYP: Three peptides from soy glycinin, modulate cholesterol metabolism in Hepg2 cells through the activation of the LDLR-SREBP2 Pathway. *Journal of Functional Foods*, *14*, 469–478.
52. Lammi, C., Zanoni, C., & Arnoldi, A., (2015). Three Peptides from soy glycinin modulate glucose metabolism in human hepatic HepG2 cells. *International Journal of Molecular Sciences, 16,* 27362–27370.

53. Lee, H. J., Lee, K. W., Kim, K. H., & Kim, H. K., (2004). Antitumor activity of peptide fraction from traditional Korean soy sauce. *Journal of Microbiology and Biotechnology, 14* (3), 628–630.
54. Lee, H. J., Namb, S. H., Seo, W. T., & Yun, H. D., (2012). The production of surfactin during the fermentation of cheonggukjang by potential probiotic *Bacillus subtilis* CSY191 and the resultant growth suppression of MCF-7 human breast cancer cells. *Food Chemistry, 131*, 1347–1354.
55. Lee, S. W., (1990). Study on the origin and interchange of dujang in ancient East Asia. *Journal of the Korean Society of Food Culture, 5*, 313–316.
56. Li, C., Matsui, T., Matsumoto, K., & Yamasaki, R., (2002). Latent Production of angiotensin I converting enzyme inhibitors from buckwheat protein. *Journal of Peptide Science, 8*, 267–274.
57. Li, F. J., Yin, L. J., & Cheng, Y. Q., (2009). Comparison of angiotensin I Converting enzyme inhibitor activities of pre-fermented douchi started with various cultures. *International Journal of Food Engineering, 5*, 1556–3758.
58. Li, L. T., Zhang, J. H., Li, Z. G., & Tatsumi, E., (2003). The comparison of natto, tempeh, and douchi. *Chinese Condiment, 5*, 3–10.
59. Lu, M. P., Wang, R., Song, X., & Chibbar, R., (2008). Dietary soy isoflavones increase insulin secretion and prevent the development of diabetic cataracts in streptozotocin-induced diabetic rats. *Nutrition Research, 28*, 464–471.
60. Lule, V. K., Garg, S., & Pophaly, S. D., (2015). Potential health benefits of lunasin: A multifaceted soy-derived bioactive peptide. *Journal of Food Science, 80*, R485–R494.
61. Masuda, S., Kudo, Y., & Kumagai, S., (1998). Reduction of *Escherichia coli* O157:H7 population in soy sauce. *Journal of Food Protection, 61*, 657–661.
62. Matsui, T., Matsufuji, H., Seki, E., & Osajima, K., (1993). Inhibition of angiotensin I converting enzyme by *Bacillus licheniformis* alkaline pro-tease hydrolyzates derived from sardine muscle. *Bioscience, Biotechnology, and Biochemistry, 57*, 922–925.
63. Matsuzaki, K., (2009). Control of cell selectivity of antimicrobial peptides. *Biochimica et Biophysica Acta, 1788*, 1687–1692.
64. Mehdad, A., Brumana, G., Souza, A. A., & Barbosa, J., (2016). Bowman-Birk inhibitor induces apoptosis in human breast adenocarcinoma through mitochondrial impairment and oxidative damage following proteasome 20S inhibition. *Cell Death Discovery, 2*, 15067–15077.
65. Mejia, E. G., & Dia, V. P., (2009). Lunasin and lunasin-like peptides inhibit inflammation through suppression of NFkB pathway in the macrophage. *Peptides, 30*, 2388–2398.
66. Mellander, O., (1950). The physiological importance of the casein phospho peptide calcium salts, II: Oral calcium dosage of infants. *Acta Society of Medical Upsaliensis, 55*, 247–255.
67. Merz-Demlow, B. E., Duncan, A. M., & Wangen, K. E., (2000). Soy isoflavones improve plasma lipids in normocholesterolemic, premenopausal women. *The American Journal of Clinical Nutrition, 71*, 1462–1469.
68. Moktan, B., Saha, J., & Sarkar, P. K., (2008). Antioxidant activities of soybean as affected by *Bacillus* fermentation to kinema. *Food Research International, 41*, 586–593.
69. Moon, G. S., & Cheigh, H. S., (1990). Separation and characteristics of antioxidative substances in fermented soybean sauce. *Korean Journal of Food Science and Technology, 22*, 461–465.

70. Moreno, M. R. F., Leisner, J. J., & Tee, L. K., (2002). Microbial analysis of Malaysian tempeh and characterization of two bacteriocins produced by isolates of *Enterococcus faecium*. *Journal of Applied Microbiology*, *92*, 147–157.
71. Mottram, D. S., Wedzicha, B. L., & Dodson, A. T., (2002). Food chemistry: Acrylamide is formed in mallard reaction. *Nature*, *419*, 448–449.
72. Nagahara, A., Benjamin, H., & Storkson, J., (1992). Inhibition of benzo [a] pyrene-induced mouse forestomach neoplasia by a principal flavor component of Japanese-style fermented soy sauce. *Cancer Research*, *52*, 1754–1756.
73. Nakahara, T., Sano, A., & Yamaguchi, H., (2010). Antihypertensive effect of peptide-enriched soy sauce-like seasoning and identification of its angiotensin I converting enzyme inhibitory substances. *Journal and Agricultural and Food Chemistry*, *58,* 821–827.
74. Namgung, H. J., Park, H. J., Cho, I. H., & Choi, H. K., (2010). Metabolite profiling of doenjang: Fermented soybean paste during fermentation. *Journal of the Science of Food and Agriculture*, *90*, 1926–1935.
75. Nordentoft, I., Jeppesen, P. B., Hong, J., Abudula, R., & Hermansen, K., (2008). Increased insulin sensitivity and changes in the expression profile of key insulin regulatory genes and beta cell transcription factors in diabetic KKAy-mice after feeding with a soybean protein rich diet high in isoflavone content. *Journal of Agricultural and Food Chemistry, 56*, 4377–4385.
76. Nunomura, M., Sasaki, M., Asao, Y., & Yokotsuka, T., (1976). Isolation and identification of 4-hydroxy-2 (or 5)-ethyl-5 (or 2)-methyl-3 (2H)-furanone, as a flavor component in shoyu. *Agricultural and Biological Chemistry*, *40*, 491–495.
77. Ohta, T., (1986). Natto. In: Reddy, N. R., Pierson, M. D., & Salunkhe, D. K., (eds.), *Legume-Based Fermented Foods* (pp. 85–93). Boca Raton, FL: CRC Press.
78. Omueti, O., (2000). Nutritional evaluation of home level prepared soy corn milk: A protein beverage. *Nutrition and Food Science*, *30*, 128–132.
79. Onda, T., Yanagidab, F., & Tsujia, M., (2003). Production and Purification of a bacteriocin peptide produced by *Lactococcus Sp*. Strain GM005, isolated from miso-paste. *International Journal of Food Microbiology*, *87*, 153–159.
80. Opara, C. C., Kuru, T., & Ezenwaka, I. B., (2012). Production of Soy yogurt from lactobacillus isolated from fermented African oil bean seed (Ugba). *Greener Journal of Agricultural Sciences*, *3*, 110–119.
81. Pabona, J. M., Dave, B., & Su, Y., (2013). The soybean peptide lunasin promotes apoptosis of mammary epithelial cells via induction of tumor suppressor PTEN: Similarities and distinct actions from soy isoflavone genistein. *Genes and Nutrition, 8*, 79–90.
82. Pak, V. V., Koo, M., Kwon, D. Y., & Yun, L., (2012). Design of a highly potent inhibitory peptide acting as a competitive inhibitor of HMG-COA reductase. *Amino Acids*, *43*, 2015–2025.
83. Park, I. Y., Cho, J. H., Kim, K. S., Kim, Y. B., Kim, M. S., & Kim, S. C., (2004). Helix stability confers salt resistance upon helical antimicrobial peptides. *The Journal of Biological Chemistry*, *279,* 13896–13901.
84. Park, J. H., Jeong, H. J., & Lumen, B. O., (2007). *In vitro* digestibility of the cancer-preventive soy peptides lunasin and BBI (Bowman-Birk protease inhibitor). *Journal of Agricultural and Food Chemistry, 55*, 10703–10706.
85. Park, J. S., Jeong, J. K., & Mo, Y. G., (2009). Impact of high dielectric on device performance of indium-gallium-zinc oxide transistors. *Applied Physics Letters*, *94* (4), 6, Article ID: 042105.

86. Park, K. Y., Jung, K. O., Rhcc, S. II., & Choi, Y. H., (2003). Antimutagenic effects of doenjang (Korean Fermented Soy paste) and its active compounds. *Mutation Research, 52*, 43–53.
87. Park, K. Y., Kong, K. R., Jung, K. O., & Rhee, S. H., (2001). Inhibitory effects of kochujang with different salt concentration. *Journal of Food Science and Nutrition, 6*, 187–191.
88. Patrzykat, A., & Douglas, S. E., (2005). Antimicrobial peptides: Cooperative approaches to protection. *Protein and Peptide Letters, 12,* 19–25.
89. Purwadaria, H. K., & Fardiaz, F., (2016). Tempe from traditional to modern practices. In: *Modernization of Traditional Food Processes and Products* (pp. 145–160). New York: Springer.
90. Rizzello, C. G., Lorusso, A., Russo, V., & Pinto, D., (2017). Improving the antioxidant properties of quinoa flour through fermentation with selected autochthonous lactic acid bacteria. *International Journal of Food Microbiology, 241*, 252–261.
91. Sachin, S., & Choubey, N. S., (2017). Literature review on soybean quality assessment and utility of neural network in seed classification. *International Journal of Current Research, 9* (5), 51160–51165.
92. Singh, B. P., Yadav, D., Vij, S., Mérillon, J. M., & Ramawat, K. G., (2019). Soybean bioactive molecules in current trend and future prospective. In: *Bioactive Molecules in Food* (pp. 267–294). Switzerland: Springer.
93. Seppo, L., Jauhiainen, T., & Poussa, T., (2003). Fermented milk high in bioactive peptides has a blood pressure-lowering effect in hypertensive subjects. *The American Journal of Clinical Nutrition, 77*, 326–330.
94. Shimakage, A., Shinbo, M., & Yamada, S., (2012). ACE inhibitory substances are derived from soy foods. *International Journal of Biological Macromolecules, 12,* 72–80.
95. Shin, H. W., Jang, E. S., & Moon, B. S., (2016). Anti-Obesity effects of gochujang products prepared using rice koji and soybean meju in rats. *Journal of Food Science and Technology, 53,* 1004–1013.
96. Shin, J. H., Kang, M. J., Yang, S. M., & Lee, S. J., (2010). Comparison of physicochemical properties and antioxidant activities of Korean traditional kanjang. *Journal of Agriculture and Life Science, 44*, 39–48.
97. Singh, B. P., Vij, S., & Hati, S., (2014). Functional significance of bioactive peptides derived from soybean. *Peptides, 54,* 171–179.
98. Smid, E. J., & Hugenholtz, J., (2010). Functional genomics for food fermentation processes. *The Annual Review of Food Science and Technology, 1,* 497–519.
99. Souza, L. C., Camargo, R., & Demasi, M., (2014). Effects of an anticarcinogenic Bowman-Birkprotease inhibitor on purified 20S proteasome and MCF-7 breast cancer cells. *PLoS One, 9,* 86600–86610.
100. Steinkraus, K., (2004). *Industrialization of Indigenous Fermented Foods Revised and Expanded* (pp. 99–112). Boca Raton, FL: CRC Press.
101. Subrota, H., Shilpa, V., & Brij, S., (2013). Antioxidative activity and polyphenol content in fermented soy milk supplemented with WPC-70 by probiotic lactobacilli. *International Food Research Journal, 20,* 2125–2131.
102. Sugimoto, S., Fujii, T., Morimiya, T., Johdo, O., & Nakamura, T., (2007). The fibrinolytic activity of a novel protease derived from a tempeh producing fungus, *Fusarium* sp. *Bioscience, Biotechnology, and Biochemistry, 71*, 2184–2189.
103. Tamang, J. P., (2015). Naturally fermented ethnic soybean foods of India. *Journal of Ethnic Foods, 2,* 8–17.

104. Tamang, J. P., (2003). Native microorganisms in fermentation of kinema. *Indian Journal of Microbiology, 43*, 127–130.
105. Tanaka, Y., Watanabe, J., & Mo, Y., (2012). Monitoring of the microbial communities involved in the soy sauce manufacturing process by PCR-denaturing gradient gel electrophoresis. *Food Microbiology, 31*, 100–106.
106. Taniguchi, A., & Yamanaka-Okumura, H., (2008). Natto and Viscous vegetables in a Japanese style meal suppress postprandial glucose and insulin responses. *Asia Pacific Journal of Clinical Nutrition, 17*, 663–668.
107. Teixeira, V., & Feio, M. J., (2012). Role of Lipids in the interaction of antimicrobial peptides with membranes. *Progress in Lipid Research, 51*, 149–177.
108. United States Department of Agriculture-Agricultural Research Service. http://www.fao.org/faostat/en/#search/soy%20bean (accessed on 4 August 2020).
109. Vasconcellos, F. C. S., & Woiciechowski, A. L., (2014). Antimicrobial and antioxidant properties of conglycinin and glycinin from soy protein isolate. *International Journal of Current Microbiology and Applied Sciences, 3*, 144–157.
110. Wah, T. T., Walaisri, S., Assavanig, A., Niamsiri, N., & Lertsiri, S., (2013). Culturing of *Pichia guilliermondii* enhanced volatile flavor compound formation by *Zygosaccharomyces Rouxii* in the model system of thai soy sauce fermentation. *International Journal of Food Microbiology, 160*, 282–289.
111. Wang, D., Wang, L. J., & Zhu, F. X., (2008). *In vitro* and *in vivo* Studies on the antioxidant activities of the aqueous extracts of douchi (a Traditional Chinese Salt-Fermented Soybean Food). *Food Chemistry, 107*, 1421–1428.
112. Wang, L. J., Yin, L. J., & Zou, L., (2007). Influences of processing and NaCl supplementation on isoflavone contents and composition during douchi manufacturing. *Food Chemistry, 101*, 1247–1253.
113. Wang, W., & De-Mejia, E. G., (2005). New frontier in soy bioactive peptides that may prevent age-related chronic diseases. *Comprehensive Reviews in Food Science and Food Safety, 4*, 63–78.
114. Wangen, K. E., & Duncan, A. M., (2001). Soy isoflavones improve plasma lipids in normocholesterolemic and mildly hypercholesterolemic postmenopausal women. *The American Journal of Clinical Nutrition, 73*, 225–231.
115. Yang, K., Han, Q., Chen, B., Zheng, Y., Zhang, K., Li, Q., & Wang, J., (2018). Antimicrobial hydrogels: Promising materials for medical application. *International Journal of Nanomedicine, 13*, 2217–2263.
116. Yang, H. J., Lee, Y. S., & Choi, I. S., (2018). Comparison of physicochemical properties and antioxidant activities of fermented soybean-based red pepper paste, gochujang, prepared with five different red pepper (*Capsicum annuum* L.) varieties. *Journal of Food Science and Technology, 55*, 792–801.
117. Yang, H. J., & Kwon, D. Y., (2012). Meju: Unsalted soybeans fermented with *Bacillus Subtilis* and *Aspergillus oryzae*, potentiates insulinotropic actions and improves hepatic insulin sensitivity in diabetic rats. *Nutrition and Metabolism, 9*, 1–12.
118. Yoshikawa, M., (2015). Bioactive peptides derived from natural proteins with respect to diversity of their receptors and physiological effects. *Peptides, 72*, 208–225.
119. Yoshikawa, M., Fujita, H., & Matoba, N., (2000). Bioactive peptides derived from food proteins preventing lifestyle-related diseases. *Biofactors, 12*, 143–146.
120. Yoshikawa, M., Takahashi, M., & Yang, S., (2003). Delta opioid peptides derived from plant proteins. *Current Pharmaceutical Design, 9*, 1325–1330.

121. Zaheer, K., (2017). Review of dietary isoflavones: Nutrition, processing, bioavailability, and impacts on human health. *Critical Reviews in Food Science and Nutrition, 57*, 1280–1293.
122. Zhang, J. H., & Tatsumi, E., (2006). Angiotensin I-converting enzyme inhibitory peptides in douchi: Chinese traditional fermented soybean product. *Food Chemistry*, *98*, 551–557.

CHAPTER 2

HONEY-BASED POLYPHENOLS: EXTRACTION, QUANTIFICATION, BIOAVAILABILITY, AND BIOLOGICAL ACTIVITIES

CSILLA BENEDEK, JOHN-LEWIS ZINIA ZAUKUU, ZSANETT BODOR, and ZOLTAN KOVACS

ABSTRACT

This chapter focuses on health-promoting phytochemicals in honey, with special emphasis on polyphenols. Polyphenol-rich foods can shield consumers from risks of several diseases, related to oxidative stress through diverse mechanisms (direct and indirect). Honey has been proven to have a variety of phytochemicals that have the potential to modulate physiological and biochemical routes and to redirect pathological processes. Polyphenols and their subclasses (especially flavonoids, phenolic acids, and their derivatives) are the most powerful nutraceuticals in honey. To explore potential benefits, one must have a clear overview of the occurrence, bioavailability, and health benefits of these phytochemicals in different honey types worldwide. Several traditional and environmentally friendly extraction methods (e.g., supercritical fluid extraction, accelerated solvent extraction) have been adapted for the extraction of polyphenols from honey. Analytical techniques (such as: high-performance liquid chromatography and capillary electrophoresis with diode array detection, ultra-high-performance liquid chromatography with mass spectrometry detection, liquid chromatography coupled with tandem mass spectrometry) are used to quantify the polyphenols from honey. Aspects of bioavailability of the polyphenols present in honey are also outlined, along with its main health-related properties.

2.1 INTRODUCTION

The honey is a complex food matrix and an outstanding natural product that is consumed mainly in its native form. It has wide range of phenolic compounds. Polyphenols are secondary plant metabolites that are characterized by multiple complex organic structures containing phenolic rings. These naturally occurring products exhibit different functions, depending on their structure. They can mainly be grouped into flavonoids and non-flavonoid polyphenols, which include phenolic acids, stilbenes, coumarins, and lignans. Flavonoids are water-soluble compounds having low molecular weight with at least two phenolic groups (OH) [12].

These are mostly present in nature as the aglycone part of different glycosides, formed with various sugars. Phenolic acids are derivatives of either hydroxybenzoic or hydroxycinnamic acids, constituting a significant portion of the total phenolic compounds [10]. The potential health benefits and natural existence of polyphenols in honey offer opportunities to many researches communities sand open a new arena in diet and nutrition, pharmaceuticals, and medicine. In botany, these compounds are mostly present as derivatives of glycosides that can be transformed into aglycone forms in honey in the presence of glucosidase (known as bee enzyme) [5]. Figure 2.1 shows some common examples of phenolic acids and flavonoids, identified in diverse types of honeys.

Polyphenols are regarded as important markers in honey because their presence in honey is dependent on many factors, such as: botanical origin, enzymatic activity from different bee species, and environmental conditions. The complexity of honey as a matrix, where the target analytes (i.e., polyphenols), are present in rather low concentrations (46–753 $\mu g\ g^{-1}$), demands multi-step analytical procedures to provide an accurate determination. A well-accepted protocol for the entire analysis process in order to gain reliable analytical results is getting importance nowadays than before. As health impacts of polyphenols are strongly linked to their occurrence and amount in food, therefore it is important to provide an inclusive knowledge on the detection and quantification of polyphenols in honey. Multi-step analytical procedures include extraction of polyphenols from honey for its quantification through advanced high-throughput equipment or device. However, its wide therapeutic application is scanty due to a lack of technologically-needed information.

This chapter provides current information on extraction, separation techniques, and detection of polyphenols in honey. The bioavailability of

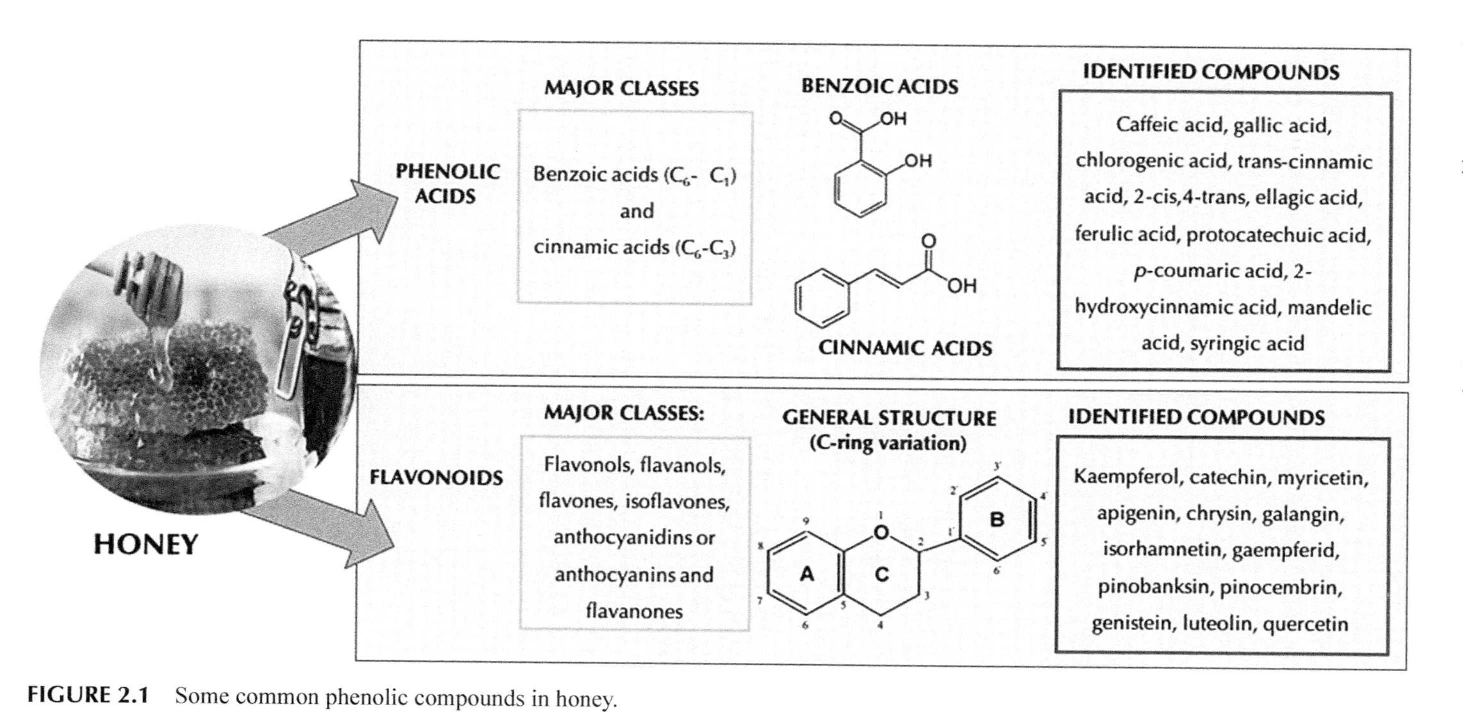

FIGURE 2.1 Some common phenolic compounds in honey.

Source: Self-developed: concepts were adapted from Refs. Al-Farsi et al. [1], Boussaid et al. [11], Biluca et al. [18], Cianciosi et al. [19], Dua et al. [27], Kavanagh et al. [49], and Ranneh et al. [74].

polyphenols is an important factor for the nutraceutical properties of honey. Therefore, information on the bioavailability of honey-based polyphenols has been represented in a judicious way. As a functional food, the health impacts and the preventive role of honey against several diseases are also outlined.

2.2 TECHNIQUES FOR EXTRACTION OF POLYPHENOLS

According to most published procedures, phenolic compounds in honey are isolated through an extraction and clean-up phase, before samples can be submitted for the separation and the identification steps. Extraction is the main sample preparation procedure necessary for identifying and quantifying polyphenols in monofloral honeys. The objective is to obtain an elevated level of the target compound and to eliminate any potential disturbing components (such as: mainly sugars and other polar compounds). Extraction and separation of phenolic compounds under different classes are crucial for achieving correct identification and quantification [23].

The choice of an extraction technique depends on the structure of the target molecules, particle size in the sample, and the presence of interfering substances. The yield of extraction is affected by temperature, solvent-to-feed ratio, number of extraction cycles, and choice of the extraction solvents (hydrophilic or hydrophobic nature). Although traditional methods (conventional liquid-liquid and solid-liquid extractions) have been successfully practiced for extraction of polyphenols from honey, yet presently they are being replaced by advanced technologies [supercritical fluid extraction (SFE), accelerated solvent extraction (ASE), or even multi-walled carbon nanotubes (MWCNTs)] [15]. In this section, different technologies for the extraction of polyphenols from honey are discussed.

2.2.1 *SOLID-LIQUID EXTRACTION*

In recent decades, the styrene-di-vinylbenzene polymeric resin known as Amberlite XAD-2 has received popularity as an adsorbent to extract phenolic compounds from honey. Samples are typically dissolved in hydrochloric acid solution (pH = 2), filtered, and subsequently passed through the resin column. Alternatively, the filtrate can be mixed with Amberlite and stirred for a specified time for better absorption prior to transferring the mixture into the column [54]. Elution is performed with aqueous hydrochloric acid (pH = 2) and then continued with water. This allows separation of less polar phenolic fractions (bound on the adsorbent of the column) from the other polar compounds, like sugars. After elution of the target molecules from the column with methanol, the extracts are evaporated under reduced pressure and then re-dissolved in either water or methanol [16].

The clean-up of the extracts is performed by an additional extraction with an appropriate organic solvent (ethyl acetate [46] or diethyl ether [66]). Finally, the obtained organic extract is evaporated and re-dissolved in methanol prior to HPLC. It has been shown that the phenolic extracts of honey using Amberlite XAD-2 had a decreased amount of phenolic target compounds. The antioxidant activity was also higher in the original honeys than in their fractionated phenolic extracts (average recovery of phenolic compounds was 20 ± 7%, whereas the average flavonoid recovery was 23 ± 5%, depending upon the method used). Compounds like kaempferol, syringic acid, and p-coumaric acid were completely adsorbed, while the recovery of phenolic acids (gallic acid and caffeic acid) and quercetin by methanolic elution was much less efficient (54% for quercetin, 80% for caffeic acid, while gallic acid was completely absent in the methanolic fraction) [16]. The extraction efficiency of polyphenols from honey is an important reason for the qualitative and quantitative differences that have been reported in several studies. Furthermore, the separated extracts are subjected to PVPP pre-treatment; and an important effect on non-phenolic reducing compounds was observed when the total content of polyphenols was measured by the Folin-Ciocalteu assay. In practice, this means aco-elution of the non-phenolic compounds with phenolic compounds, when methanolic elution is applied on Amberlite XAD-2 [16].

Some researchers have simplified this phase by implementing SPE method, in which polyphenols are bound by hydrophobic interactivity with a solid sorbent. Thus, the combination of the extraction and clean-up phases results in higher analyte recovery and substantial savings in time and solvent. Similarly, for extraction with Amberlite XAD-2, the sample is usually solubilized in acidified water. Before use, the SPE cartridge (e.g., C18, Strata-XSPE) should be washed and activated using an appropriate solvent mixture that is dependent on the type of the sorbent, e.g., methanol, and water [18].

During elution, polar interfering substances are removed first, followed by elution of target analytes, which is usually achieved with methanol [23]. Dimitrova et al. proposed a two-stage sample conditioning, adjusted to the acid character of the sample: the C18 SPE column is conditioned first with methanol containing sodium hydroxide, then with the methanol containing hydrochloric acid. Finally, elution is performed using acetonitrile:

tetrahydrofuran (1:1); and subsequently, reversed-phase HPLC is applied for separation [26].

The performance of different sorbents (Bond Elut C18, Strata-X, Oasis HLB, and Amberlite XAD-2) was compared during the extraction and pre-concentration of some phenolic acids and flavonoids from honey samples. Best results were accomplished when Oasis HLB was used as the sorbent, washed with acidified water (pH = 2), and then eluted with methanol [62]. An entirely new adsorbent, mesoporous silica-coated with nano-Al_2O_3 (Al_2O_3/SiO_2), was proposed for the isolation of flavonoids. Its performance was tested on myricetin, quercetin, luteolin, and kaempferol; and the efficiency of the extraction was better than for a commercial SPE cartridge [35].

The performance of the extraction of nine known polyphenols (including phenolic acids and flavonoids) was compared by Istasse et al. [46]. Solid-phase extraction (SPE) was adapted for isolating polyphenols from an artificial honey solution. They compared two extraction methods using Amberlite XAD-2 polymer and a C18 cartridge. The C18 cartridges showed an average recovery of 74.2%, while a much lower efficiency (43.7%) was obtained with Amberlite XAD-2. Recoveries were excellent (>90%) for naringin, vanillic acid, and rutin; though Gallic acid was lost using C18 cartridge. It seems that Gallic acid has a stronger affinity for Amberlite XAD-2, which allows not only its adsorption to the resin, but also its consequent elution in the last phase of the extraction.

Therefore using Amberlite XAD-2, all nine polyphenols under study were successfully extracted. However, the recycling experiments of Amberlite XAD-2 showed that the average polyphenol recovery dropped from 43.7% (in the first extraction) to 29.3% (in the third extraction cycle). As a conclusion, although extraction using Amberlite XAD-2 resulted in more extracted compounds, yet this may be compensated by a better extraction recovery on the C18 cartridge. On the other hand, the high cost of Amberlite XAD-2 resin is not counterweighed by its multiple uses due to its continuous decay and loosing extraction capacity [84].

Recovery problems for Gallic acid were also reported using Amberlite XAD-2. This may be due to the fact that Gallic acid is too polar and remain in resin-type adsorbent after the initial washing-step with water. The validation experiments were performed with artificial honey. Polyphenol recoveries showed variable performances ranging from 102.15% ± 10.49% for kaempferol to 54.65% ± 14.04% for p-coumaric acid. The loss of some polyphenols might be due to their stronger interaction with the adsorbent. It was concluded that none of the commonly known adsorbents are able to extract all phenolic compounds with different polarities.

Although there are resins (like Oasis HLB) that are more efficient for polar compounds (gallic acid can be retained), yet their performance is generally lower for less polar substances (like quercetin or kaempferol). Istasse et al. [46] confirmed the formation of *cis* isomers from the *trans* polyphenol standards. The *cis* isomers were identified by comparing their spectra with the spectra of the standard solutions in HPLC-DAD and LC-UV-MS. The authors concluded that *cis* isomers can be formed even in the absence of light, even though the light was previously considered to be essential for this transformation. They confirmed that the causes are not yet clear, as isomerization can be induced by simple exposure to ambient temperature in methanolic solution or an unknown reaction with the adsorbent [43].

MS were used as sorbents for phenolics by a Saudi Arabian research group by adding MWCNTs to an acidified honey solution. The mixture was stirred to facilitate adsorption, then the sorbent containing target compounds were removed by vacuum filtration. First, it was washed with water, then it was treated with methanol to release the adsorbed phenolic compounds. After evaporation of the entire amount of methanol, the solid residue was cleaned-up by re-dissolving in water and extraction with diethyl ether. The ether extract was reduced to dryness by evaporation, then the residue was re-dissolved in methanol and subsequently subjected to HPLC measurements.

The HPLC method has the advantage of simultaneous extraction of a large number of phenolic compounds having various structures (phenolic acids, flavonoids, and derivatives of these) with good yields. Additionally, the MWCNTs were efficiently regenerated [46].

2.2.2 LIQUID-LIQUID EXTRACTION

Liquid-liquid extraction (LLE) has been suggested for extraction of honey polyphenols. First honey is dissolved in water and then LLE extraction is performed with an organic solvent, typically ethyl acetate [22, 52].

LLE was adapted for sixty Turkish honey samples. Initially, honey was dissolved in ultrapure water, then polyphenols were extracted from the water phase with ethyl acetate (three times shaking for 30 min). Ethyl acetate was completely evaporated and the residue was dissolved in methanol and was filtered prior to quantification. Other solvents, such as chloroform and dichloromethane showed lower performances during extraction process [51].

Extraction of polyphenols from Chinese honey was performed with mild acidic water, followed by centrifugation and concentration by evaporation.

Four different types of polyphenols were extracted from different types of honeys [80].

Four main polyphenols (such as: ferulic acid, kaempferol, apiolin, and luteolin)were extracted with n-butanol from Chinese honeys. Samples were simply dissolved in water, and then followed by extraction with n-butanol [86].

A special LLE, called dispersive liquid-liquid microextraction (DLLME), was employed for the extraction of phenolic substances, such as flavonoid aglycones from honey. Recovery of polyphenols was generally higher than 70% [23].

2.2.3 OTHER EXTRACTION METHODS

Conventional extraction procedures usually require high processing temperature and long working time. Therefore, alternative extraction techniques are being sought continuously for recovering phenolic compounds from honey. Among these, extraction, assisted by either ultrasound or microwave or both of them were reported for recovering phenolic compounds from honey [87]. Furthermore, environmentally benign processes (such as: SFE and ASE) were used for this purpose [13, 15].

Ultrasound-assisted LLE has been used to solubilize honey samples. Honey samples are dissolved in water or in the HPLC eluent; and sonication is employed for 10 min. The procedure results in excellent recoveries (up to 98% for all analytes) [23].

The stability of phenolic compounds was checked by Biesaga and Pyrzynska during ultrasonic- or microwave-assisted extraction (USE or MAE). The authors confirmed that the extraction assisted by ultrasound delivers generally higher yields than the conventional LLE. It has been proven that phenolic acids and glycosides are highly stable during MAE and USE treatments. Aglycones of flavonols (e.g., quercetin) are unstable under such conditions. Therefore, a low amount of yields are obtained for extraction of Aglycones of flavonols from honey with MAE and USE treatments [13].

ASE has been used to extract the polyphenols from honey. Honey samples are dissolved in acidified water (HCl is used for acidification at pH = 2) at room temperature; and ASE is employed with four different static cycles. Phenolics compounds are eluted with methanol and subsequently, solvent is removed. Residue is re-suspended in distilled water and cleaned-up by extracting with diethyl-ether for three times. Subsequently, ether is allowed to evaporate from the extract and to dissolve in a methanol/water mixture [23].

SFE may be a good choice for extraction of polyphenols from natural matrixes, because it is eco-friendly and mitigates solvent wastage. It helps to prevent the oxidation and/or decomposition of polyphenols. In this case, supercritical CO_2 acts as a solvent. Several other supercritical fluids have been recently proposed for extraction, e.g., propane, argon (Ar), and SF_6 [15], which may offer promising alternatives for extraction of polyphenol compounds from honey.

2.2.4 HYDROLYSIS

To extract aglycones from honey, hydrolysis is performed in most cases with hydrochloric or formic acid at relatively high temperatures (80–100°C). Recovery of polyphenols is a function of acid concentration, duration of the hydrolysis, and operating temperature. Generally, the objective of hydrolysis is to accomplish a complete release of free aglycones and to keep polyphenol degradation to a minimum level [23].

Unlike most food matrixes, honey contains mainly aglycones instead of their glycosides, because honey bees hydrolyze the glycosidic bonds during the elaboration/ ripening process [7, 86]. Rutin (quercetin-3-rutinoside) and hesperidin (hesperetin 7-rhamnoglucoside) are the most frequently reported glycosides that have been detected in honey [71].

2.3 TECHNIQUES FOR QUANTIFICATION OF POLYPHENOLS

Identification and quantification of individual polyphenol are usually performed by chromatographic or electrophoretic techniques. The choice of the analytical method and the working conditions strongly depend on the objective of the analysis (qualitative and/or quantitative). Costs, analysis time, and selectivity are also critical issues in this context.

2.3.1 CHROMATOGRAPHIC TECHNIQUES

Generally, separation of phenolic acids and flavonoids is performed by liquid chromatography (LC), using C18 or RP-C18 column [35]. Pyrzynska and his co-workers provide detailed information on methods for the identification of different polyphenols. Gradient elution is prevalently used. Binary solvent systems, including a water-based component combined with a less polar organic solvent (e.g., acetonitrile or methanol), have been used in several cases. To ensure a suitable pH level during gradient runs, several organic acids (formic acid, acetic acid, phosphoric acid) are usually added [84]. Isocratic elution has been reported for the measurement of phenolic acids (homogentisic acid) from strawberry-tree honey [71].

Zhang et al. used a unique multivariate HPLC-DAD second-order calibration method (based on a trilinear decomposition algorithm) for the quantification of nine polyphenols. Their calibration technique uses mathematical models for the overlapped chromatographic peaks, obtaining thus the neat profile for each analyte. Short processing and evaluation period, linearity, and excellent recoveries (90%–110%) are the main advantages of this method [87].

Although several research groups still prefer to use the HPLC-UV detection, yet the visible trend for polyphenol analyses is to use ultra-high-performance liquid chromatography (UHPLC), which has several advantages, such as: high resolution, sensitivity, and speed of analysis. Thirty-two phenolic compounds in 60 Turkish honeys have been identified by UHPLC with electrospray ionization (ESI) coupled to tandem mass spectrometry (UHPLC-ESI-MS/MS). The methods provide fast and reliable results using only LLE as a preparatory step. The Serbian group used successfully UHPLC coupled to ion trap mass analyzer (UHPLC-LTQ), equipped with a heated ESI probe [52]. GC has also been applied for the measurement of phenolics (mainly phenolic acids), however, this technique has its drawbacks. The derivatization step in GC mainly produces methylated or trimethylsilyl derivatives of the non-volatile compounds [61].

Recently, high-performance thin-layer chromatography (HPTLC) is a fingerprinting method to detect phenolic compounds. It is a suitable and authenticated to detect monofloral honeys. HPTLC may be considered as a convenient method for the clarification of the botanical origin of honey samples based on their polyphenol profile. It has several advantages, e.g., low instrumentation costs, short running time, a wide choice of adsorbents and eluents, a small sample size for analysis, good precision and accuracy [55, 76].

The diode-array detectors (DAD) or similar photodiode array detectors are still popular [54, 84]. Usually, wavelengths of 280 nm or 290 nm are used (mainly for phenolic acids), whereas 340 nm is used for the detection of flavonoids [66]. However, the employment of conventional analytical methods based on UV spectra is often restricted for the presence of similar compounds in the samples.

This is the main reason of the increasing popularity of a method, which is based on mass spectrometry (MS or multiple MS^n), because they accomplish significantly higher sensitivity and can deliver information on molecular mass and structural features. They are used in combination with UV detection or as self-standing detectors, without any other simultaneously used detectors [43].

However, various MS ionization techniques (e.g., electrospray ionization (ESI), atmospheric pressure chemical ionization (APCI), fast atom bombardment (FAB), and thermo-spray (TSP)) can be used for the analysis of flavonoids [71]; and the ESI has been used for detection of polyphenols in honey [43, 46, 52].

2.3.2 ELECTROPHORETIC TECHNIQUES

Capillary electrophoresis (CE) has received popularity as an alternative to LC in the analysis of various foodstuffs. The main advantages of CE over LC are shorter running time, low sample size, and low cost due to the fact that organic solvent is not needed for elution. The technique is applied for the estimation of polyphenols from both honeys and propolis. The predominant CE modes in use are capillary-zone electrophoresis (CZE) [80] and micellar electro-kinetic chromatography (MEKC) [86].

The separation selectivity of MEKC is superior compared with CZE, because electrophoretic behavior in CZE is affected by factors, such as: stereochemistry of the C-ring in flavonoids and number, position, and substitution of the OH groups attached to the rings. The method of validation with MEKC is also in use. A hybrid method, called capillary electrochromatography (CEC), has also been developed by combining CZE and HPLC. However, the considerable time, labor, and cost demands of this method prevent its wider use [71].

2.4 BIOAVAILABILITY OF HONEY POLYPHENOLS

Bioavailability generally refers to the accessible portion of a nutrient or non-nutrient component for maintaining good health, specifically their adsorption, controlling, and modulating physiological systems and storage. According to our present knowledge, the bioavailability of phenolic compounds includes the following main digestive routes [17]:

- Elimination through the kidney or re-excretion into the gut via bile and pancreatic juices;

- Microbiological fermentation of phenolics that were not absorbed or were re-eliminated via bile or the pancreas, form extra metabolite products;
- Modifications of polyphenols during digestion in gastric/small-intestine;
- Primary/secondary phase enzyme changes of the polyphenols in the gut;
- Release of phenolic compounds from the food matrix (e.g., honey);
- Transport in the bloodstream and successive relocation in tissues;
- Up-take of polyphenols by enterocytes, either in their free aglycone forms or in some conjugated forms.

In case of flavonoids, bioavailability is determined by series of factors, such as: food matrix, accompanying drinks and actual contents in the stomach; and individually variable factors like intestinal peristalsis, pH of the gastrointestinal tract, specificity of carrier (transporter protein), blood, and lymph flow [69]. As regards to the first category, dietary fibers (e.g., hemicellulose), divalent metal ions, viscous, and protein-rich food matrixes are known to hinder polyphenol bioaccessibility, while digestible carbohydrates, lipids, and other antioxidants may enhance the bioavailability of phenolics [17].

If the flavonoid is present as a glycoside, then its structure plays an additional role for determining bioavailability, depending on the type of the glycosylic sugar and the type of the glycosidic bond, the glycosylation sites and the number of sugar backbones linked to the flavonoid. The type of flavonoid compound itself has a significant influence on absorption in the gastrointestinal tract. In a clinical trial, it has been proven that a single oral dose of 50 mg of flavonoid, isoflavones were best absorbed, while flavan-3-ols, flavanones, and flavonols had much lower absorption efficiencies. Nevertheless, differences in the corresponding molar amounts and possible metabolites were not considered in this study. The administration mode of polyphenols is also a relevant factor. Intravenous administration of polyphenols is fully absorbed in the gastrointestinal tract [69].

Upon uptake in the gastrointestinal epithelium, flavonoids can moderate phase II metabolism and excretion, thus improving the bioavailability of polyphenols. Moreover, as a result of their impacts on efflux transporters (i.e., p-glycoprotein), phenolics may act in a synergistic way [17]. It is well known that flavonoid glycosides are not readily absorbed, unless they are first hydrolyzed. This is performed either by bacterial enzymes in the intestine or catalyzed by two β-endo-glucosidase enzymes capable of hydrolyzing flavonoid glycosides. Lactase phlorizin hydrolase (LPH) acts in the brush border of the epithelial cells of the small intestine and cytosolic β-glucosidase (CBG) takes part in an alternative hydrolytic step within the epithelial cells [5]. Microbiota seems to have a key role in flavonoid absorption in the gastrointestinal tract [81].

Compared to their native glycosidic forms, the phenolic aglycones formed are more easily absorbed through the gut barrier by passive diffusion. In an animal experiment, it was found that concentration of aglycone in plasma is more compared to glycoside, which indicates that absorption of the glycoside is higher in distal parts of the intestine, probably in the colon. It has been also proposed that after the release of glycosides from aglycone, about 15% of the flavonoid aglycones are absorbed into the epithelial cells with bile micelles, being then transported by the lymph. While vast studies are available on the bioavailability of flavonoids in different food matrixes, yet only limited information was found on specific studies dealing with honey [5].

Several studies have been published on *in vivo* animal and human metabolism of classes of flavanones (such as: hesperetin), naringenin, and their glycosides using MS for identification of metabolites during biotransformation. Glucuronidation seems to be principal biotransformation of hesperetin and major urinary metabolites are sulfo glucuronides. Identified metabolites were glucuronide and sulfate conjugates, and the phase II conjugation of aglycones taking place in the intestine [81]. Flavones and flavonols are also exposed to the action of phase-II enzymes, forming in the same way glucuronide, sulfate, and/or methylated metabolites [69].

Absorption of polyphenols into the systemic circulation is restricted due to an efflux, which promotes and returns these conjugates into the intestinal lumen. Practically, no free flavanone aglycones were detected post-digestion. However, this may be dose-dependent. Flavanone conjugates that are not absorbed in the intestine are further transported to the colon. Here a microbial metabolism converts to phenolic acids, such as: propionic, hydroxyl-phenylacetic, hydroxycinnamic, and hydroxybenzoic acid derivatives. Information on the distribution of polyphenols in tissues is scanty; and no data are available for humans. After absorption, flavanones circulate in the bloodstream, where they are bound with proteins and delivered rapidly to organs, such as: brain, liver, heart, lung, kidney, and spleen.

In animal experiments, it was found that phenolic aglycones were the predominant form in this stage of distribution in organs. They were detected in the liver and heart. Hesperetin and naringenin were able to cross the blood-brain barrier (BBB), and were detected in many brain regions [81]. Quercetin was also accumulated in low concentrations in the brain, although the mechanism of passing the BBB is not yet understood [69]. Absorption of hydroxycinnamic acids in their free form takes place through the entire gastrointestinal tract, where the jejunum is apparently the main site of their absorption. It is mainly a transcellular mechanism to facilitate the permeation of free hydroxycinnamic acids across the intestinal epithelium. Their metabolism occurs mainly in the liver, but also in the intestine and the colon, thus forming glucuronic, sulfated, methylated, hydrogenated/dehydrogenated conjugates.

Caffeic acid has been more intensively metabolized than ferulic acid (its methylated derivative). However, caffeic acid is less absorbed and transported than ferulic acid. Metabolites of chlorogenic acid are eliminated by urinary excretion @ 29% of the total intake of chlorogenic acid, and only 0.29–4.9% of the original amount is excreted in its intact form, accompanied by the lower amount of phenolic acids (phenyl-acetic and hydroxybenzoic acids) [77].

The research study on bioavailability of polyphenols in honey indicates that after an intake of two honey types in amounts of 1.5 g/kg body mass, the total-phenolic content in plasma of the 40 subjects under study was increased ($p < 0.05$). At the same time, plasma antioxidant and reducing capacities were increased correspondingly ($p < 0.05$).

These results supported the theory that honey polyphenols are readily bio-available and these boosts plasmatic antioxidant capacity, which improves the defense against oxidative stress. However, the honey in this experiment delivered mg amounts of p-hydroxybenzoic and p-hydroxycinnamic acids per kg of body weight; and the subsequent HPLC analysis did not confirm their concentration in plasma. This may be due to [5]: (i) less than one-third of these compounds were absorbed; (ii) these compounds may have been distributed quickly into parts of the body other than plasma; and (iii) presystemic metabolism of the monophenols took place in the human organism.

Methyl syringate 4-O-β-D-gentiobioside (leptosperin) and its aglycone in Manuka honey were used to determine their biological potential. The absorption and metabolism of the phenolic acid and its glycosides were monitored in humans and animals. In humans fed with Manuka honey, methyl syringate was detected in both plasma and urine in its glucuronide, sulfate, and the aglycone forms. The levels of these substances in plasma reached a maximum value within 0.5–1 h after intake; and then most metabolites disappeared within 3 h. Simultaneously, a significant quantity of metabolites and traces of leptosperin were found in the urine excreted within 4 h. To clarify the metabolic pathways of leptosperin and methyl syringate, the compounds were administered separately to mice. In each case, the presence of glucuronide, sulfate, and the aglycone was confirmed in both plasma and urine [42].

The P-gp (permeability glycoprotein) is an important protein of the cell membrane that eliminates many foreign substances out of cells through the ATP-dependent efflux pump, member of the ATP-binding cassette transporters (ABC transporters) family [5]. It is also known as multidrug resistance protein 1 (MRP1), being responsible for both drug resistance and decreased bioavailability of different medicines during cancer chemotherapy, by lowering their intracellular concentrations. On the other hand, both MRPs and breast cancer resistance protein (BCRP) are involved in the cellular uptake and transport of conjugated flavonoid metabolites [69].

Due to the high concentration of bio-available flavonoids in honey, it can be an efficient inhibitor of the P-gp. Flavonoids (such as: kaempferol, genistein, chrysin, biochanin, quercetin, and naringenin) in most of the honeys are able to interact with P-gp transporters. Thus, flavonoids are bi-functional in overturning the multi-drug resistance (MDR) through their transport mechanism and their interaction with P-gp transporters. As a consequence, an enhancement of the therapeutic index can occur, so that honey may be a potential nutraceutical for patients suffering from cancer [44].

2.5 BIOLOGICAL ACTIVITIES OF HONEY-BASED POLYPHENOLS

Health and diet have always a symbiotic relationship that partly continues to be unraveled: diet has a direct impact on health. Aristotle noted that pale honey has "*beneficial for sore eyes and skin lesions*" [58]. Many similar hypotheses represent the roots of apitherapy, a modern-day alternative liaison of medicine whose primary interest is the potencies of honey including diverse bee products for treatment of diseases and illnesses.

In spite of high interest in honey, its composition is nonetheless still difficult to define (depends on several factors), however phytochemicals (such as: enzymes, vitamins, and polyphenols) in honey play vital roles in our health. Figure 2.2 gives a general overview of the therapeutic benefits of honey.

2.5.1 *ANTIMICROBIAL PROPERTIES OF HONEY*

It is not only prudent, but essential to channel significant research efforts into natural remedies to tackle microbial infections. A plethora of reports have ascertained and detailed the antimicrobial activity (*in vitro*) of honey from diverse sources. It has been reported that honey from *Leptospermum scoparium* offers inhibitory activity against almost 60 different bacterial species, such as: anaerobic-aerobic, gram-negative, and gram-positive [57].

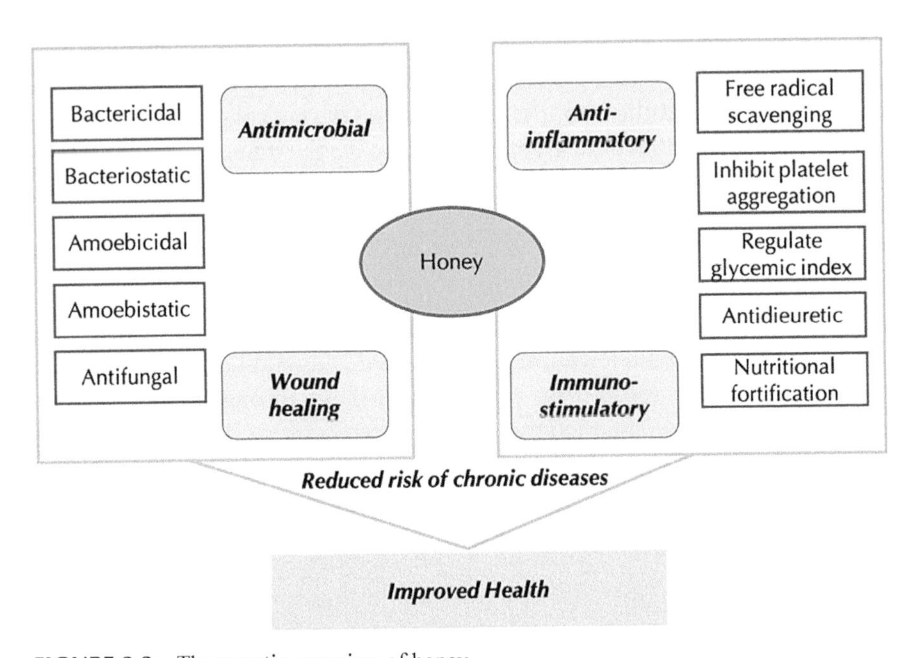

FIGURE 2.2 Therapeutic overview of honey.
Source: Self-developed: Concepts were adapted from Refs. [8, 26, 30, 45, 58, 61, 64, 80].

Manuka honey has an antibacterial effect against *Pseudomonas aeruginosa* (Imipenem resistant and sensitive), *Staphylococcus aureus* (methicillin-resistant and sensitive), *Escherichia coli* (penicillin-resistant and sensitive) and *Streptococcus pyogenes* (penicillin-resistant and sensitive), while honey from *Nigella sativa* in Kingdom of Saudi Arabia only offers antibacterial activity against *Pseudomonas aeruginosa* [2].

The antibiotic properties of honey were further confirmed when *S. aureus, E. coli,* and *P. Aeruginosa* were isolated during the treatment of diabetic foot ulcers with honey, which proved effective [3]. *P. aeruginosa* is a major causative organism for nosocomial illness including bacteremia, urinary tract infections (UTIs), and pneumonia; whereas *S. pyogenes* is the leading causative organism for skin diseases, such as: necrotizing fasciitis, impetigo, and cellulitis. In addition to respiratory problems, *S. aureus* can cause an array of disease-causing proteins, not limited to epidermolytic toxins, catalase, hemolysins (alpha, beta, gamma), and enterotoxins [59].

Antimicrobial property of honey, extracted from *G. thoracica* and *H. itama*, has also been reported against *Staphylococcus xylosus* (gram-positive bacteria), *Pseudomonas aeruginosa* (gram-negative bacteria) and *Vibrio parahaemolyticus* (gram-negative bacteria) in Malaysia [79]. Amoebistatic and amoebicidal potencies were exhibited in a dose-dependent manner when honey was used to treat acanthopodia and resulted in a detached and shrank amoebae (*Acanthamoeba castellanii)* [75]. In *the in-vitro* experiment, antimicrobial activity of honey was confirmed against *H. pylori* isolates, a major pathogen for gastritis [78]. Table 2.1 shows some phenolic compounds and their inhibitory effects on certain microbes [25].

TABLE 2.1 Antimicrobial Activity of Some Phenolic Compounds [25, 79, 26]

Phenolic Compound	Antimicrobial Activity	Susceptible Microbes
Flavan-3-ol	Antibacterial	*B. cereus; C. pneumonia. Campylobacter jejuni; Clostridium perfringens; Escherichia coli; F. nucleatum; H. pylori; L. acidophilus, A. naeslundii; P. gingivalis; P. melaninogenica; P. oralis; S. aureus; Streptococcus mutans; Vibrio cholera.*
	Antiviral	Adenovirus; Enterovirus; Flu virus.
	Antifungal	*Candida albicans; Microsporum gypseum; Trichophyton mentagrophytes; Trichophyton rubrum.*
Flavonol	Antibacterial	*E. coli; S. aureus. S. mutans.*
	Antiviral	Herpes simplex virus (HSV) type 1
Condensed tannin	Antibacterial	*Bacillus clostridium; Campylobacter lysteria;* Different *Salmonella strains; Escherichia coli; Staphylococcus helicobacter.*
	Antiviral	*Epstein-Barr virus Herpes virus, HSV-1 and HSV-2.*
	Antifungal	*Candida parapsilosis.*
Hydrolyzable tannins	Antibacterial	*Escherichia coli; L. monocytogenes; P. aeruginosa; Staphylococcus aureus.*
Neolignan	Antibacterial	Different strains of Mycobacterium tuberculosis.

The antimicrobial properties of honey have been largely attributed to the formation of hydrogen peroxide by the bee-derived enzyme glucose oxidase [58], however other factors are suspected to contribute synergistically. Glucose oxidase enzyme activity has been observed to increase with mild thermal treatment [14] and may explain why ancient Chinese traditional medicines and some plant-based drugs were sometimes roasted with honey to improve health treatments [68]. However, heating of honey is a seemingly double-bind and controversial topic, as it should always be done with caution to limit the formation of toxicants. Elevated concentration of sugars in honey also eliminates microbes, especially bacteria that are sensitive to high osmotic pressure and stop the proliferation of more osmotolerant microorganisms.

It has been reported that the antioxidant mechanism, associated with honey, contributes to the healing of sores [31] by limiting ROS induction and creating a condition favorable to cell growth by enhancing hypoxia-inducible factor 1 (HIF-1) marker. Flavonols, tannins, and flavan-3-ols are the main recognized polyphenols in antimicrobial studies due to their wide spectrum and their ability to suppress number of microbial virulence factors, such as: prevention of biofilm, counterbalance of harmful bacterial secretions and a decrease in the clinging of ligands [25, 45]. These studies portray that honey may be a therapeutic agent for some skin diseases, respiratory-, and gastrointestinal-infections. In microorganisms, antibiotic-resistance potentiality offered by honey is still not reported [78].

2.5.2 IMMUNO-STIMULATORY EFFECT OF HONEY

A weak immune system is one of the major reasons behind many deadly infections and illnesses. Biofortification is an alternative way of improving the immune response. Daily doses of 40 g and 60 g of honey can improve both physical- and psychological quality of life (QOL) and CD4 levels in human immunodeficiency virus (HIV) (asymptomatic) patients, who are not on highly active antiretroviral therapy (HAART) [83]. A good QOL mediates disease progression to AIDS. Asymptomatic HIV subjects, who were administered Tualang honey, showed a decrease in viral load, thus underlining the potential of honey in boosting the immune system in HIV patients.

Heated honey can be harmful due to the release of 5-hydroxymethylfurfural (5-HMF), a toxic compound produced through the Maillard reaction [6]. However, some research reports have suggested potential benefits of this practice. Heated honey is believed to increase the immune-stimulatory effect [68] for the management of various types of diseases. Granulocyte colony-stimulating factor (G-CSF) is the glycoprotein that is responsible for stimulating bone marrow to produce and release stem cells into the bloodstream. Its secretion from enterocytes is a combination of temperature and time-dependent processes that can be induced by heating the honey [68]. Melanoidins from temperature treated honey have exhibited peroxyl radical-scavenging activity [50].

Manuka honey contains a polymyxin B-insensitive molecule that is capable of stimulating monocytoid cells in humans to produce a signaling protein (TNF)-α through appropriate receptors [50]. Other compounds in the honey could also stimulate chemotactic activity in the neutrophils. Arabinogalactan proteins in Kanuka honey are highly glycosylated in nature with a molecular weight of 110 kDa; and these were identified as precursors for TNF-alpha [38]. Honey flavonoids (apigenin and kaempferol) showed inhibitory functions towards TNF-alpha in keratinocytes in a clinical trial [68]. Although the exact mechanism behind the immune-stimulatory ability of honey is yet to be elucidated, yet the product contributes to tissue regeneration, reduces inflammation, and clears infections [78].

2.5.3 EFFECTS OF HONEY ON CARDIOVASCULAR DISEASES

The influx of Cardiovascular diseases (CVDs) is a global concern that can be prevented by assessing the ramifications of obesity, hypertension, high lipid levels, diabetes, and inadequate diet, etc., [37, 38]. As a matter of fact, increased intake of dietary polyphenols and antioxidants may play a strategically vital role in overcoming this challenge [4, 11, 12, 20, 70] with a major role in chemical and pathogenic defense in humans [71].

A strong correlation has been established between regular consumption of polyphenols and flavonoids with relatively reduced occurrence of CVDs in middle-aged and low-risk Mediterranean subjects [37]. Several flavonoids (such as: anthocyanidins, flavanones, flavones, flavanols, and isoflavones [9]) have been identified in honey [1, 24, 39, 71]. These present diverse potencies, including anti-inflammatory and anti-thrombotic activities by curbing unstable oxygen molecules, inhibiting enzymes, superoxide anions, and per-oxidation process through reduction of alkoxyl-and peroxyl-radicals [24].

Dietary polyphenol acts as an antioxidant and reduces the risk of CVDs [27]. Polyphenols provide ancillaries, such as: free-radical scavenging activities, triggering the production of nitric oxide (NO), correcting inflammation-related cells and their molecular targets, synthesis of vascular-relaxing structures (like prostacyclin (PGI2)), chelating trace metal ions, restoration of α-tocopherol (membrane-based antioxidant) and neutralization of production of vasoconstrictor endothelin-1 (ET-1) [37].

Antioxidants play a key role against unstable molecules in the body [53]. Honey from *G. thoracica* exhibited phenolic compound concentrations as high as 99.04 ± 5.14 mg mL^{-1} and 17.67 ± 0.75 mg mL^{-1} for flavonoids [79]. Kelulut honey had gallic acid (phenolic compound) (concentration 228.09 ±

7.9 to 235.28 ± 0.6 mg kg^{-1}) and flavonoid catechin (concentration 97.88 ± 10.1 mg kg^{-1} to 101.5 ± 11.4 mg kg^{-1}) [72]. Honey from *Ceratonia silique* in Morocco contains phenolic compounds and flavonoids that are correlated to their biochemical constituents [30]. Orange honey from Tunisia has an antioxidant concentration of 4.72 mg of carotene kg^{-1} [83].

For the management of hypertension, pulmonary edema, and heart diseases, the diuretic property of honey is greatly acknowledged. Diuretic property of honey is largely attributed to the antioxidant functions of flavonoids. Carob honey in Morocco was noted for activities, related to the excretion of ions (sodium, potassium) and other diuretic functions that presented no negative outcome for hypokalemia as with other popular clinical diuretics like furosemide. The sodium excretion activity of carob honey was mainly due to the presence of Na-K-2Cl membrane transport proteins in Henle's loop, as a result of the release of o-AQP2 and AVP on the renin-angiotensin-aldosterone system [33].

Honey in combination with propolis is sometimes used for the non-clinical treatment of diverse diseases. In recent studies, a combined delivery of honey-propolis offered significant diuresis: honey alone remarkably augmented the amount of urine and induced riddance of urine electrolyte, but propolis when solely administered showed less contribution [64]. However, ions in plasma did not change when honey was administered alone or in conjunction with propolis [29]. Urinary NO was also elevated but urinary prostaglandins were reduced when honey was orally ingested in a clinical trial [39].

Clumping of platelets or aggregation is a similarly important factor for developing CVDs, particularly the appearance of plaques inside the arteries (atherosclerosis) [85]. Honey polyphenols (such as: catechin, quercetin, luteolin, rutin, and apigenin) inhibit the clumping of platelets through the thromboxane A2 site, *in vitro* understudies [39]. Quercetin is also responsible for the reduction of systolic blood pressure (BP) and plasma oxidized LDL levels in persons with obesity [31]. It has been proven that when honey, beebread, and royal jelly are mixed, there is a beneficial effect of maintaining a good body mass index (BMI). Tualang honey was also remarkably better for controlling arterial pressure and blood glucose in postmenopausal women [41].

2.5.4 ANTI-DIABETIC ACTIVITY OF HONEY

In nutritional chemistry, the metabolic effect of carbohydrate is frequently demonstrated by its glycemic index (GI). Foods with carbohydrates contribute to high and low blood glucose levels, respectively. The recent focus on honey as a natural food supplement takes into account that monofloral honeys have different sugar ratios, especially fructose and glucose ratio [47, 67, 74].

Acacia types of honey have a comparatively higher concentration of fructose with lower GI [9]. Depending on the origin of floral honey, the ratio of fructose-glucose may range from 21–43.5% and 0.46–1.62% [31]. Honeys with low GI's are more valued compared to honeys with high GI because fructose adsorption in the liver is independent of insulin. It rather induces liver enzymes (such as: glucokinase) that boost glucose absorption in the liver [19], so that it can be stored as glycogen. This mechanism is crucial for the reduction of hypoglycemia by fructose. For glucose intolerance or diabetes, honey stimulates plasma insulin and C-peptide to compensate for insulin deficiency. Honey consumption shields the pancreas from oxidative stress and cellular damage [31]. With its rich phenolic and antioxidant contents, honey limits the presence of oxidized low-density lipoprotein (ox-LDL), prevents oxidation in cell membrane lipids, and strengthens intracellular antioxidant defense to control inflammation in several tissues, organs, and endothelial cells [32, 36].

2.5.5 ANTI-MUTAGENIC AND ANTI-CANCER ACTIVITIES OF HONEY

Cancer appears due to the proliferation of abnormal cells that do not correspond to normal regulatory mechanisms through developmental stages known as carcinogenesis [82, 88]. However, the exact cause and mechanism of this chronic disease is an empirical question and it is still under investigation. It is believed that cancer appears as a result of cell genome destruction due to a genetic disorder or improper lifestyle [34].

According to Badolato et al. [10], hydrogen peroxide plays a significant role in the cancer gamut. Elevated levels by somatic cancer cells promote self-alterations that are implicated in cancer development. This was substantiated by López-Lázaro [56], who found that hydrogen peroxide can selectively trigger the apoptosis in cancer cells and reduce the activity of some anticancer drugs, commonly used in the clinical practice. Oxidative stress has also been noted to promote the carcinogenesis through free radicals [40].

Cancers which are developed due to diet and lifestyle are more prominent in developed countries, while cancers as a result of infection are reported in developing countries. This has driven much focus on the chemopreventive measures to curb this epidemic [56]. Honey is considered as a potential chemopreventive agent [82]. Figure 2.3 shows the nutraceutical relationship between honey and cancer.

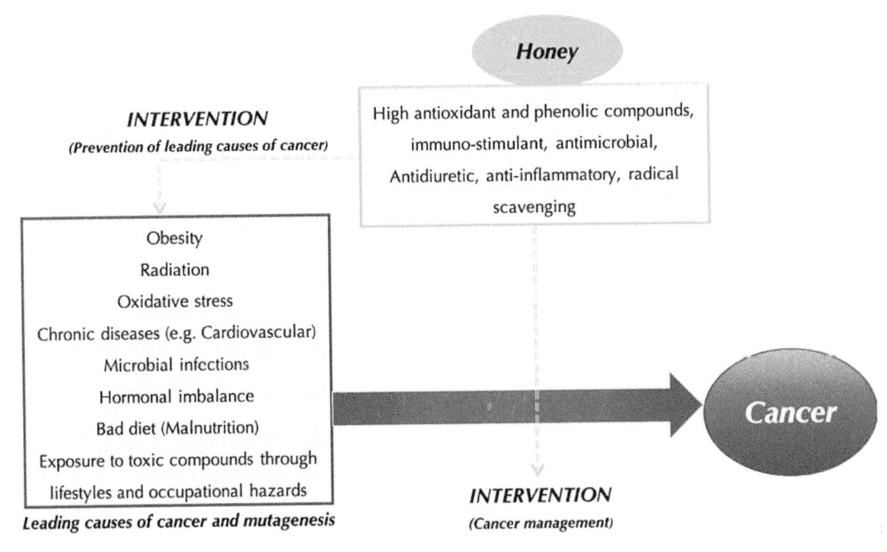

FIGURE 2.3 Combined effect of honey constituents in cancer prevention.
Source: Self-developed: Concepts were adapted from Refs. [10, 24, 30, 31, 50, 66, 75, 89].

The anticancer activity of honey has been investigated and confirmed in different groups of tissues and cell lines, such as: colorectal-, breast-, endometrial-, prostate-, renal-, oral-, and cervical-cancer [82]. Tualang honey with different concentrations (ranging from 1–20%)showed effectiveness in controlling cancer cell proliferation with dose- and time-dependent manners [73]. It has been reported that treatment with Tualang honey (dose 20 $gday^{-1}$ for 12 weeks) significantly promoted the white blood cell and platelet counts, but also creatinine concentration in postmenopausal breast cancer in women [88].

Conventional honey is a viable option for the prophylaxis and treatment of radiotherapy-induced oral mucositis [65]. According to Zaida et al. [88], the combination of Tualang honey and anastrozole presented higher impact than sole administration of anastrozole to decrease the breast background parenchymal enhancement (BPE) in breast cancer postmenopausal female patients having a positive-receptor (ER+) for estrogen.

2.6 SUMMARY

The procedures to purify the phenolic compounds from honey are different types of extraction (traditional solid-liquid, liquid-liquid or alternative methods) methods before chromatographic separation, identification, and quantification. The yield of these compounds, however, depends on the influencing factor. A combination of LC techniques with UV or MS detection is the most frequent method for the detection of phenolics compounds in honey.

As biologically active compounds, flavonoids in aglycone form are present in honey due to their hydrolysis during honey production by bees. Aglycones are more favored in terms of absorption in the gut compared to their glycosides, which imply an optimum bioavailability of flavonoids in honey. There are recent encouraging reports, revealing that flavonoids can be incorporated in lipoprotein domains and plasma membranes. As these sites are usually targets of lipid peroxidation, flavonoids are supposed to have a protective interaction with lipid bilayers. They can also accumulate in the nucleus and mitochondria, affecting diverse cell metabolic functions.

Generally, thorough studies on the bio-accessibility and pharmacokinetics of polyphenols with the human model are necessary because many aspects of the fate of polyphenols upon ingestion are still unknown. In addition, data reported on absorption are often inconsistent and variable. Polyphenols in honey have been associated to anti-inflammatory, anti-diabetic, and antimicrobial activities. Furthermore, honey-based polyphenols reduce risks of metabolic disorder, abnormalities in cardiovascular, respiratory, and immune systems. Some of these health potentials have already been proven clinically, but a more in-depth investigation is needed. Extrapolation of the results in a more precise and judicious way may encourage the community to accept honey as a therapeutic against some chronic diseases.

ACKNOWLEDGMENTS

Zoltan Kovacs acknowledges the support New National Excellence Program of the Ministry for Innovation and Technology (ÚNKP-19-4-SZIE-27) and the Bolyai János Scholarship from the Hungarian Academy of Sciences. Zsanett Bodor acknowledges the support of the New National Excellence Program of the Ministry for Innovation and Technology (ÚNKP-19-3-I-SZIE-71; Szent István University). Zsanett Bodor and John-Lewis Zinia

Zaukuu acknowledge the support from the Doctoral School of Food Science, Szent István University. The project was supported by the European Union and co-financed by the European Social Fund (Grant Agreement no. EFOP-3.6.3-VEKOP-16-2017-00005).

KEYWORDS

- **accelerated solvent extraction**
- **atmospheric pressure chemical ionization**
- **background parenchymal enhancement**
- **bioavailability**
- **blood-brain barrier**
- **polyphenols**

REFERENCES

1. Al-Farsi, M., Al-Amri, A., Al-Hadhrami, A., & Al-Belushi, S., (2018). Color, flavonoids, phenolics, and antioxidants of Omani honey. *Heliyon, 4*, 1–14.
2. Al-Nahari, A. A. M., Almasaudi, S. B., El-Ghany, E. S. M. A., Barbour, E., Jaouni, S. K. A., & Harakeh, S., (2015). Antimicrobial activities of Saudi honey against *Pseudomonas aeruginosa*. *Saudi Journal of Biological Sciences, 22*, 521–525.
3. Augustine, A. J., Baliga, S., Adhikari, P., Kateel, R., Ullal, S., & Bhat, G., (2017). Antibacterial action of tropical honey on various bacteria obtained from diabetic foot ulcers. *Complementary Therapies in Clinical Practice, 30*, 29–32.
4. Ávila, S., Hornung, P. S., Teixeira, G. L., Malunga, L. N., Apea-Bah, F. B., Beux, M. R., Beta, T., & Ribani, R. H., (2019). Bioactive compounds and biological properties of Brazilian stingless bee honey have a strong relationship with the pollen floral origin. *Food Research International, 123*, 1–10.
5. Alvarez-Suarez, J. M., Giampieri, F., & Battino, M., (2013). Honey as a source of dietary antioxidants: structures, bioavailability, and evidence of protective effects against human chronic diseases. *Current Medicinal Chemistry, 20*, 621–638.
6. Al-diab, D., & Jarkas, B., (2015). Effect of storage and thermal treatment on the quality of some local brands of honey from Latakia markets. *Journal of Entomology and Zoology Studies, 3*, 328–334.
7. Anklam, E., (1998). Review of the analytical methods to determine the geographical and botanical origin of honey. *Food Chemistry, 63*, 549–562.
8. Bucekova, M., Buriova, M., Pekarik, L., Majtan, V., & Majtan, J., (2018). Phytochemicals: Mediated production of hydrogen peroxide is crucial for high antibacterial activity of honeydew honey. *Scientific Reports, 8*, 1–9.

9. Blasa, M., Candiracci, M., Accorsi, A., Piacentini, M. P., Albertini, M. C., & Piatti, E., (2006). Raw millefiori honey is packed full of antioxidants. *Food Chemistry, 97*, 217–222.
10. Badolato, M., Carullo, G., Cione, E., Aiello, F., & Caroleo, M. C., (2017). From the hive: Honey a novel weapon against cancer. *European Journal of Medicinal Chemistry, 142*, 290–299.
11. Boussaid, A., Chouaibi, M., Rezig, L., Hellal, R., Donsì, F., Ferrari, G., & Hamdi, S., (2018). Physicochemical and bioactive properties of six honey samples from various floral origins from Tunisia. *Arabian Journal of Chemistry, 11*, 265–274.
12. Belščak-Cvitanović, A., Durgo, K., Huđek, A., Bačun-Družina, V., & Komes, D., (2018). Overview of polyphenols and their properties. In: Galanakis, C., (ed.), *Polyphenols: Properties, Recovery, and Applications* (pp. 3–44). Woodhead Publishing, Vienna, Austria.
13. Biesaga, M., & Pyrzyńska, K., (2013). Stability of bioactive polyphenols from honey during different extraction methods. *Food Chemistry, 136*, 46–54.
14. Bucekova, M., Juricova, V., Di Marco, G., Gismondi, A., Leonardi, D., Canini, A., & Majtan, J., (2018). Effect of thermal liquefying of crystallized honeys on their antibacterial activities. *Food Chemistry, 269*, 335–341.
15. Brglez, M. E., Knez, H. M., Škerget, M., Knez, Ž., & Bren, U., (2016). Polyphenols: Extraction methods, antioxidative action, bioavailability, and anticarcinogenic effects. *Molecules (Basel, Switzerland), 21*, 1–38.
16. Bridi, R., Martínez, P., Montenegro, G., Aguilar, P., Giordano, A., Nuñez-Quijada, G., & Lissi, E., (2018). Differences between phenolic content and antioxidant capacity of quillay Chilean honeys and their separated phenolic extracts. *Ciencia e Investigación Agraria (Agricultural Science and Research), 44*, 251–250.
17. Bohn, T., (2014). Dietary factors affecting polyphenol bioavailability. *Nutrition Reviews, 72*, 429–452.
18. Brudzynski, K., Abubaker, K., & Miotto, D., (2012). Unraveling a mechanism of honey antibacterial action: Polyphenol/H_2O_2-induced oxidative effect on bacterial cell growth and on DNA degradation. *Food Chemistry, 133*, 329–336.
19. Brunswick, N., (2002). Small amounts of dietary fructose dramatically increase hepatic glucose uptake through a novel mechanism of glucokinase activation. *Nutrition Reviews, 2002*, 253–257.
20. Biluca, F. C., Schulz, M., Braghini, F., Vitali, L., Costa, A. C. O., Gonzaga, L. V., Micke, G. A., & Rodrigues, E., (2017). Phenolic compounds: Antioxidant capacity and bio-accessibility of minerals of stingless bee honey. *Journal of Food Composition and Analysis, 63*, 89–97.
21. Cianciosi, D., Forbes-Hernández, T. Y., Afrin, S., Gasparrini, M., Reboredo-Rodriguez, P., Manna, P. P., Zhang, J., & Lamas, L. B., (2018). Phenolic compounds in honey and their associated health benefits: A review. *Molecules, 23*, 1–20.
22. Chau, T., Owusu-Apenten, R., & Nigam, P., (2017). Total phenols, antioxidant capacity and antibacterial activity of manuka honey extract. *Journal of Advances in Biology and Biotechnology, 15*, 1–6.
23. Ciulu, M., Spano, N., Pilo, M. I., & Sanna, G., (2016). Recent advances in the analysis of phenolic compounds in unifloral honeys. *Molecules, 21*, 1–32.
24. Can, Z., Yildiz, O., Sahin, H., Akyuz, T. E., Silici, S., & Kolayli, S., (2015). An investigation of Turkish honeys: Their physicochemical properties, antioxidant capacities and phenolic profiles. *Food Chemistry, 180*, 133–141.

25. Daglia, M., (2012). Polyphenols as antimicrobial agents. *Current Opinion in Biotechnology, 23*, 174–181.
26. Dimitrova, B., Gevrenova, R., & Anklam, E., (2007). Analysis of phenolic acids in honeys of different floral origin by solid-phase extraction and high-performance liquid chromatography. *Phytochemical Analysis, 18*, 24–32.
27. Deng, J., Liu, R., Lu, Q., Hao, P., Xu, A., Zhang, J., & Tan, J., (2018). Biochemical properties, antibacterial, and cellular antioxidant activities of buckwheat honey in comparison to manuka honey. *Food Chemistry, 252*, 243–249.
28. Dua, A., Garg, G., & Mahajan, R., (2013). Polyphenols, flavonoids, and antimicrobial properties of methanolic extract of fennel (*Foeniculum vulgare* Miller). *European Journal of Experimental Biology, 3*, 203–208.
29. El-Guendouz, S., Al-Waili, N., Aazza, S., Elamine, Y., Zizi, S., Al-Waili, T., Al-Waili, A., & Lyoussi, B., (2017). Antioxidant and diuretic activity of co-administration of *Capparis spinosa* honey and propolis in comparison to furosemide. *Asian Pacific Journal of Tropical Medicine, 10*, 974–980.
30. El-Haskoury, R., Kriaa, W., Lyoussi, B., & Makni, M., (2018). *Ceratonia siliqua* honeys from morocco: Physicochemical properties, mineral contents and antioxidant activities. *Journal of Food and Drug Analysis, 26*, 67–73.
31. Erejuwa, O. O., (2019). Honey: Profile and features: Applications to diabetes. In: Watson, R., & Preedy, V., (eds.), *Bioactive Food as Dietary Interventions for Diabetes* (2nd edn. pp. 461–494). New York: Elsevier Inc.
32. Erejuwa, O. O., & Sulaiman, S. A., (2012). Honey: Novel antioxidant. *Molecules, 17*, 4400–4423.
33. El-haskoury, R., Zizi, S., Touzani, S., Al-waili, N., Al-ghamdi, A., Abdallah, M. B., York, N., & Care, M., (2015). Diuretic activity of carob (*Ceratonia silique* L.) honey: Comparison with furosemide. *African Journal of Traditional, Complementary, and Alternative Medicines, 12*, 128–133.
34. Flytkjær, V. L., Møller, H., & Vedsted, P., (2019). Cancer diagnostic delays and travel distance to health services: A nationwide study in Denmark. *Cancer Epidemiology, 59*, 115–122.
35. Gašić, U. M., Milojković-Opsenica, D. M., & Tešić, Ž. L., (2017). Polyphenols as possible markers of botanical origin of honey. *Journal of AOAC International, 100*, 852–861.
36. Gheldof, N., & Engeseth, N. J., (2002). Antioxidant capacity of honeys from various floral sources based on the determination of oxygen radical absorbance capacity and inhibition of in vitro lipoprotein oxidation in human serum samples. *Journal of Agricultural and Food Chemistry, 50*, 3050–3055.
37. Gea, A., Martin-Moreno, J. M., Martinez-Gonzalez, M. A., Pimenta, A. M., Carvalho, N. C., Mendonça, R. D., Bes-Rastrollo, M., & Lopes, A. C. S., (2018). Total polyphenol intake, polyphenol subtypes and incidence of cardiovascular disease. *Nutrition, Metabolism, and Cardiovascular Diseases, 29*, 69–78.
38. Vallianou, G. N., (2014). Honey and its anti-inflammatory, anti-bacterial, and anti-oxidant properties. *General Medicine: Open Access, 2*, 100–132.
39. Hossen, M. S., Ali, M. Y., Jahurul, M. H. A., Abdel-Daim, M. M., Gan, S. H., & Khalil, M. I., (2017). Beneficial roles of honey polyphenols against some human degenerative diseases: A review. *Pharmacological Reports, 69*, 1194–1205.
40. Hassan, M. I., Mabrouk, G. M., Shehata, H. H., & Aboelhussein, M. M., (2012). Antineoplastic effects of bee honey and *Nigella sativa* on hepatocellular carcinoma cells. *Integrative Cancer Therapies, 11*, 354–363.

41. Hassan, I. I., Tohit, N. M., Zakaria, R., Wahab, S. Z. A., Abdul, K. A., Mohamed, N., Norhayati, M. N., & Nik, H. N. H., (2018). Long-term effects of honey on cardiovascular parameters and anthropometric measurements of postmenopausal women. *Complementary Therapies in Medicine, 41*, 154–160.
42. Ishisaka, A., Ikushiro, S., Takeuchi, M., Araki, Y., Juri, M., Yoshiki, Y., Kawai, Y., & Niwa, T., (2017). *In vivo* absorption and metabolism of leptosperin and methyl syringate, abundantly present in manuka honey. *Molecular Nutrition and Food Research, 61*, 17–22.
43. Istasse, T., Jacquet, N., Berchem, T., Haubruge, E., Nguyen, B. K., & Richel, A., (2016). Extraction of honey polyphenols: Methodology and evidence of Cis isomerization. *Analytical Chemistry Insights, 2016*, 49–57.
44. Jaganathan, S. K., (2011). Can flavonoids from honey alter multidrug resistance? *Medical Hypotheses, 76*, 535–537.
45. Joerg, E., & Sontag, G., (1993). Multichannel colorimetric detection coupled with liquid chromatography for determination of phenolic esters in honey. *Journal of Chromatography A, 635*, 137–142.
46. Kassim, M., Achoui, M., Mustafa, M. R., Mohd, M. A., & Yusoff, K. M., (2010). Ellagic acid, phenolic acids, and flavonoids in Malaysian honey extracts demonstrate in vitro anti-inflammatory activity. *Nutrition Research, 30*, 650–659.
47. Karuranga, S., Cho, N. H., Ohlrogge, A. W., Shaw, J. E., Da Rocha, F. J. D., Huang, Y., & Malanda, B., (2018). IDF Diabetes Atlas: Global estimates of diabetes prevalence for 2017 and projections for 2045. *Diabetes Research and Clinical Practice, 138*, 271–281.
48. Kavanagh, S., Gunnoo, J., Marques, P. T., Stout, J. C., & White, B., (2019). Physicochemical properties and phenolic content of honey from different floral origins and from rural versus urban landscapes. *Food Chemistry, 272*, 66–75.
49. Khalil, M. I., & Sulaiman, S. A., (2010). The potential role of honey and its polyphenols in preventing heart diseases: A review. *African Journal of Traditional, Complementary, and Alternative Medicines, 7*, 315–321.
50. Kim, L., & Brudzynski, K., (2018). Identification of menaquinones as novel constituents of honey. *Food Chemistry, 249*, 184–192.
51. Kıvrak, Ş., & Kıvrak, İ., (2017). Assessment of phenolic profile of Turkish honeys. *International Journal of Food Properties, 20*, 864–876.
52. Kečkeš, S., Tešić, Ž., Gašić, U., Dabić, D., Trifković, J., Natić, M., & Milojković-Opsenica, D., (2013). Phenolic profile and antioxidant activity of Serbian polyfloral honeys. *Food Chemistry, 145*, 599–607.
53. Liang, N., & Kitts, D. D., (2015). Role of chlorogenic acids in controlling oxidative and inflammatory stress conditions. *Nutrients, 8*, 1–20.
54. Lachman, J., Orsák, M., Hejtmánková, A., & Kovářová, E., (2010). Evaluation of antioxidant activity and total phenolics of selected Czech honeys. *LWT-Food Science and Technology, 43*, 52–58.
55. Locher, C., Neumann, J., & Sostaric, T., (2017). Authentication of honeys of different floral origins via high-performance thin-layer chromatographic fingerprinting. *JPC-Journal of Planar Chromatography-Modern TLC, 30*, 57–62.
56. López-Lázaro, M., (2007). Dual role of hydrogen peroxide in cancer: Possible relevance to cancer chemoprevention and therapy. *Cancer Letters, 252*, 1–8.
57. Mandal, M. D., & Mandal, S., (2011). Honey: Its medicinal property and antibacterial activity. *Asian Pacific Journal of Tropical Biomedicine, 1*, 154–160.

58. Meo, S. A., Al-Asiri, S. A., Mahesar, A. L., & Ansari, M. J., (2017). Role of honey in modern medicine. *Saudi Journal of Biological Sciences, 24*, 975–978.
59. McLoone, P., Warnock, M., & Fyfe, L., (2016). Honey: A realistic antimicrobial for disorders of the skin. *Journal of Microbiology, Immunology, and Infection, 49*, 161–167.
60. Mandal, S., DebMandal, M., Pal, N. K., & Saha, K., (2010). Antibacterial activity of honey against clinical isolates of *Escherichia coli, Pseudomonas aeruginosa* and *Salmonella enterica* serovar typhi. *Asian Pacific Journal of Tropical Medicine, 3*, 961–964.
61. Mato, I., Huidobro, J. F., Simal-Lozano, J., & Sancho, M. T., (2006). Analytical methods for the determination of organic acids in honey. *Critical Reviews in Analytical Chemistry, 36*, 3–11.
62. Michalkiewicz, A., Biesaga, M., & Pyrzynska, K., (2008). Solid-phase extraction procedure for determination of phenolic acids and some flavonols in honey. *Journal of Chromatography A, 1187*, 18–24.
63. Moussa, A., Noureddine, D., Mohamed, H. S., Abdelmelek, M., & Saad, A., (2012). Antibacterial activity of various honey types of Algeria against *Staphylococcus aureus* and *Streptococcus pyogenes*. *Asian Pacific Journal of Tropical Medicine, 5*, 773–776.
64. Mouhoubi-Tafinine, Z., Ouchemoukh, S., & Tamendjari, A., (2016). Antioxydant activity of some Algerian honey and propolis. *Industrial Crops and Products, 88*, 85–90.
65. Münstedt, K., Momm, F., & Hübner, J., (2019). Honey in the management of side effects of radiotherapy-or radio/chemotherapy-induced oral mucositis. A systematic review. *Complementary Therapies in Clinical Practice, 34*, 145–152.
66. Nousias, P., Karabagias, I. K., Kontakos, S., & Riganakos, K. A., (2017). Characterization and differentiation of Greek commercial thyme honeys according to geographical origin based on quality and some bioactivity parameters using chemometrics. *Journal of Food Processing and Preservation, 41*, 13–61.
67. Nedeljković, N. M., Pezo, L. L., Sakač, M. B., Marić, A. Z., Jovanov, P. T., Novaković, A. R., & Kevrešan, Ž. S., (2018). Physicochemical properties and mineral content of honey samples from Vojvodina (Republic of Serbia). *Food Chemistry, 276*, 15–21.
68. Ota, M., Ishiuchi, K., Xu, X., Minami, M., Nagachi, Y., Yagi-Utsumi, M., Tabuchi, Y., Cai, S. Q., et al., (2019). The immunostimulatory effects and chemical characteristics of heated honey. *Journal of Ethnopharmacology, 228*, 11–17.
69. Piskula, M., Murota, K., & Terao, J., (2012). Bioavailability of flavonols and flavones. In: Spencer, J. P. E., & Crozier, A., (eds.), *Flavonoids*, and *Related Compounds: Bioavailability and Function* (pp. 93–108). CRC Press, Taylor and Francis Group, Boca Raton.
70. Pita-Calvo, C., & Vázquez, M., (2017). Differences between honeydew and blossom honeys: A review. *Trends in Food Science and Technology, 59*, 79–87.
71. Pyrzynska, K., & Biesaga, M., (2009). Analysis of phenolic acids and flavonoids in honey. *TrAC-Trends in Analytical Chemistry, 28*, 893–902.
72. Ranneh, Y., Ali, F., Zarei, M., Akim, A. M., Hamid, H. A., & Khazaai, H., (2018). Malaysian stingless bee and Tualang honeys: A comparative characterization of total antioxidant capacity and phenolic profile using liquid chromatography-mass spectrometry. *LWT-Food Science and Technology, 89*, 1–9.
73. Ramsay, E. I., Rao, S., Madathil, L., Hegde, S. K., Baliga-Rao, M. P., George, T., & Baliga, M. S., (2019). Honey in oral health and care: A mini review. *Journal of Oral Biosciences, 61*, 32–36.
74. Serra, B. J., Ventura, C. F., & Orantes, B. J. F., (2019). Characterization of avocado honey (*Persea americana* Mill.) produced in Southern Spain. *Food Chemistry, 287*, 214–221.

75. Siddiqui, R., Khan, N. A., Mehmood, M. H., Malik, A., & Yousuf, F. A., (2016). Antiacanthamoebic properties of natural and marketed honey in Pakistan. *Asian Pacific Journal of Tropical Biomedicine, 6*, 967–972.
76. Stanek, N., & Jasicka-Misiak, I., (2018). HPTLC phenolic profiles as useful tools for the authentication of honey. *Food Analytical Methods, 11*, 2979–2989.
77. Stalmach, A., Williamson, G., & Clifford, M. N., (2016). In: Spencer, J., & Crozier, A., (eds.), *J. Dietary Hydroxycinnamates and Their Bioavailability* (pp. 123–156). CRC Press, Taylor and Francis Group, Boca Raton.
78. Tahereh, E. O., & Moslem, N., (2013). Traditional and modern uses of natural honey in human diseases: A review. *Iranian Journal of Basic Medical Sciences, 16*, 731–742.
79. Tuksitha, L., Chen, Y. L. S., Chen, Y. L., Wong, K. Y., & Peng, C. C., (2018). Antioxidant and antibacterial capacity of stingless bee honey from Borneo (Sarawak). *Journal of Asia-Pacific Entomology, 21*, 563–570.
80. Tu, J. Q., Zhang, Z. Y., Cui, C. X., Yang, M., Li, Y., & Zhang, Y. P., (2017). Fast separation and determination of flavonoids in honey samples by capillary zone electrophoresis. *Kemija u Industriji (Chemistry and Industry),* 66, 129–134.
81. Urpi-Sarda, M., Rothwell, J., Morand, C., & Manach, C., (2016). Bioavailability of flavanones. In: Spencer, J., & Crozier, A., (eds.), *Flavonoids*, and *Related Compounds. Bioavailability and Function* (pp. 1–44). CRC Press, Taylor and Francis Group, Boca Raton.
82. Waheed, M., Hussain, M. B., Javed, A., Mushtaq, Z., Hassan, S., Shariati, M. A., Khan, M. U., & Majeed, M., (2019). Honey and cancer: A mechanistic review. *Clinical Nutrition*, 1–5.
83. Wan-Mohammad, W. M. Z., Sulaiman, S. A., Abd-Aziz, C. B., Gan, S. H., Wan-Yusuf, W. N., & Mustafa, M., (2018). Tualang honey ameliorates viral load, CD4 counts and improves quality of life in asymptomatic human immunodeficiency virus infected patients. *Journal of Traditional and Complementary Medicine, 2018*, 1–8.
84. Yung, A. C., Hossain, M. M., Alam, F., Islam, M. A., Khalil, M. I., Alam, N., & Gan, S. H., (2016). Efficiency of polyphenol extraction from artificial honey using C-18 cartridges and Amberlite® XAD-2 resin: A comparative study. *Journal of Chemistry, 2016*, 1–6.
85. Zaverio, M. R., (2002). Ruggeri-platelets in atherothrombosis. *Nature Medicine, 8*, 1227–1234.
86. Zhou, X. J., Chen, J., & Shi, Y. P., (2015). Rapid and sensitive determination of polyphenols composition of unifloral honey samples with their antioxidant capacities. *Cogent Chemistry, 1*, 1–10.
87. Zhang, X. H., Wu, H. L., Wang, J. Y., Tu, D. Z., Kang, C., Zhao, J., Chen, Y., & Miu, X. X., (2013). Fast HPLC-DAD quantification of nine polyphenols in honey by using second-order calibration method based on trilinear decomposition algorithm. *Food Chemistry, 138*, 62–69.
88. Zakaria, Z., Zainal, A. Z. F., Gan, S. H., Wan, A. H. W. Z., & Mohamed, M., (2018). Effects of honey supplementation on safety profiles among postmenopausal breast cancer patients. *Journal of Taibah University Medical Sciences, 13*, 535–540.

CHAPTER 3

TROPICAL HERBS AND SPICES AS FUNCTIONAL FOODS WITH ANTIDIABETIC ACTIVITIES

ARNIA SARI MUKAROMAH and FITRIA SUSILOWATI

ABSTRACT

The high biodiversity of herbs and spices are concomitant with the ethnobotanical medicines that are used as natural food preservatives, herbal medicine, cosmetics, crop protection, etc. Herbs and spices (such as: celery, clove, cumin, garlic, ginger, nutmeg, onion, red pepper, tamarind, and turmeric) as functional foods can reduce the risk factors related to *diabetes mellitus* that is associated with cardiovascular illness, stroke, diabetic retinopathy (eye damage), diabetic nephropathy (kidney damage), diabetic neuropathy (nerve damage) and diabetic foot (ulcers). These plants are rich in bioactive compounds, such as, phenolics, terpenoids, flavonoids, and xyloglucan. This chapter highlights phytochemicals in herbs and spices for anti-diabetic activity.

3.1 INTRODUCTION

Diabetes is one of the main metabolic disorders of the endocrine system, which is usually attributed by increasing blood sugar levels [109]. According to the WHO report, 1.6 million deaths in 2016 were due to diabetes. International Diabetes Federation Atlas (IDFA) indicated that approximately, about 8.8% of adults (20–79 years) were living with diabetes in 2017 [40]. Traditional natural herbs have an important role to control diabetes.

Tropical herbs and spices belong to several families, namely: *Achariaceae, Amaryllidaceae, Apiaceae, Asteraceae, Brassicaceae, Euphorbiaceae, Fabaceae, Lamiaceae, Myrtaceae, Oxalidaceae, Poaceae, Rubiaceae, Rutaceae,*

Solanaceae, and *Zingiberaceae.* Numerous investigations have been performed on the isolation, structure identification, production, and biological activities of herbs and spices with anti-diabetic activities [64, 49, 64, 66, 99].

This chapter represents comprehensive on common tropical herbs and spices with anti-diabetic activities. The chapter also includes biochemical mechanisms of biomolecules.

3.2 DIABETES MELLITUS

Diabetes has been dramatically rising during 1980 till date. According to WHO, 422 million adults suffer from type-2 diabetes. The common sign of diabetes is indicated by enhancing of blood glucose level and abnormality of glucose metabolism [109]. Diabetes is classified as:

- **Type-1 Diabetes:** It referred to insulin-dependent *diabetes mellitus* (IDDM) and appears in the early human life and symptoms are quickly detected. Patients with Type-1 diabetes loose ability to produce adequate insulin because of the destruction of pancreatic β-cells. Diabetic therapy is required to maintain blood glucose level. The preliminary therapies are control of dietary intake and insulin therapy.
- **Type-2 Diabetes:** It is referred to non-insulin-dependent *diabetes mellitus* (NIDDM), which arises slowly in older and obese individuals with exhibited unrecognized symptoms. In this case, insulin is produced from the pancreatic islets but several features of the insulin-response system are failed. Characteristics in a type-2 diabetic patients are: (a)inability to uptake of glucose in an efficient way from the blood, (b) incomplete fatty acid oxidation in the liver, (c) accumulation of acetyl co-A, (d) overproduction of ketone bodies, acetoacetate, and β-hydroxybutyrate in the blood, (e) lowering blood pH, and (f) insulin resistance in type-2 diabetes.

Diabetic characteristic symptoms are: (a) excessive thirst and frequent urination (polyuria), (b) intake of a large volume of water (polydipsia), and excretion of glucose in the urine (glucosuria). Health management for type-2 diabetes patients involves [86]: (a) dietary restriction, (b) perform regular body exercise, and (c) avoid drugs that enhance insulin sensitivity or insulin production.

According to fasting plasma glucose test (FPG) values: subjects are categorized as: (1) normal fasting glucose (FPG<100 mgdL^{-1} (5.6 mmol L^{-1})]); (2)

impaired fasting glucose (IFG) (FPG 100–125 mgdL^{-1} (5.6–6.9 mmol L^{-1})); and (3) provisional diagnosis of diabetes (FPG ≥ 126 mgdL^{-1} (7.0 mmol L^{-1})).

Furthermore, oral glucose tolerance test (OGTT) also can be used to categorize subjects as: (1) normal glucose tolerance (2-h post-load glucose, 140 mg dL^{-1} (7.8 mmol L^{-1})), (2) impaired glucose tolerance (IGT) (2-h post-load glucose 140–199 mg dL^{-1} (7.8–11.1 mmol L^{-1})), and (3) provisional diagnosis of diabetes 2 h post-load glucose ≥ 200 mg dL^{-1} (11.1 mmol L^{-1})).

According to the previously mentioned information, patients with IFG and/or IGT are referred to the pre-diabetes group. Pre-diabetes patients have a high risk of encounter by *diabetes mellitus* (DM). In addition, IFG, and IGT are related to other metabolic syndromes, such as: obesity, dyslipidemia by abnormal behavior of high-triglyceride and/or low-high-density lipoprotein, and hypertension [7].

Obesity is a life-threatening metabolic disease, which consumes more calories from the diet than under normal conditions. It triggers several diseases, such as, type-2 diabetes, heart attack, stroke, and cancers in the colon, breast, endometrium, and prostate [69]. Therefore, obesity, and diabetes are positively correlated as the main target of metabolic disorder research. In humans, the prevention of diabetes can be achieved by the consumption of antioxidant-rich foods, (such as: herbs and spices) [12].

3.3 TROPICAL HERBS AND SPICES

Historical records indicate that the uses of herbs and spices have an important role in human nutrition, in food preservation and flavoring agents [62, 75]. In 19th century, "Silk Road (a trade route connecting east and west civilization" promoted the use of herbs and spices at an affordable cost [106]. Also, herbs, and spices were utilized as the source of medicinal therapy in ancient Egypt and Assyria; and for culinary purposes [75].

The Merriam-Webster dictionary defines "*herbs as any plant or parts of plants, which are valuable for its medicinal, savory or aromatic qualities.*" Often spices belong to vegetable products (such as: pepper or nutmeg) and are often used as seasonal functional or flavor foods [63]. The Food and Agriculture Organization (FAO) indicated that around 50,000 plants were used as medicinal plants in 2002 and most of them are also presently used as herbs and spices [85].

In 2016, Royal Botanic Gardens in Kew reported that 17,810 species were used for medicinal purposes among 30,000 plants [82]. A tropical environment with high humidity and rainfall has become home for millions of herbs and spices, such as: cinnamon, clove, nutmeg, pepper, ginger, citrus, saffron, turmeric, etc.

The herbs and spices have been used to cure fever, nausea, scars, colds, headaches, arthritis, etc. Ephedra (*Ma huang*: Chinese herb) is used to relieve asthma, whereas willow tree bark and *Salix alba* helped to relieve fever [22]. Herbal knowledge has been passed down from one generation to the next through oral conversation and daily rituals.

Research continues on usefulness of these traditional remedies, such as: (a) ginger is for relieves from nausea, (b) turmeric to help arthritis symptoms, and (c) cinnamon and saffron for reducing blood sugar levels. These days, people have faith in traditional natural remedies because of several health benefits. The World Health Organization (WHO) indicates that around 80% of the world population entrust on traditional medicine to maintain good health [31, 95].

3.4 FUNCTIONAL FOODS

The term of functional food was first coined as Foods for Specified Health Used (FOSHU) in Japan (1991). Today, functional foods are used for specific health benefits. For decades, numerous research studies on: plant-based bioactive compounds; development of functional foods in a smart way. Evidences portray that many herbs and spices show specific biological activities besides their nutritional values. Phytochemicals in herbal medicine interact with the human body, metabolic, and immune systems against several diseases. Herbs and spices are considered as functional foods [16].

3.5 HERBS AND SPICES AS ANTI-DIABETIC AGENTS

In tropical regions, herbs, and spices have received importance as therapeutic agents. These plants offer anti-diabetic activities. Figure 3.1 shows tropical herbs and spices that have anti-diabetic activities.

3.5.1 CELERY

Celery (*Apium graveolens*) is a commercial seed spice for use in flavoring and seasoning of the food. Celery leaves are ordinarily used in cooking to add a balmy flavor to foods. Celery leaves are commonly dried and are sprinkled in soups, stews, and baking purposes, fried, and roasted foods, and even the uncooked celery is an edible item. It is introduced in a salad or as a garnish. This plant possesses many health benefits, such as: antioxidant, anti-diuretic, hypolipidemic, hypoglycemic, etc.

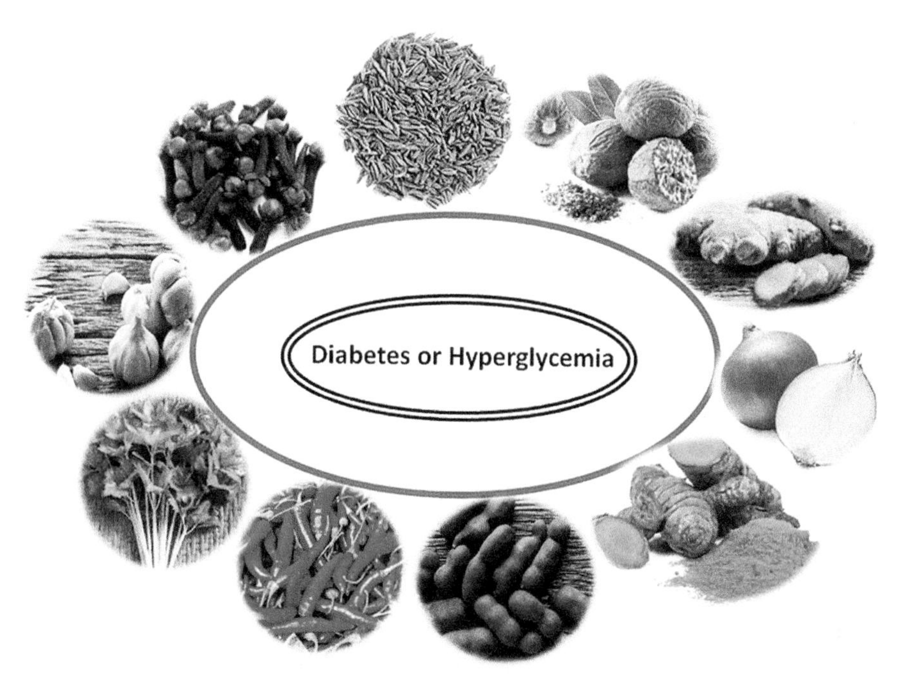

FIGURE 3.1 Tropical herbs and spices with anti-diabetic activities.
Source: Self-developed with concepts from Refs. [64, 105].

Gelodar et al. [34] reported that celery reduced blood glucose in diabetic mice. For the purpose of the experiment, 20 out of 25 mice were induced by alloxan (170 mg kg^{-1}) and divided into 4 groups (5 mice per group). Another 5 mice were chosen as control group receiving neither alloxan nor celery-based diet. Following this, 3 groups of alloxan-induced mice were treated with a celery-based diet (6.25% of the bodyweight) for 15 days. The remaining 5 mice were chosen as a negative control group receiving an only intraperitoneal injection of alloxan. As a result, the concentration of glucose, cholesterol, and creatinine and the activities of ALT, AST, and ALP were lower in the celery-treated group ($P<0.05$) [34].

A research study indicated that leaf extract of celery has the potential to reduce blood glucose levels in elderly pre-diabetic patients. The experiment was performed with 16 elderly pre-diabetic patients (6 males and 10 females: older than 60 years). Two groups were formed: a control (placebo) group and a treated (celery leaf extract) group. The treated group was given a celery leaf extract in a capsule @ 250 mg (30 minutes before a meal, three times per day) for 12 days. This study showed that the celery capsule was able to lower the blood glucose level [113].

Celery contains alkaloids, carbohydrates, flavonoids (apigenin, apiin, isoquercitrin), glycosides, steroids, vitamin A and vitamin C [4, 53]. Other phytochemical compounds in celery leaves are: apigenin, bergapten, chrysoeriol 7-glucosides, furanocoumarin, luteolin, isopimpinellin, phenols, phthalide, psoralen, and xanthotoxin [18, 53]. Celery has also been documented to have several hypoglycemic compounds (such as, essential oils, flavonoids (kaempferol, quercetin, triterpenes, and luteolin), phenolics, triterpenes with antidiabetic [37] antioxidant properties.

The seed extract of celery has the ability to decrease the blood glucose level and to increase insulin levels in diabetes-induced mice [27]. The celery seed is rich in flavonoids, such as: apigenin, luteolin, and phenolics [4, 51, 71]. Apigenin inhibits the aldose reductase enzyme activity [27, 42]. A diabetic patient can suffer from cataracts, retinopathy, and neuropathy due to increasing levels of sorbitol [51]. Another anti-diabetic mechanism explains the ability of celery seed in stimulating the secretion of insulin from pancreatic β-cells and decreasing the liver gluconeogenesis pathway [5, 18, 71]. Based on the histology test, celery seeds promoted the integrity of pancreatic β-cells [5, 71].

Flavonoid is one of the major antioxidant compounds in celery [18]. Flavonoid is able to overcome the free radicals and preclude devastation to pancreatic β-cells [2, 5, 18, 53]. Flavonoids can regulate the absorption of glucose in the intestine, carbohydrate metabolism, and glucose uptake, especially in the regulation of the cell-signaling AMP-activated protein kinase (AMPK) pathways. Flavonoid is also able to help glucose uptake in the skeletal muscle cells [5, 18, 29, 53]. Gutierrez et al. [37] reported that flavonoids as an anti-diabetic agent can diminish apoptosis, enhance pancreatic β-cell proliferation, promote the secretion of insulin, manage glucose digest, and reduce hyperglycemia [37]. Flavonoid is also able to suppress insulin resistance, give relief from inflammation in adiposity cells, and suppress oxidative stress chain in the skeletal muscles [35]. Flavonoid plays a key role in the up-regulation of Glucose Transporter-1 (GLUT-1) expression level, which is useful in the treatment of type-2 DM (T2DM).

Kaempferol, a natural flavonol, is an anti-diabetic agent due to its activity in pancreatic β-cell protection, which is correlated to T2DM [5]. It has been reported that kaempferol has the ability to reduce hyperglycemia and increase the glucose uptake through the PI3K and protein kinase C (PKC) pathways in rat's muscle. An oral dose of kaempferol was also able to increase the fasting blood glucose (FBG), glucose tolerance, and HbA1c concentration, and depress insulin resistance [29].

Gutierrez et al. [37] reported that quercetin can reduce the levels of plasma glucose in alloxan-induced diabetic mice. Quercetin commits to GLUT-4 mRNA rearrangement in cell membranes of adipocytes and skeletal muscle cells. Also, it up-regulates GLUT-4 mRNA level, which reduces the blood glucose level. Quercetin increases the liver glucose uptake and promotes the secretion of insulin from pancreatic β-cells.

Quercetin and naringenin can protect β-cells from the toxicity of cytokine via Phosphatidylinositol-3-kinase (PI3K) pathway. Apigenin has anti-hyperglycemic property [39]. Apigenin-treated diabetic mice displayed an improvement in hyperglycemia levels and its antioxidant status. Apigenin treatment in alloxan-induced diabetic mice decreased the glucose blood levels and repaired pancreatic β-cells.

Luteolin was investigated for its ability to raise the insulin action and to promote GLUT-4 activity in diabetic mice and it increased antioxidant activity in diabetic nephropathy in diabetic mice. Its antioxidant property promotes the secretion of insulin via the NF-κB and iNOS-NO signaling pathways [107].

Apigenin and luteolin act as Sodium-glucose Cotransporter-2 (SGLT-2) inhibitors in neuropathic diabetes [39]. Several scientific facts indicate that oxidative stress affects DM pathogenesis and diabetic complications. The elaboration of glucose oxidation, glycation of proteins, and oxidative deterioration of proteins can lead to the formation and accumulation of free radicals in DM patients. This peculiarity of free radicals and chronic deterioration in endogenous antioxidants cause harm to cell organelles and oxidative enzymes [57].

3.5.2 CLOVES

Syzygium aromaticum L. (Cloves) is a native herb in Indonesia, Malaysia, Brazil, Mexico, India, Haiti, Madagascar, Sri Lanka, and Tanzania. Cloves are commonly used as spices for an aromatic flavor. Cloves contain essential oil, phenolic compounds, and hydrolyzable tannins [61]. Tahir et al. reported that the yield of essential oil in cloves was 8.5% with pale yellow color. It is soluble in alcohol and has a refractive index of 1.23 [103].

Quantity percentage of clove essential oil is determined by phenological state and agro-climatic conditions. Essential oils in cloves contain 2.93% of caryophyllene oxide, 3.36% of eucalyptol, 3.72% of citronellyl acetate, 3.84% of γ-terpinene, 4.51% of methyl benzoate, 5.79% of geraniol, 8.12% of *n*-hexane, 8.84% of acetaldehyde, 15.3% of eugenyl acetate and 18.7% of eugenol. In addition, clove contains acetophenone, chromone glycosides, phenylpropanoids, 2,4,6-trihydroxy-3-methyl acetophenone-2-O-β-D-glucoside, sesquiterpenoids, triterpenoids, sterol, and tannin. Those mentioned compounds showed anticancer activity in human ovarian cancer cells (A2780) using MTT assay [63].

Clove oil reveals anti-diabetic benefit by reducing high blood sugar levels and lipid peroxidation, and recovering the antioxidant enzymes [90].

Alpha-amylase inhibition assay is commonly used to determine anti-diabetic activity. According to the research, clove essential oil exhibits maximum anti-diabetic activity at 100 μg mL^{-1}, whereas clove essential oil emulsion (essential oil (25%) + tween 80 (75%) + ethanol (25%) + water (25%)) shows a maximum anti-diabetic capability (as much as 95.30% inhibition of α-amylase activity) [103]. In addition, clove is an excellent functional food and is appropriate for alternative therapeutic of type-2 *DM*, because it can prevent activity of key enzymes in type-2 diabetic patients, such as: α-amylase and α-glucosidase [1].

Glycation is an attachment of excess sugar to the protein or lipid without enzymatic process. The free aldehyde or ketone group from this reducing sugar is able to adduct the alpha-amino group from N-terminal or E-amino amine as on lysine, arginine, cysteine, and histidine. It forms a cross-linking protein AGEs with other dipeptides having an N-terminal free side group (mainly on arginine and tryptophan). Advanced glycation end-products (AGEs) can induce pathogenesis and atherosclerosis in diabetic complications. A study reported that the antiglycation effect of some medicinal plants can be detected using the SDS-PAGE method [76].

Clove extracts demonstrated strong inhibition protein cross-linking formation at 25 μg mL^{-1} concentration. Therefore, clove has the ability as a glycation agent and to induce protein cross-linking inhibitory activity [76]. Another study reported that the aqueous extract of clove suppresses atherosclerosis and diabetes due to its strong antioxidant activity, anti-apolipoprotein A-I (apoA-I) glycation and phagocytosis prevention through inhibition of low-density lipoprotein (LDL) and cholesteryl ester transfer protein (CETP), and reducing hypolipidemic activity [44, 76].

3.5.3 CUMIN

Cumin (*Cuminum cyminum* L.) originated from Southwest Asia and the Eastern Mediterranean region. It is cultivated in Europe, Egypt, the Middle East, India, and Iran. Cumin powder is used as a spice in various cuisines because of its unique flavor, strong, and warm aroma. The essential oil of cumin is also popular among traditional healers [56]. The different parts (roots, stems, leaves, and flowers) of the cumin plant have various bioactive compounds (phenolics, flavonoids, and tannins) with unique biological activities. The essential oil is produced from flowers (1.7%), leaves, and stems (0.1%), and roots (0.03%).

The main constituent in essential oil from roots is bornyl acetate (23%). However, α-terpinene is a major compound in the stem and leaves; and γ-terpinene is abundant in the flower. Quercetin is a primary phenolic compound in the roots, while vanillic acid is present in a higher amount in the flowers. However, *p*-coumaric, rosmarinic, *trans*-2-dihydrocinnamic acids, and resorcinol are found predominantly in stems and leaves. Antioxidant activity assessment of the cumin plant parts has been checked using four types of assays, such as: 1,1-diphenyl-2-picrylhydrazyl (DPPH), β-carotene/linoleic acid, reducing power and chelating power assays. The cumin flowers acetone extract has strong antioxidant activity, lipid peroxidation inhibitor, and reducing agent, whereas acetone extract of stem and leaves exhibits the highest chelating power. Nevertheless, the essential oil offers moderate antioxidant activity as shown by antioxidant assay [20].

Chemical distribution and cumin fruit antioxidant activity were determined by correlating harvesting time with plant maturity. Cumin fruit antioxidant activity in various growth levels is still lower than butylated hydroxytoluene (BHT) using both DPPH and FRAP assays. The IC_{50} value of BHT is about 27 times higher than the 13.59 mg mL^{-1} of IC_{50} in immature cumin. On the other hand, intermediate, and premature cumin parts indicate the highest radical scavenging activity than the other growth stages, although their antioxidant activity is still lower than BHT.

FRAP assay can quantify the antioxidant compounds' capability to diminish the ferric ion (Fe^{3+}) to ferrous (Fe^{2+}). FRAP assay studies reveal that BHT reducing power (602 mol Fe^{2+} per mass) is greater than immature, intermediate, premature, and fully mature cumin. The BHT values were 132, 121, 106, and 89 moles of Fe^{2+} per mg of essential oil for immature, intermediate, premature, and fully mature cumin, respectively. In addition, phenolic contents (which enhance at intermediate and premature stages) have a positive correlation with their antioxidant capacity (AOC) of essential oils [65].

The essential oil from cumin can repair the metabolic condition and prevent the progression of diabetes. These biological activities are unique to cumin as an alternative adjuvant therapeutics for pre-diabetic subjects and support lifestyle alteration [41].

Dietary cumin seeds are known as an anti-diabetic agent to reduce blood glucose content and to decrease excretions of urea and creatinine in diabetic rats [108]. Cumin seed-derived peptides have radical scavenging activity; and therefore can be used as a substance in the functional food and pharmaceutical studies. Three novel cumin-seed peptides (CSPs) have successfully been extracted and identified as "α-amylase inhibitory peptide," which plays an important role in retardation of carbohydrate digestion and glucose absorption. This process reduces the risks of *DM* development [93, 94]. Cumin improves metabolic index and anthropometric in overweight and/or type-2 diabetic subjects [42]. Furthermore, essential oil from cumin exhibits maximum anti-diabetic activity at a concentration of 100 $\mu g\ mL^{-1}$. Cumin essential oil emulsion (essential oil (25%) + tween 80 (75%) + ethanol (25%) + water (25%)) demonstrates maximum anti-diabetic activity (95%) and α-amylase inhibition (83%). Inhibition of α-amylase activity delays the carbohydrate complex hydrolysis and glucose absorption in the intestinal tract, which reduces blood glucose levels [103].

The ethanolic seed extract of cumin exhibits good anti-hyperglycemic and anti-dyslipidemic activities. These have been validated with type-2 *DM* rat and male Syrian golden hamsters. Furthermore, the anti-dyslipidemic activity can be shown by declining of triglycerides, cholesterol, LDL-C in the serum, and enhancing of serum HDL-C. The ethanolic extract of cumin seeds also has aldose inhibitory activity. It can reduce secondary diabetic complications in the later stage [100].

3.5.4 GARLIC

Garlic (*Allium sativum* L.) is used as a common seasoning in almost every cuisine around the world. It is also used as a traditional therapeutic agent.

Scientific reports have documented that garlic can diminish the level of blood glucose (hypoglycemic activities) in diabetic animals. Jain and Vyas [43] reported the hypoglycemic effects of garlic extracts (ethanol, 40–60° petroleum ether and diethyl ether) in diabetic rabbits induced by alloxan. The garlic ethyl extract offered maximum hypoglycemic activity ($P < 0.001$) [43].

Garlic extract offers anti-diabetic and hypoglycemic activities due to the enhancement of insulin-like activity in plasma. It influences the secretion of insulin from β-cells, the release of bound insulin or increases insulin sensitivity. These processes liberate fixed insulin. The effect of garlic oil from fresh cloves has been studied on diabetic rats and mice (streptozotocin-induced and alloxan-induced) [73]. Treated animals group and control were administrated (intra-gastrically) @ 50 mg per kg of body weight daily of garlic oil supplement for 28 days. Treated animal group exhibited a significant alleviation in red cell acid phosphatase ($p<0.001$), serum acid phosphatase and alkaline phosphatase ($p<0.001$), aminotransferases ($p<0.001$) and amylase ($p>0.002$) levels compared to the diabetic control rats.

Yang et al. [112] found that garlic diminishes concentrations of LDL cholesterol and total serum cholesterol in diabetic patients in a significant way as compared to a placebo, possibly due to the inhibition of reactive oxygen species (ROS) through the modulation of NADPH oxidase subunit expression and can restore erectile function in diabetic patients. Sheela et al. have reported that S-allyl cysteine sulfoxide from garlic provoked a reversal of diabetic conditions, such as, the depletion of liver glycogen, glucose tolerance, weight loss, etc., in alloxan-induced diabetic rats [87]. S-Allyl cysteine sulfoxide (isolated from garlic) rectified diabetic conditions in alloxan-induced diabetic rats and it was as effective as glibenclamide- and insulin-therapy [88]. S-Allyl cysteine sulfoxide can also control lipid peroxidation better than the glibenclamide and insulin. It also increased insulin secretion from pancreatic β-cells [99].

Pandiya and Banerjee indicated that the beneficial effect of garlic in *DM* is attributed to the presence of volatile sulfur compounds, such as, alliin, allicin, diallyl disulfide, diallyl trisulfide, diallyl sulfide, S-allyl cysteine, ajoene, and allyl mercaptan. Allicin can enhance serum insulin by effectively combining with cysteine, which would spare insulin from SH group that is common cause of insulin inactivation [74].

3.5.5 GINGER

Ginger (*Zingiber officinale* Roscoe) is similar to turmeric and it is a popular ingredient in cooking due to a peculiar flavor, aroma, and pungent odors. Besides cooking, ginger is also well-known for its medicinal properties against gastrointestinal disorder, motion sickness, nausea relief, cold, and flu relief, pain, and inflammation (reduce muscle pain, knee, and elbow). Ginger contains several biochemical constituents, such as: gingerol, shogaol, paradol, and zingerone [23]. It may also reduce the risks of cardiovascular diseases (CVDs), blood clotting, and hyperglycemia.

A clinical trial was conducted by Shidfar et al. [89], who reported the impact of ginger on glycemic indices in patients with type 2-diabetes. The 45 patients with T2DM (20–60 years old), who did not receive insulin, were divided into control and treated groups. Patients in the treated group received 3 g of ginger supplement (capsule) daily for 3 months. The results of this study showed that ginger supplementation increased the glycemic indices, TAC, and PON-1 activity in patients with T2DM. It is reported that the phenolic compounds (such as: gingerol and shogaol) act as α-amylase and α-glucosidase inhibitors in carbohydrate metabolism and hyperglycemia. Ginger supplementation can enhance the homeostasis of glucose in T2DM patients by lowering insulin resistance and improving glucose tolerance. This study also reported that ginger supplementation significantly reduces the C-reactive protein (CRP) as an inflammation marker [89].

Khandouzi et al. [50] reported the effects of ginger on fasting blood sugar (FBS). Out of 41 patients with T2DM, 22 patients were assigned to the ginger group compared to 19 in the control group. The treated group received 2 g of ginger powder daily for 12 weeks. The results of this study exhibited that oral administration of ginger powder significantly reduced the levels of blood sugar, Hemoglobin A1c (HbA1c), Apolipoprotein B, Apolipoprotein B/Apolipoprotein A-I, and Malondialdehyde (MDA) [50]. Therefore, it is concluded that oral administration of ginger powder may relieve the risks of some chronic complications related to diabetes. These evidences are in agreement with results by Azimi et al. [13], who also reported the ability of ginger supplementation for decreasing the CRP concentration and reduction of the risk of Type 2 diabetes Mellitus (T2DM).

3.5.6 NUTMEG

Nutmeg (*Myristica fragrans* Houtt.) is a dried kernel of the seed, which is grown in Indonesia and Grenada (West Indies). Nutmeg has a slightly sweet flavor and is used in the preparation of various foods [81].

Acetone extract of nutmeg seed contains 28.16% of sabinene, 10.26% of β-pinene, 9.72% of α-pinene, 4.30% of myristicin, 2.72% of isoeugenol, 1.81% of p-cymene, 1.54% of carvacrol, 0.89% of eugenol and 0.82% of β-caryophyllene with antimicrobial and antioxidant activities [36]. Furthermore, AMP-activated protein kinase (AMPK) enzyme can be induced by nutmeg extract. Bioactive components of nutmeg extract include tetrahydrofuroguaiacin B, nectandrin B, and galbacin, which offer strong AMPK stimulation in mouse C2C12 skeletal my oblast. Therefore, nutmeg, and its active components have become popular as a T2DM therapeutic agent and to develop obesity medication [70].

3.5.7 ONION

Onion (*Allium cepa* L.) is popular as food seasoning and natural medicine [101]. Besides fresh onions, all parts of onion can be utilized and processed for extract, juice, paste, bagasse, powder, etc. In addition, onion contains rutin, quercetin [33], isorhamnetin, kaempferol, luteolin, fructan, coumarin, steroids, S-methyl-L-cysteine [84], S-methylthio-L-cysteine, S-propylthio-L-cysteine, and (S- (1-propenyllthio)-L-cysteine, and isoalliin (S- (1-propenyl-S-oxo-L-cysteine) [101]. According to Ko et al. [52], the methanol extract of onion bulbs contains quercetin-4'-*O*-monoglucoside, quercetin-3,4'-*O*-diglucoside and quercetin [52]. Several investigations with onion have been performed under *in vivo* and *in vitro* conditions to study anti-diabetic activities.

3.5.7.1 ONION JUICE

Onion juice is an antioxidant and anti-hyperglycemic agent to relieve liver and kidney damages, that are caused by alloxan-induced diabetes [30]. Furthermore, the water solution of onion showed the hypolipidemic and hypoglycemic activities in STZ diabetic rats. For the preparation of water solution of onion, 40 g of minced onion bulbs were frozen for 48 h and were infused in the 100 mL of water. This solution was able to reduce triacylglycerol, serum cholesterol, and LDL-cholesterol in STZ diabetic rats. Furthermore, onion juice can be used for hypoglycemic and hypolipidemic conditions. As hypolipidemic, onion juice reduces the concentration of lipoperoxide, lipid hydroperoxide, and activity of superoxide dismutase (SOD) [25].

3.5.7.2 ONION EXTRACT

According to Bang et al. [17], onion powder can prevent the weight loss of STZ-induced diabetic rats. Furthermore, onion prevents diabetic complications by maintaining and reducing blood glucose. It is correlated with lowering serum lipids. Meanwhile, oxidative stress in kidneys of diabetic animals can also be prevented by onion extract. It was found that polyphenol compounds (such as: rutin and quercetin) in ethanolic extract of onion have antihyperglycemic and antidyslipidemic activities by enhancing glucose transporter type-4 (GLUT4) translocation, glucose uptake and insulin activity in skeletal muscle cells (L6 myotubes) and adipocytes (3T3L1 cells) due to activation of phosphatidylinositol-4,5-biphosphate 3-kinase/Akt dependent pathway [33].

Activities of both α-amylase and α-glucosidase are inhibited by aqueous extract of white onion and purple onion in the concentration range of 0–4 mg mL^{-1} under *in vitro* conditions. Furthermore, inhibition of α-amylase and α-glucosidase activities can reduce hyperglycemia, prevent ROS yield, and prevent Fe^{2+}-induced lipid peroxidation [72].

3.5.7.3 ONION PEEL EXTRACT

Jung et al. [45] have reported that onion peel extract (OPE) can reduce insulin resistance and hyperglycemia in diabetic rats. Anti-diabetic activity of OPE is attributed to: glucose metabolism; uptake in peripheral tissues through insulin receptor (INSR) and GLUT4 expressions in skeletal muscle; reduction in plasma free fatty acid (FFA) level; and suppression of inflammatory and oxidative stresses in the liver.

3.5.7.4 ONION WASTE

Onion waste contains high dietary fiber with anti-diabetic activity. Furthermore, several *in vitro* studies have revealed that fiber concentrates (FCs) offer glucose absorption, delay glucose diffusion, increase amylase activity, and reduction of starch digestibility. Moreover, FCs from bagasse (onion wastes were triturated and pressed) have shown a higher ability to absorb glucose because they contain pectic polysaccharides in higher concentrations. Bagasse Fiber Concentrates (BFCs) can retard and delay diffusion and absorption of glucose in the gastrointestinal tract [19].

3.5.8 RED PEPPER

Red pepper (*Capsicum annuum* L.) contains capsaicinoids, β-carotene, phenols, flavonoid, squalene, vitamin C, vitamin E, calcium, phosphor, potassium, phytosterols, and fatty acid. Harvesting time determines the phytochemical composition and antioxidant activity of red pepper through the stable accumulation of vitamin E, phytosterols, and fatty acid [21]. Concentrations of quercetin and catechin differ among colored bell pepper at maturation stages [35]. Furthermore, peel, and seed extracts of red sweet pepper inhibit pancreatic α-amylase activity but it does not influence α-glucosidase activity. They contain high concentration of phenolic compounds (such as: rutin, gallic acid, carvacrol, and 4-OH benzoic acid) with antioxidant activity [98].

Capsaicinoids are major components in the chili pepper, which offers anti-diabetic activity. Moderate consumption of 4.4 mg of capsaicinoids daily induces glucose level, LDL cholesterol and CRP formation [48]. In addition, Korean red pepper prevented *DM* by activation of PPAR-γ and AMPK [111]. The hypoglycemic effect of capsaicin and dihydrocapsaicin does not exhibit a synergistic effect. Natural capsaicinoid (capsaicin (66%) and dihydrocapsaicin (29%)) and capsaicin (95%) significantly showed a hypoglycemic effect. Capsaicinoids can elevate insulin level and reduces blood glucose levels, inhibit glucose absorption in the ileum [114].

3.5.9 TAMARIND

Tamarind (*Tamarindus indica* L.) is an indigenous plant in Africa, India, Thailand, Bangladesh, Sri Lanka, Indonesia, Mexico, and Costa Rica. Tamarind has been used as a food seasoning and flavoring agent for food and drinks. Tamarind fruit pulp contains tartaric acid, reducing sugar, vitamin B, vitamin C, carotene; and this pulp has been used as a food component, curries, sauces, confectionery flavoring and the major component in beverages. Moreover, tamarind leaves, flowers, and seeds are also served as vegetables, salad, curries, and soup [24].

Tamarind is an anti-diabetic agent for type-1 *DM* and can control oxidative stress [58, 68]. Ethanolic extract of tamarind barks can decrease blood glucose levels and prevent oxidative stress by antioxidant activity [3]. Potential of tamarind seed as an antioxidant and anti-diabetic activity was confirmed *in vivo* and *in vitro* studies. Tamarind seed aqueous extract exhibits hyperglycemic attenuation, as shown in several indicators, such as, FBG level, glycogen content, and glucose-6-phosphatase action. The anti-hyperglycemic mechanism appears through the regulation of intracellular glucose utilization in target organs [59].

Furthermore, the contribution of anti-diabetic activity of tamarind seed aqueous extract is related to its anti-inflammatory activity on β-cell cells of islets and cytokines by complex mechanisms, involving calcium [Ca^{2+}] andSREBP-1c gene in the liver [97].

Tamarind seed also contains edible mucoadhesive polysaccharide (such as: tamarind xyloglucan (TGX)), which has beneficial as dietary fiber and as thickener or stabilizer in the food industry. Xyloglucan enhanced lipid metabolism in the rat by reducing total cholesterol, β-lipoprotein, total lipid, phospholipid, and adipose tissue weight [110]. Xyloglucan has an important function to attenuate inflammation and healing of ulcerative colitis by the mucoadhesive process and it has antioxidant properties [78, 79]. Therefore, the polyphenol-rich seeds of tamarind and their aqueous extract can be developed as new herbal drug for diabetes therapy [96].

Hyperlipidemia can be reduced by the utilization of tamarind fruit pulp. Tamarind fruit pulp extract diminishes the potential risk of atherosclerosis development in humans by reducing triglycerides, non-HDL cholesterol, total cholesterol, increasing HDL cholesterol in serum, and enhancing of antioxidant defense system [60]. Moreover, tamarind fruit pulp extract is a potential hypocholesterolemic agent. It enhances cholesterol efflux, preventing cholesterol biosynthesis, increasing uptake and removal of LDL-C from peripheral tissues, and suppressing the accumulation of triglyceride in the liver. In addition, tamarind fruit pulp contains phenolic compounds and flavonoids that can prevent oxidative damage, especially caused by LDL-C oxidation [55]. Consumption of tamarind fruit pulp is safe for health, because it does not show the cytotoxic effect on DNA damage in the cells of *Mus musculus* and *Rattus norvegicus* [92].

3.5.10 TURMERIC

Turmeric (*Curcuma longa* L.) is popular as a spice, her band food coloring. Traditionally turmeric is used as a home remedy for abdominal pains [10], cough, rheumatism, and sinusitis [8]. The major phytochemical is curcuminoid, which is responsible for its deep orange-yellow color and anti-diabetic agent [9, 14, 54].

Arun and Nalini [11] have reported the efficacy of turmeric and curcumin on diabetic rats. The results indicated that the injection of turmeric or curcumin reduced the blood glucose levels, Hb, and glycosylated hemoglobin levels. The supplementation of turmeric and curcumin also diminished the oxidative stress chain in diabetes-induced mice, possibly due to a decrease in the influx of blood glucose into the polyol pathway. It is associated with an increase of NADPH/NADP ratio and enhanced activity of the antioxidant enzyme, i.e., glucose peroxidase. Additionally, the activity of succinate dehydrogenase was reduced significantly due to turmeric or curcumin therapy [11].

Turmeric is beneficial for suppressing hyperglycemia. Mohamed et al. [67] investigated the effects of aqueous, methanol, and n-hexane root extracts of *Curcuma longa* on alloxan-induced diabetic rats. Total 36 rats were grouped into 6 groups with 6 rats per group:

The group I rats were considered as normal control:

- Group II rats were considered as diabetic control.
- Group III rats were induced with diabetes and standard drug (metformin, 100 mg kg^{-1}) was administered.
- Rats in group IV were induced with diabetes and administered with crude aqueous extract with a dose of 400 mg kg body $weight^{-1}$ for 4 weeks.
- Rats in group V were induced with diabetes and administered with methanol with a dose of 400 mg kg body $weight^{-1}$ for 4 weeks.
- Rats in group VI were induced with diabetes and administered with n-hexane extraction fractions with a dose of 400 mg kg body $weight^{-1}$ for 4 weeks.

This study by Mohamed et al. concluded that the methanol fraction had the highest potency for reducing blood glucose levels in a significant way ($p<0.05$) compared to diabetic control after few days of administration of fractions. Furthermore, hypoglycemic activity was observed after a week of administration of the aqueous extract. The fraction of n-hexane extraction had a remarkable hypoglycemic activity only after 2 weeks of administration [67].

Curcumin has the ability to enhance wound repair in diabetic patients. Sidhu et al. [91] evaluated the efficacy of curcumin for the treatment of wounds in diabetic rats. Wounds of rats treated with curcumin showed earlier re-epithelialization, accelerated neovascularization process, increased movement of various cells, including dermal myofibroblasts, fibroblasts, macrophages into the wound bed, and higher collagen content. For its immune-histochemical, the results showed an increase in transforming growth factor-beta 1 in curcumin-treated wounds compared to the control group. A delay in the apoptosis patterns was shown in diabetic wounds compared to curcumin-treated wounds [91].

3.6 CHALLENGES AND FUTURE PERSPECTIVES

Utilizations of herbs and spices as a functional food for diabetic treatment have several challenges. In order to maximize the beneficial health effects through the utilization of herbs and spices, the following issues have been discussed.

3.6.1 SYNTHESIS AND RIPENING STAGES

The quantity of the phytochemicals from selected herbs and spices is mostly affected by the growth stage and its origin of a plant. Rebey et al. [80] found that the ripening stage of cumin seed significantly influenced the contents of polyphenol, flavonoids, tannins, and their antioxidant activities. During three ripening stages, cumin seed contained a high phenolic constituent with significant antioxidant activity. The study exhibited that full ripe cumin seeds are enriched with polyphenol and condensed tannin than unripe seeds. Moreover, unripe cumin seeds contain higher total flavonoid content (TFC) that half-ripe and fully-ripe seeds. AOC is not only represented by the high value of the total phenolic compound but also on the phenolic composition [80].

The origin of the synthesis of the phytochemical compounds is varied in leaves, stem, bark, flowers, roots, and seeds. Rebey et al. found that rosmarinic acid is a phenolic acid while p-coumaric acid is a constituent in half-ripe and full-ripe seeds [80].

Bettaieb [20] reported that the differences in the chemical composition of cumin depend on the plant parts. Bornyl acetate in roots, γ-pinene, and vanillic acid in flower, thymol/ R-terpinene/ p-coumaric in stems, and leaves are abundant. Therefore, certain health-promoting antioxidants development in the food industry can be enhanced by the arrangement of breeding strategies, selection of resources of herbs and spices, and type of solvent for extraction [20].

3.6.2 RECOVERY METHOD FOR PHYTOCHEMICALS

Selected extraction method plays a significant role for determining the quantity and the beneficial effect of a phytochemical compound. Chemical property and characteristics of each compound depend on the isolation method. Su et al. [102] found a significant correlation between the determination of solvent extraction of herbs and spices and antioxidant activity. Supercritical extraction is an excellent technique to extract oil from nutmeg because it can produce high quality of botanical extract, free of solvent, a noncytotoxic compound with significant cost saving [7]. Furthermore, Rebey et al. [80] reported that the soxhlet extract of cumin seed consisted of the higher content of polyphenols and flavonoids than the maceration extract. Also, maceration extract revealed higher antiradical and bleaching power capacity than the soxhlet extracts [80]. Therefore, the recovery of phytochemicals must consider the type of extraction method.

3.6.3 CULTIVATION AND PROCESSING OF HERBS AND SPICES

Several problems related with raw material and oil production of nutmeg have been reported in the literature. Such problems are related with cultivation, plant protection measures, processing, shortage of production of nutmeg, and recovery of valuable compounds. Therefore, enhancing cultivation, processing, and awareness shall increase the productivity, product quality, and sustainability of the nutmeg processing industry [46]. In the rural area, the sustainability of nutmeg production depends on local government support, traditional culture, and training about cultivation, and processing to farmers [47].

3.6.4 PROTECTION OF HERBS AND SPICES

During cultivation, the management of pests and diseases are the major challenges. Shahnazzi et al. [86] explained that identification and determination of plant diseases, pesticide resistance plant disease, and development of environmental-benign pesticides shall improve breeding strategy and productivity of nutmeg.

3.6.5 IDENTIFICATION OF HERBS AND SPICES

The identification of spices is usually based on morphological characters using microscopic instruments or magnifying glasses. However, molecular detection of spices has been developed using DNA isolation, whole genome amplification with PCR, and sequencing. Furthermore, the utilization of molecular detection should be supported by excellent DNA isolation method, primer design, and operating conditions of PCR [32].

3.6.6 SUSTAINABILITY OF GENETIC DIVERSITY

Genetic diversity of herbs and spices may be damaged by the modernization in the agricultural sector, urbanization, deforestation, climate factor, and changing land uses. Genetic diversity reduction itself may occur in the crop, variety, and allele of the plants. Therefore, conservation strategies should consider both conventional and modern technologies to maintain the valuable genetic diversity of herbs and spices, and improving social education for sustainable development [16].

3.7 SUMMARY

DM (type-1 diabetes and type-2 diabetes) is a metabolic condition. Diabetic therapies include insulin therapy or exercise, and the use of herbs and spices. Herbs and spices contain significant amounts of phytochemicals, such as, essential oil, phenolic compounds, terpenoids, flavonoids, xyloglucan with anti-diabetic activity. The utilization of herbs and spices as functional foods involves several challenges, such as: (1) quantity of the phytochemical versus ripening stage and extraction method; (2) cultivation and sortation of raw materials; (3) availability and security of herbs and spices; (4) identification methods of phytochemicals; and (5) sustainability of herbs and spices.

ACKNOWLEDGMENTS

Authors express their gratitude to: Ardhiani Kurnia Hidayanti, K. Rusmadi, and Miswan, M. Hum for their cooperation and support in this project.

KEYWORDS

- **advanced glycation end-products**
- **antidiabetic activity**
- **bagasse Fiber Concentrates**
- **butylated hydroxytoluene**
- **cholesteryl ester transfer protein**
- **cumin-seed peptides**
- **diabetes mellitus**

REFERENCES

1. Adefegha, S. A., & Oboh, G., (2012). *In Vitro* inhibition activity of polyphenol-rich extracts from *Syzygium aromaticum* (Clove) buds against carbohydrate hydrolyzing enzymes linked to type 2 diabetes and Fe^{2+}-induces lipid peroxidation in rat pancreas. *Asian Pacific Journal of Tropical Biomedicine*, 774–781.

2. Aditama, T. Y., (2015). *Jamu Dan Kesehatan (Herbalism and Health)* (2nd edn., pp. 20–30). Jakarta, Indonesia: Lembaga Penerbit Badan Penelitian dan Pengembangan Kesehatan (Institute for Health Research and Development Agency).
3. Agnihotri, A., & Singh, V., (2013). Effect of *Tamarindus indica* Linn. and *Cassia fistula* Linn. stem bark extracts on oxidative stress and diabetic conditions. *Acta Poloniae Pharmaceutica-Drug Research, 70* (6), 1011–1019.
4. Al-Rawi, S., Ibrahim, A. H., & Rahman, N. N. A., (2011). The Effect of supercritical fluid extraction parameters on the nutmeg oil extraction and its cytotoxic and antiangiogenic properties. *Procedia Food Science, 1*, 1946–1952.
5. Al-Sa'aidi, J. A. A., & Alrodhan, M. N. A., (2012). Antioxidant activity of N-butanol extract of celery (*Apium Graveolens*) seed in streptozotocin-induced diabetic male rats. *Research in Pharmaceutical Biotechnology, 4* (2), 24–29.
6. Al-Sa'aidi, J. A., & Al-shihmani, B. A., (2013). Anti-hyperglycaemic and pancreatic regenerative effect of n-butanol extract of celery (*Apium Graveolens*) seed in STZ-induced diabetic male rats. *Suez Canal Veterinary Medicine Journal, 18* (1), 24–29.
7. American Diabetes Association (ADA), (2005). Diagnosis and classification of diabetes mellitus. *Diabetes Care, 28* (Suppl. 1), 537–542.
8. Ammon, H. P. T., Anazodo, M. I., & Safayhi, H., (1992). Curcumin: Potent inhibitor of leukotriene B4 formation in rat peritoneal polymorphonuclear neutrophils (PMNL). *Planta Medica, 58* (2), 226–230.
9. Arafa, H. M., (2005). Curcumin attenuates diet-induced hypercholesterolemia in rats. *Medical Science Monitor, 11* (7), BR228–BR234.
10. Araujo, C. A. C., & Leon, L. L., (2001). Biological activities of *Curcuma Longa* L. *Memórias do Instituto Oswaldo Cruz, 96*, 723–728.
11. Arun, N., & Nalini, N., (2002). Efficacy of turmeric on blood sugar and polyol pathway in diabetic albino rats. *Plant Foods for Human Nutrition, 57*, 41–52.
12. Assefa, A. D., Keum, Y. S., & Saini, R. K., (2018). Comprehensive study of polyphenols contents and antioxidant potential of 39 widely used spices and food condiments. *Journal of Food Measurement and Characterization, 12* (3), 1548–1555.
13. Azimi, P., Ghiasvand, R., Feizi, A., Hariri, M., & Abbasi, B., (2014). Effects of cinnamon, cardamom, saffron, and ginger consumption on markers of glycemic control, lipid profile, oxidative stress, and inflammation in type 2 diabetes patients. *The Review of Diabetic Studies*, 11 (3), 258–266.
14. Babu, K. N., Sastry, E. V. D., Saji, K. V., & Divakaran, M., (2015). Diversity and erosion in genetic resources of spices. Chapter 9, In: Ahuja, M. R., & Jain, S. M., (eds.), *Genetic Diversity and Erosion in Plants* (pp. 225, 226) Switzerland: Springer International Publishing.
15. Babu, P. S., & Srinivasan, K., (1997). Hypolipidemic action of curcumin: The active principle of turmeric (*Curcuma Longa*) in streptozotocin-induced diabetic rats. *Molecular and Cellular Biochemistry, 166* (1, 2), 169–175.
16. Bailey, R., (2005). Functional foods in Japan: FOSHU (Foods for Specified Health Uses). Chapter 15; In: Hasler, C. M., (eds.), *Regulation of Functional Foods and Nutraceuticals: A Global Perspectives* (p. 249). London, UK: Blackwell Publishing Ltd.
17. Bang, M. A., & Kim, H. A., (2009). Alterations in the blood glucose, serum lipids and renal oxidative stress in diabetic rats by supplementation of onion (*Allium cepa.* Linn). *Nutrition Research and Practice, 3* (3), 242–246.

18. Barnes, J., Anderson, L. A., & Phillipson, J. D., (2007). *Herbal Medicines* (3rd edn., p. 146). New York: Pharmaceutical Press.
19. Benítez, V., & Mollá, E., (2017). Physicochemical properties and *In Vitro* antidiabetic potential of fiber concentrate from onion by-products. *Journal of Functional Foods, 36*, 34–42.
20. Bettaieb, I., Bourgou, S., & Wannes, W. A., (2010). Essential Oils, phenolics, antioxidant activities of different parts of cumin (*Cuminum cyminum* L.). *Journal of Agricultural and Food Chemistry, 58*, 10410–10418.
21. Bhandari, S. R., Bashyal, U., & Lee, Y. S., (2016). Variation in proximate nutrients, phytochemicals, and antioxidant activity of field-cultivated red pepper fruits at different harvest times. *Horticulture, Environment, and Biotechnology, 57* (5), 493–503.
22. Bruneton, J., (1995). *Pharmacognosy, Phytochemistry, Medicinal Plants* (p. 265). Paris, France: Lavoisier Publishers.
23. Butt, M. S., & Sultan, M. T., (2011). Ginger and its health claims: molecular aspects. *Critical Reviews in Food Science and Nutrition, 51*, 383–393.
24. Caluwé, E. D., Halamová, K., & Damme, P. V., (2010). Tamarind (*Tamarindus indica* L.): A review of traditional uses, phytochemistry, and pharmacology. chapter 5; In: Juliani, H. R., Simon, J. E., & Ho, C. T., (eds.), *African Natural Plant Products: New Discoveries and Challenges in Chemistry and Quality* (pp. 85–100). New York: Oxford University Press (ACS Symposium Series).
25. Campos, K. E., Diniz, Y. S., & Cataneo, A. C., (2003). Hypoglychemic and antioxidant effects of onion, *Allium Cepa*: Dietary onion addition, antioxidant activity and hypoglycaemic effects on diabetic rats. *International Journal of Food Sciences and Nutrition, 3*, 241–246.
26. Chen, Z. B., Liu, J. Q., Chen, B. H., Xu, G. Y., & Xie, R. F., (2007). Content Mensuration of red pigment in flesh of *Symplocos paniculata* and Elementary study on its basic physical and chemical properties. *Journal of Zhangzhou Normal University, 1*, 108–112.
27. Chentli, F., Azzoug, S., & Mahgoun, S., (2015). *Diabetes mellitus* in elderly. *Indian Journal of Endocrinol Metabolism, 19*, 744–752.
28. Craig, W. J., (1999). Health-promoting properties of common herbs. *The American Journal of Clinical Nutrition, 70* (Suppl. 3), 491S–499S.
29. Ebrahimi, E., Shirali, S., & Afrisham, R., (2016). Effect and mechanism of herbal ingredients in improving *Diabetes mellitus* complications. *Jundishapur Journal of Natural Pharmaceutical Product, 2016*, 1–9.
30. El-Demerdash, F. M., Yousef, M. I., & El-Naga, N. I. A., (2005). Biochemical study on the hypoglycemic effects of onion and garlic in alloxan-induced diabetic rats. *Food and Chemical Toxicology, 43*, 57–63.
31. Farnsworth, N. R., Akerele, O., Bingel, A. S., Soejarto, D. D., & Guo, Z., (1985). Medicinal plants in therapy. *Bulletin of the World Health Organization, 63* (6), 965–981.
32. Focke, F., Haase, I., & Fischer, M., (2011). DNA-based identification of spices: DNA isolation, whole genome amplification, and polymerase chain reaction. *Journal Agricultural and Food Chemistry, 59*, 513–520.
33. Gautam, S., Pal, S., Maurya, R., & Srivastava, A. K., (2015). Ethanolic extract of *Allium cepa* stimulates glucose transporter: 4-mediated glucose uptake by the activation of insulin signaling. *Planta Medica, 81,* 208–214.
34. Gelodar, G., & Nazify, H. A. S., (1997). Effect of celery, apple tart and carrots on some biochemical parameters in diabetic rats. *Journal of Kerman University Medical Science, 3* (40), 114–119.

35. Ghasemnezhad, M., Sherafati, M., & Payvast, G. A., (2011). Variation in phenolic compounds, ascorbic acid and antioxidant activity of five colored bell pepper (*Capsicum Annum*) fruits at two different harvest times. *Journal of Functional Foods, 3*, 44–49.
36. Gupta, A. D., Bansal, V. K., Babu, V., & Maithil, N., (2013). Chemistry, antioxidant, and antimicrobial potential of nutmeg (*Myristica fragrans* Hott). *Journal of Genetic Engineering and Biotechnology, 11*, 25–31.
37. Gutierrez, R. M., Juarez, V. A., Sauceda, J. V., & Sosa, I. A., (2014). *In Vitro* and *In Vivo* antidiabetic and antiglycation properties of *Apium Graveolens* in type 1 and 2 diabetic rats. *International Journal of Pharmacology, 10*, 368–379.
38. Hajiaghaalipour, F., Khalilpourfarshbafi, M., & Arya, A., (2015). Modulation of glucose transporter protein by dietary flavonoids in type 2 *Diabetes mellitus*. *International Journal of Biological Science, 11*, 508–524.
39. Hossain, M. K., Dayem, A. A., & Han, J., (2016). Molecular Mechanism of the anti-obesity and anti-diabetic properties of flavonoids. *International Journal of Molecular Sciences, 17*, 569–573.
40. International Diabetes Federation (IDF), (2017). *IDF Diabetes Atlas*. (8th edn., pp. 14–106). https://www.idf.org/e-library/epidemiologyresearch/diabetes-atlas.html (accessed on 4 August 2020).
41. Jafari, T., Mahmoodnia, L., & Tahmasebi, P., (2018). Effect of cumin (*Cuminum Cyminum*) essential oil supplementation on metabolic profile and serum leptin in pre-diabetic subjects: A randomized double-blind placebo-controlled clinical trial. *Journal of Functional Foods, 47*, 416–422.
42. Jafarnejad, S., Tsang, C., Taghizadeh, M., & Asemi, Z., (2018). Meta-analysis of cumin (*Cuminum Cyminum* L.) consumption on metabolic and anthropometric indices in overweight and type 2 diabetics. *Journal of Functional Foods, 44*, 313–321.
43. Jain, R. C., & Vyas, C. R., (1975). Garlic in alloxan induced diabetic rabbits. *American Journal of Clinical Nutrition, 28*, 684–685.
44. Jin, S., & Cho, K. H., (2011). Water extracts of cinnamon and clove exhibits potent inhibition of protein glycation and anti-atherosclerotic activity *in vitro* and *in vivo* hypolipidemic activity in zebrafish. *Food and Chemical Toxicology, 49*, 1521–1529.
45. Jung, J. Y., Lim, Y., Moon, M. S., Kim, J. Y., & Kwon, O., (2011). Onion peel extracts ameliorate hyperglycemia and insulin resistance in high-fat diet/streptozotocin-induced diabetic rats. *Nutrition and Metabolism, 8* (18), 1–8.
46. Juwita, R., & Tsuchida, S., (2017). Current conditions and profitability of the nutmeg industry in Bogor regency, Indonesia. *Journal of the International Society for Southeast Asian Agricultural Sciences, 23* (2), 33–44.
47. Juwita, R., Tsuchida, S., & Munarso, S. J., (2018). Local government support for the development of nutmeg industry in Fakfak regency, West Papua, Indonesia. *International Journal of Environmental and Rural Development, 9* (1), 63–70.
48. Kenig, S., Baruca-Arbeiter, A., & Mohorko, N., (2018). Moderate but not high daily intake of chili pepper sauce improves serum glucose and cholesterol levels. *Journal of Functional Foods, 44*, 209–217.
49. Khan, V., Najmi, A. K., & Akhtar, M., (2012). Pharmacological appraisal of medicinal plants with antidiabetic potential. *Journal of Pharmacy and Bioallied Sciences, 4*, 27–42.
50. Khandouzi, N., Shidfar, F., & Rajab, A., (2015). The effects of ginger on fasting blood sugar, hemoglobin A1C, apolipoprotein B, apolipoprotein A-I and malondialdehyde in type 2 diabetic patients. *Iranian Journal of Pharmaceutical Research, 14* (1), 131–140.

51. Kirkman, M. S., Briscoe, V. J., Clark, N., Florez, H., & Haas, L. B., (2012). Diabetes in older adults. *Diabetes Care, 35*, 2650–2664.
52. Ko, E. Y., Nile, S. H., Jung, Y. S., & Keum, Y. E., (2018). Antioxidant and antiplatelet potential of different methanol fractions and flavonols extracted from onion (*Allium cepa* L.). *Journal Biotech, 8*, 155.
53. Kooti, W., Ali-Akbari, S., & Asadi-Samani, M., (2014). Review on medicinal plant of *Apium graveolens. Advanced Herbal Medicine, 1* (1), 48–59.
54. Kuroda, M., Mimaki, Y., Nishiyama, T., & Mae, T., (2005). Hypoglycemic effects of turmeric (*Curcuma Longa* L. Rhizomes) on genetically diabetic KK-A^y mice. *Biological and Pharmaceutical Bulletin, 28* (5), 937–939.
55. Lim, C. Y., Junit, S. M., Abdulla, M. A., & Aziz, A. A., (2013). *In vivo* biochemical and gene expression analyses of antioxidant activities and hypocholesterolemic properties of *Tamarindus indica* fruit pulp extract. *PLoS ONE, 8* (7), 1–12.
56. Lim, T. K., (2013). *Edible Medicinal and Non-Medicinal Plants, Volume 5: Fruits* (pp. 19–27) Dordrecht, Netherlands: Springer Science Business Media.
57. Lukačínová, A., Mojžiš, J., & Beňačka, R., (2008). Preventive effects of flavonoids on alloxan-induced *Diabetes mellitus* in rats. *Acta Veterinaria Brno, 77*, 175–182.
58. Maiti, R., Das, U. K., & Ghosh, D., (2005). Attenuation of hyperglycemia and hyperlipidemia in streptozotocin-induced diabetic rats by aqueous extract of seed on *Tamarindus indica. Biological and Pharmaceutical Bulletin, 28* (7), 1172–1176.
59. Maiti, R., Jana, D., Das, U. K., & Ghosh, D., (2004). Anitidiabetic effect of aqueous extract of seed of *Tamarindus indica* in streptozotocin-induced diabetic rats. *Journal of Ethnopharmacology, 92*, 85–91.
60. Martinello, F., Soares, S. M., Franco, J. J., & Santos, A. C., (2006). Hypolidemic and antioxidant activities from *Tamarindus indica* L. pulp fruit extract in hypercholesterolemic hamsters. *Food and Chemical Toxicology, 44*, 810–818.
61. Mbaveng, A. T., & Kuete, V., (2017). *Syzygium aromaticum*. Chapter 29; In: Kuete, V., (ed.), *Medicinal Spices and Vegetables from Africa* (pp. 611–613). London, UK: Academic Press.
62. McCormick Research Institute (2006). *The History of Spices*; https://www.mccormick-scienceinstitute.com/resources/history-of-spices; (accessed on 24 August 2020). (accessed on 4 August 2020).
63. Merriam, C., (1964). *Merriam-Webster Dictionary.* Online https://www.merriam-webster.com/ (accessed on 4 August 2020).
64. Modak, M., Dixit, P., Londhe, J., Ghaskadbi, S., & Devasagayam, T. P. A., (2007). Indian herbs and herbal drugs used for the treatment of diabetes. *Journal of Clinical Biochemistry and Nutrition, 40*, 163–173.
65. Moghaddam, M., Miran, S. N. K., & Pirbalouti, A. G., (2015). Variation in essential oil composition and antioxidant activity of cumin (*Cuminum Cyminum* L.) fruits during stages of maturity. *Industrial Crops and Products, 70*, 163–169.
66. Mohammed, A., & Islam, M. S., (2018). Spice-derived bioactive ingredients: Potential agents or food adjuvant in the management of *Diabetes Mellitus. Frontiers in Pharmacology, 9*, 893–898.
67. Mohammed, A., Wudil, A. M., & Alhassan, A. J., (2017). Hypoglycemic activity of curcuma longa linn root extracts on alloxan-induced diabetic rats. *Haya: The Saudi Journal of Life Sciences, 2* (2), 43–49.

68. Nahar, L., Nasrin, F., Zahan, R., Haque, A., Haque, E., & Mosaddikm, A., (2014). Comparative study of antidiabetic activity of *Cajanus Cajan* and *Tamarindus Indica* in alloxan-induced diabetic mice with reference to *In Vitro* antioxidant activity. *Pharmacognosy Research, 6* (2), 180–187.
69. Nelson, D. L., & Cox, M. M., (2008). *Principles of Biochemistry.* 5th Edition; New York, USA: W. H. Freeman and Company; pp. 929–939.
70. Nguyen, P. H., Le, T. V. T., & Kang, H. W., (2010). AMP-activated protein kinase (AMPK) activators from *Myristica Fragrans* (Nutmeg) and their anti-obesity effects. *Bioorganic and Medicinal Chemistry Letters, 20*, 4128–4131.
71. Niaz, K., Gull, S., & Zia, M. A., (2013). Antihyperglycemic/hypoglycemic effect of celery seeds in streptozotocin-induced diabetic rats. *Journal of Rawalpindi Medical College (Pakistan), 17*, 134–137.
72. Oboh, G., & Ademiluyi, A. O., (2018). Inhibition effect of garlic, purple onion, and white onion on key enzymes linked with type 2 diabetes and hypertension. *Journal of Dietary Supplements, 2018*, 1–15.
73. Ohaeri, O. C., (2001). Effect of garlic oil on the levels of various enzymes in the serum and tissue of streptozotocin-diabetic rats. *Bioscience Reports, 21* (1), 19–24.
74. Pandiya, R., & Banerjee, S. K., (2013). Garlic as an antidiabetic agent: Recent progress and patent reviews. *Recent Patents on Food, Nutrition, and Agriculture, 5*, 105–127.
75. Parry, J. W., (1953). *The Story of Spices* (pp. 190–207). New York, USA: Chemical Publishing Co, Inc.
76. Perera, H. K. I., & Handuwalage, C. S., (2015). Analysis of glycation induced protein cross-linking inhibitory effects of some antidiabetic plants and spices. *BMC Complementary and Alternative Medicine, 15* (175), 1–9.
77. Perez-Torres, I., Ruiz-Ramirez, A., Banos, G., & El-Hafidi, M., (2013). *Hibiscus sabdariffa* linnaeus, curcumin, and resveratrol as alternative medicinal agents against metabolic syndrome. *Cardiovascular and Hematological Agents in Medicinal Chemistry, 11* (1), 25–37.
78. Periasamy, S., Lin, C. H., & Nagarajan, B., (2018). Tamarind xyloglucan attenuates dextran sodium sulfate induced ulcerative colitis: Role of antioxidation. *Journal of Functional Foods, 4*, 327–338.
79. Periasamy, S., Lin, C. H., Nagarajan, B., & Sankaranarayanan, N. V., (2018). Mucoadhesive role of tamarind xyloglucan on inflammation attenuates *Ulcerative Colitis*. *Journal of Functional Foods, 47*, 1–10.
80. Rebey, I. B., Kefi, S., & Bourgou, S., (2014). Ripening stage and extraction method effects on physical properties, polyphenol composition and antioxidant activities of cumin (*Cuminum Cyminum*) seeds. *Plant Foods for Human Nutrition, 69* (4), 358–364.
81. Rema, J., & Krishnamoorthy, B., (2012). Nutmeg and mace. Chapter 22; In: Peter, K. V., (eds.); *Handbook of Herbs and Spices* (pp. 399–400). New York, USA: Woodhead Publishing.
82. Royal Botanic Garden Kew, (2017). *State of the World's Plants*. https://stateoftheworld-splants.org/ (accessed on 4 August 2020).
83. Ryu, B., Kim, H. M., & Woo, J. H., (2016). New acetophenone glycoside from the flower buds of *Syzygium aromaticum* (Cloves). *Fitoterapia, 115*, 46–51.
84. Sablu, S., Madende, M., & Ajao, A. A., (2019). The genus *Allium*: Features, phytoconstituents, and mechanisms of antidiabetic potential of *Allium cepa* and *Allium sativum.*

Chapter 9; In: Watson, R., & Preedy, V., (eds.), *Bioactive Food as Dietary Interventions for Diabetes* (pp. 137–154). New York: Academic Press.
85. Schippmann, U., Leaman, D. J., & Cunningham, A. B., (2002). Impact of cultivation and gathering of medicinal plants on biodiversity: Global trends and issues. In: *Biodiversity and the Ecosystem Approach in Agriculture, Forestry, and Fisheries* (pp. 1–21). FAO; Rome: Inter-Departmental Working Group on Biological Diversity for Food and Agriculture.
86. Shahnazi, S., Meon, S., & Vadamalai, G., (2012). Morphological and molecular characterization of *Fusarium* spp. Associated with yellowing disease of black pepper (*Piper nigrum* L.) in Malaysia. *Journal of General Plant Pathology, 78*, 160–169.
87. Sheela, C. G., & Augusti, K. T., (1992). Antidiabetic effects of s-allyl cysteine sulfoxide isolated from garlic (*Allium sativum* Linn). *Indian Journal of Experimental Biology, 30*, 523–526.
88. Sheela, C. G., Kumari, K., & Augusti, K. T., (1995). Anti-diabetic effects of onion and Garlic sulfoxide amino acids in rats. *Planta Medica, 61*, 356–367.
89. Shidfar, F., Rajab, A., & Rahideh, T., (2015). The effect of ginger (*Zingiber officinale*) on glycemic markers in patients with type 2 diabetes. *Journal of Complementary and Integrative Medicine, 12* (2), 165–170.
90. Shukri, R., Mohamed, S., & Musapha, N. M., (2010). Cloves protect the heart, liver, and lens of diabetic rats. *Food Chemistry, 122*, 116–1121.
91. Sidhu, G. S., Mani, H., Gaddipati, J. P., & Singh, A. K., (1999). Curcumin enhances wound healing in stretozotocin induced diabetic rats and genetically diabetic mice. *Wound Repair Regen., 7* (5), 362–374.
92. Silva, F. M. V., Spadaro, A. C. C., Uyemura, S. A., & Maistro, E. L., (2009). Assessment of potential genotoxic risk of medicinal *Tamarindus Indica* fruit pulp extract using *in vivo* assay. *Genetics and Molecular Research, 8* (3), 1085–1092.
93. Siow, H. L., & Gan, C. Y., (2016). Extraction, identification, and structure-activity relationship of antioxidative and α-amylase inhibitory peptides from cumin seeds (*Cuminum cyminum*). *Journal of Functional Foods, 22*, 1–12.
94. Siow, H. L., Tye, G. J., & Gan, C. Y., (2017). Pre-clinical evidence for the efficacy and safety of A-amylase inhibitory peptides from cumin (*Cuminum cyminum*) seed. *Journal of Functional Foods, 35*, 216–223.
95. Smith-Hall, C., Larsen, H. O., & Pouliot, M., (2012). People, plants, and health: A conceptual framework for assessing changes in medicinal plant consumption. *Journal Ethnobiology and Ethnomedicine, 8*, 43–47.
96. Sole, S. S., & Srinivasan, B. P., (2012). Aqueous extract of tamarind seeds selectivity increases glucose transporter-2, glucose transporter-4, and islets' intracellular calcium levels and stimulates β-cell proliferation resulting in improved glucose homeostasis in rats with streptozotocin-induced *Diabetes mellitus*. *Nutrition Research, 32*, 626–636.
97. Sole, S. S., Srinivasan, B. P., & Akarte, A. S., (2013). Anti-inflammatory action of tamarind seeds reduces hyperglycemic excursion by repressing pancreatic B-cell damage and normalizing SREBP-1C concentration. *Pharmaceutical Biology, 51* (3), 350–360.
98. Sotto, A. D., Vecchiato, M., & Abete, L., (2018). *Capsicum Annuum* L. var. cornetto di pontecorvo PDO: Polypenolic profile and *in vitro* biological activities. *Journal of Functional Foods, 40*, 679–691.
99. Srinivasan, K., (2005). Plant foods in the management of diabetes mellitus: Spices as beneficial antidiabetic food adjuncts. *International Journal of Food Sciences and Nutrition, 56* (6), 399–414.

100. Srivastava, R., & Srivastava, S. P., (2011). Antidiabetic and antidyslipidemic activities of *Cuminum cyminum* in validated animal models. *Medical Chemistry Research, 20*, 1656–1666.
101. Starkenmann, C., Niclass, Y., & Troccaz, M., (2011). Nonvolatile S-Alk (en)ylthio-L-cysteine derivates in fresh onion (*Allium cepa* L.). *Journal Agricultural and Food Chemistry, 59*, 9457–9465.
102. Su, L., Yin, J. J., Charles, D., Zhou, K., Moore, J., & Yu, L., (2007). Total Phenolic contents, chelating capacities, and radical-scavenging properties of black peppercorn, nutmeg, rosehip, cinnamon, and oregano leaf. *Food Chemistry, 100*, 990–997.
103. Tahir, H. U., Sardraz, R. A., Ashraf, A., & Adil, S., (2016). Chemical composition and antidiabetic activity of essential oils obtained from *Syzygium aromaticum* and *Cuminum cyminum*. *International Journal of Food Properties, 19*, 2156–2164.
104. Tang, W. H., Martin, K. A., & Hwa, J., (2012). Aldose reductase, oxidative stress, and *Diabetic mellitus*. *Frontier Pharmacology, 3* (87), 1–8.
105. Upasani, S. V., Ingle, P. V., & Patil, P. H., (2014). Traditional Indian spices useful in *Diabetes mellitus:* An updated review. *Journal of Pharmaceutical and Biosciences, 4*, 157–161.
106. USC-UCLA Joint East Asian Studies Center. (1993). *Along the Silk Road: People, Interaction, and Cultural Exchange*. https://festival.si.edu/2002/the-silk-road/the-silk-road-connecting-peoples-and-cultures/smithsonian (accessed on 4 August 2020).
107. Vinayagam, R., & Xu, B., (2015). Antidiabetic properties of dietary flavonoids: Cellular mechanism review. *Nutrition and Metabolism, 12*, 60–65.
108. Willatgamuwa, S. A., & Platel, K., (1998). Antidiabetic influence of dietary *Cuminum cyminum* in streptozotocin induces diabetic rats. *Nutrition Research, 18* (1), 131–142.
109. World Health Organization (WHO), (2016). *Global Report Diabetes* (pp. 15–31). https://www.who.int/diabetes/global-report/en/ (accessed on 4 August 2020).
110. Yamatoya, K., & Yamazaki, K., (2011). Effects of xyloglucan with partial removal of galactose on plasma lipid concentration. *Journal of Functional Foods, 3*, 275–279.
111. Yang, H. J., Jang, D. J., & Hwang, J. T., (2012). Anti-diabetic effects of Korean red pepper via AMPK and PPAR-Γ activation in C2C12 myo-tubes, *Journal of Functional Foods, 4*, 552–558.
112. Yang, J., Wang, T., Yang, J., & Rao, K., (2013). S-allile cysteine restores erectile function through inhibition of reactive oxygen species generation in diabetic rats. *Andrology, 1*, 487–494.
113. Yusni, Y., Zufry, H., Meutia, F., & Sucipto, K. W., (2018). The effects of celery (*Apium Graveolens* L.) leaves treatment on blood glucose and insulin levels in elderly pre-diabetics. *Saudi Medical Journal, 39* (2), 154–160.
114. Zhang, S., You, Y., Liu, J., Qin, X., & Liu, X., (2018). Hypoglycemic effect of capsaicinoids via elevation of insulin level and inhibition of glucose absorption in streptozotocin-induced diabetic rats. *Journal of Functional Foods, 51*, 94–103.

PART II

Role of Phytochemicals in Traditional Ethnomedicines

CHAPTER 4

VALUE-ADDED PRODUCTS AND BIOACTIVE COMPOUNDS FROM FRUIT WASTES

RANJAY KUMAR THAKUR, RAHEL SUCHINTITA DAS, PRASHANT K. BISWAS, and MUKESH SINGH

ABSTRACT

Fruits and vegetables produce a hefty amount of wastes that contribute to grave nutritional loss, negative economic impact, and environmental pollution. These wastes have enormous potential for composting, producing novel efficient, and eco-friendly products. Several research studies have been conducted for extraction, isolation, and characterization techniques of several bioactive compounds from these wastes, through conventional extraction, microwave-supported extraction, supercritical fluid-enhanced extraction, ultrasound-assisted, high hydrostatic pressure extraction, pulsed electric field extraction, and enzyme-assisted extraction. In this chapter, the extraction, and application of bioactive and functional compounds from fruit (pineapple, pomegranate, mango, grape, banana, apple, and citrus fruit) wastes are discussed.

4.1 INTRODUCTION

Fruits and vegetables have a significant role in the national economy, controlling malnutrition, and solving the hunger issue. Bananas, apples, citrus fruits, and grapes are traded worldwide. Latin America is a leading global export force and China has a huge and increasing import market [181].

Fruits are rich in nutrients due to the presence of numerous bioactive compounds, in both their edible and non-edible parts, such as, peels, and seeds. Developing nations face major challenges in their practice of handling post-harvest losses and causes of resulting agro-waste can be traced through every stage, from harvesting to processing, packaging, transporting till the final stage of consumption. The United Nations-Food and Agriculture Organization (FAO) has projected that losses in fruits and vegetables are highest reaching up to 60% amongst all types of food wastes [147].

Agro-waste is one of the principal sources of municipal solid wastes that lead to an enormous environmental burden. Traditional solid waste handling methods of land filing results in harmful methane and carbon dioxide emissions, if not properly controlled. Incineration of the agro-waste leads to the emission of dangerous pollutants, such as: dioxins, furans, and particulate matters. Therefore, there is an urgent and critical necessity to hunt for value-added utilization from these agro-wastes to cater to environmental protection, sustainability, and economic viability. Fruit waste mainly comprises of: peels, rinds, seeds, pomace, etc. These waste components are rich in phytochemicals and bioactive compounds, such as: polyphenols, carotenoids, enzymes, vitamins, and dietary fibers with several health benefits [153].

Sugars, amino acids, fatty acids, glycerol, and phosphates ensuing from food waste are chief nutrients for the growth of bacteria, fungi, and algae to generate biomass and a range of primary and secondary metabolites, which can be utilized for the production of chemicals, materials, and energy [94]. Therefore, the utilization of unique biomolecules and bioactive molecules from food, vegetable, and fruit wastes are considered as a major attempt to reduce environment contamination in the food, agricultural, textile, cosmetic, chemical, and pharmaceutical industries.

Different technologies (such as: conventional extraction (e.g., soxhlet extraction, maceration, hydro-distillation) and emerging technologies (such as: ultrasound-assisted extraction (UAE), supercritical fluid-applied extraction, microwave-assisted extraction (MAE), high hydrostatic pressure (HHP)-applied extraction, pulsed electric field-dependent extraction and enzyme-assisted extraction) have been adopted for extraction and isolation of valuable compounds from food, vegetable, and fruit wastes [151].

This chapter focuses on the identification and application of bioactive and functional compounds from fruit (pineapple, pomegranate, mango, grape, banana, apple, and citrus fruit) wastes.

4.2 DIFFERENT TYPES OF WASTES FROM FRUITS

4.2.1 PINEAPPLE

Pineapple by-products include residual pulp, peels, stems, and leaves. Processing of pineapple (*Ananas comosus*) generates peels, core, pulps, top, and final-product amounting to 14%, 9%, 15%, 15%, 48%, respectively [40]. Peel is the major biowaste generated during pineapple processing [74].

A substantial amount of sugars is present in pineapple peel that can be utilized as substrates in fermentation and also as a potential substrate for methane, ethanol, and hydrogen generation [33, 80]. Bromelain (EC 3.4.22.32), a commercially available proteolytic enzyme, is often derived from the stem of pineapple. It is used in a meat tenderizer, a bread dough improver, to prevent browning, as a clarifying agent in beer, animal feed, cosmetic, and textile industries [13]. Pineapple waste can also be used for the production of organic acids, such as, lactic acid [3] and acetic acid [162].

The increase in demand for naturally derived flavors has encouraged the production of natural vanillin from substrates through the process of microbial biotransformation. Vanillin (4 hydroxy-3-methoxy-benzaldehyde) is produced from vanillic acid. Pineapple peel waste harbors ferulic acid, which is the precursor of vanillic acid [169]. Therefore, vanillin can be synthesized from pineapple peels from a series of biochemical reactions [98].

The pineapple by-products also contain a significant amount of dietary fiber, mainly insoluble dietary fiber, and can be used for the development of food-grade low-calorie fiber-enriched food products [68]. Crown of pineapple was used as the carbon source for the media for the growth of three different strains, namely, *Rhodopseudomonas Saccharomyces cerevisiae* (baker's yeast), and *Trichoderma harzianum.* These strains were used for microbial single-cell protein production [118].

4.2.2 POMEGRANATE

Pomegranate (*Punica granatum*) is rich fruit with bioactive phytochemicals. Pomegranates contains aril (ranging from 50% to 70% of entire fruit); and it consists of 78% juice along with 22% of seeds [114].

Residues arising from pomegranate processing (such as: pomegranate pee land pomegranate seeds) have numerous bioactive compounds with great potential to produce value-added products. El-Fallah et al. [47] reported that the peel of pomegranate has higher antioxidant activity than its seeds [47]. Malviya et al. [102] reported the antibacterial activity of extracts of the peel of pomegranate against *Staphylococcus aureus, Salmonella typhi, Enterobacter aerogenes,* and *Klebsiella pneumonia* [102]. Khan et al. [76] evaluated the potential production of single-cell protein using pomegranate rind as substrate in the fermentation process using *Saccharomyces cerevisiae.* The yield was estimated at 54.28% of crude protein from 100 g of pomegranate rind [75, 76]. Combination of pomegranate peel extract with chitosan and polyvinyl alcohol was used to develop for bicomposite-polymer with antimicrobial activity against Gram-positive bacteria [11].

4.2.3 MANGO

Mango (*Mangifera indica* L.) a seasonal fruit, wide ranges of products (such as: pickles, nectar, puree, chutney, and canned leather slices from mango) have become popular around the world [52]. Processing of mangoes generates 18% of inoperable pulp, 13%of seeds, 11% of peel, and 58% of total finished product [88]. The mango processing waste is derived mainly from the epicarp and endocarp and has been projected at 75,000 MT in India [45]. There is very low commercial use of mango seed kernel, which is discarded mostly as a fruit waste in the processing industries [100].

Mango waste is enriched with antioxidant compounds for several health benefits [108]. Carotenoids present in mango waste are used to manufacture natural colorants in the food industry [121]. The carotenoid in raw mango peels is about 4–8 times higher than in the ripe mangoes [8]. Different varieties of the dietary fiber in mango peels have been isolated. The total amount of dietary fiber varied from 45% to 78% in the dry peel. The content of soluble dietary fiber content, in raw and ripened mango peels, constitutes more than 35% of total dietary fiber [127].

It has been reported that mango seed kernel oil contains polyunsaturated fatty acids, such as, linoleic, and oleic acids [79]. Antimicrobial activity of mango kernel against *Staphylococcus aureus*, *Pseudomonas aeruginosa*, *Bacillus subtilis*, *Candida albicans,* and *Escherichia coli* has been identified [6]. Also, mango kernel contains several phytochemicals, such as, tannins, flavonoids, coumarins, and terpenes [122]. Mango kernel extract with soy protein isolate and fish gelatin was used to develop a biopolymer with antioxidant activity [46]. *Saccharomyces cerevisiae* was grown on mango waste for the production of single-cell protein through submerged fermentation. It was found that single-cell protein production was 40%, when mango peel waste was used in fermentation medium [5].

4.2.4 GRAPES

Grapes (*Vitis spp.*) can be consumed raw and in formulating products, such as, wine, raisins, juice, jam, jelly, vinegar, and seed oil [58]. According to the Statistical Report of International Organization of Vine and Wine (OIV) in 2016, globally 75.8 million tons of grapes were produced and the major (57%) amount of grape was used for the production of wine [124]. In the wineries, tons of residues in the form of grape pomace are produced after the fermentation process [55]. Globally grape and wine industries produce approximately 5 to 9 million MT of solid waste annually globally [147].

Grape stems, seeds, and skins are collective parts of grape pomace, which is utilized for oil extraction and their extracts are employed to formulate antioxidant and antibacterial agent preparation [187]. Grape pomace-derived dietary fibers and phenolics are mainly concentrated in the fruit skins, seed, and pulp. These dietary fibers are phenolic-rich dietary fiber and can be used as a dietary supplement and antioxidant. The phenolic rich grape pomace dietary fiber can be isolated by solvent extraction at 90°C [4].

Grape pomace-derived dietary fibers are considered as a potential functional ingredient in bakery products, to remove red wine tannins, to increase the dietary fiber [148], to increase the total phenolic content, and in delaying lipid oxidation in salad dressings and yogurt [62], seafood [171], to reduce rancidity in red wines [53]. The extract of grape pomace has also emerged into hydrophobic and hydrophilic chitosan edible film to increase the shelf-life and offer antioxidant properties [155].

The amount of phenolic compounds is higher in the seeds than the other parts of grapes and its concentration is about 10% in the pulp, 20% in the skin, and 70% in the seeds [30]. Grape seeds contain 8–15% of oils, mainly comprising of oleic and linoleic acids [39], which amounts to more than 89% of the total oil content [31]. Grape seed oil can be used as a functional ingredient to modify and formulate healthier food products in the meat industry. According to the studies by Ismail et al. [69], grape seed oil is neuroprotective, hepatoprotective, and can reduce cholesterol in the liver.

Soy protein isolate with grape seed extract was used to develop edible film, which offer antimicrobial activity against *Salmonella typhimurium* and *Escherichia coli* O157:H7 [7]. Furthermore, grape fruit seed extract with barley bran protein and gelatin were used to develop packing material for salmon fish. It was found that this biocomposite can inhibit formation of *Listeria monocytogenes* and *Escherichia coli* O157:H [106]. Wine grape pomace was used as a cross-linking agent with minerals (Mg^{2+}, Ca^{2+}, and Fe^{3+}), proteins, organic acids, amino acids and phenolic acids [72].

4.2.5 BANANA

Banana (*Musa acuminate* and *Musa balbisiana*) peel are the major by-product, which amounts to 30% (nearly) of the fruit [60]. The ratio of banana waste and product is 2:1. The lignocellulosic mass is discarded or transported to open dumping grounds. The process helps to preserve the soil moisture and supply organic matter, but has a latent risk of transmission of diseases. The lignocellulosic biomass produces greenhouse gases during decomposition [18].

Banana peel has abundant phytochemicals, such as, phenolic antioxidant (gallocatechin), carotenoids (β-carotenoids, xanthophylls, and α-carotenoid), catecholamine's, anthocyanin (delphinidin and cyanidin), sterols, and triterpenes [73, 160]. Banana peel is rich in dietary fiber, which can be incorporated into biscuits in the form of powder. In biscuit, it cannot change overall aroma, taste, and color and but maintains the calorie (low calorie food products) [71].

On the other hand, banana peel is potential to remove heavy metals, such as chromium (III) [129] and chromium (IV) [111, 129] through sorption process. Banana peels can also be used to produce silver nanoparticles, as these are rich in pectin, lignin, and hemicellulose. These nanoparticles have shown antimicrobial activity against pathogenic fungi, such as, *Candida albicans (BX and BH)* and *Candida lipolytica* (National collection of industrial microorganisms (NCIM) 3589) and bacterial cultures, such as, *Proteus valgaris* (microbial type culture collection and gene bank (MTCC) 426*)*, *Citrobacter kosari* (MTCC 1657), *Enterobacter aerogenes* (MTCC 111), *Klebsiella sp.*, *Pseudomonas aeruginosa* (MTCC 728) and *Escherichia coli* (MTCC 728) [19].

The by-products of banana include: leaves, stalk, and inflorescence. Furthermore, banana waste has non-food applications, such as, natural fiber, thickening agent, alternative source of micro- and macro-nutrients, bioactive compounds and bio-fertilizer [126]. Banana waste has been used as growth medium for *Aspergillus niger* and it was found that after 6 days of fermentation, concentration of single-cell protein was 0.57 ± 0.01 g L^{-1} in fermentation medium [130]. In an investigation, *Saccharomyces cerevisiae* was grown on waste of banana for the production of single-cell protein and the yield was 58.62% [5].

4.2.6 APPLE

Processing of apples generates stems, skin, and residual flesh [97, 179]. The by-product of apple (*Malus domestica Borkh.*) is apple cider and apple pomace, which is generated from 25% of the original fruit with 85% (wet basis) moisture content [166]. On slicing, apples produce 11% of seeds with pulp as by-products, which account for about 89% of the final product [143].

It is a source of dietary fiber, especially pectin with concentration about 10–15% (dry basis) depending upon the variety [24]. According to Younis et al. [184], apple pomace has versatile functional properties, such as: diffusion retardation index in glucose, emulsifying activity, holding capacity of water-oil, and antimicrobial activity. The most reasonable utilization for apple pomace is the pectin production [21, 42]. Apple pomace also contains large amount of non-starch polysaccharides, high amounts of insoluble fiber, and soluble fiber [28, 57, 177, 164]. By the addition of dried apple pomace powder, fiber-enhanced bakery products were formulated with wheat flour [28, 164]. The cake is prepared by powder of apple pomace that contains high amount of dietary fiber and phenolic content [109]. Apple pomace contains phloretin glycosides, chlorogenic acid and quercetin glycosides [26, 96]. Catechins and procyanidins are also present in waste of apples [56].

Apple waste, generated from processing of juice, was tested for colon tumor-cell lines, such as, HT29, HT115 and CaCo-2. Results indicate that compounds from apple waste can bestow protection against DNA injury, augment barrier function and restrain cell invasion [110]. In another study, inhibitory effects of non-extractable antioxidants and extractable antioxidants from apple waste in freeze-dried form on HeLa, HepG2 and HT-29 human cancer cells were studied and the non-extractable antioxidants were found to be more effective to reduce the growth of cancer cells [170].

Biocomposite with antimicrobial activity and enriched with polyphenols has been developed with the apple skin waste. Such biocomposite showed antimicrobial activity against *Salmonella enteric*, *Escherichia coli O157:H* and *Listeria monocytogenes* [86]. Single-cell protein was produced with *Saccharomyces cerevisiae* with apple waste in fermentation medium. The yield of single-cell protein was 50.86% [5].

4.2.7 CITRUS

The peel of citrus (*Citrus L.*) is rejected as a waste, which has diverse array of secondary biocomponents with significant antioxidant activity compared to other fruit parts [105]. The mandarins generate 16% of peels and 84% of finished product [143]. About 50% of citrus fruit is wasted in the form of rag, peel, and seeds [63, 65].

Food industry utilizes citrus peel for production of pectin, oil, molasses, and limone. It has been considered for investigation for its high content of bioactive compounds, such as, flavanones, flavonols, polymethoxylated flavones and phenolic acids. These compounds are used as natural antioxidants for biotechnological, pharmaceutical, and food industries [22].

Residues of orange juice processing industry create large proportions of by-products that are used as source of bioactive compounds [82]. The total phenolic content in peels of oranges, lemons, and grapefruit was 15% higher than those in the pulp of the fruits [61]. Lime and lemon peel oils are used as aroma and flavor compounds in non-alcoholic and alcoholic beverages and foods; and as flavoring agents in pharmaceutical industries to mask distasteful drugs. In cosmetics, they are used to form the base of lot of compositions [95]. Different antimicrobial packaging arrangements have been developed by including lemon extracts for protection of mozzarella cheese [35].

Dry powder of the pomace with orange (*Citrus sinensis)* peels was used to prepare bio-composite with good visual appearance, such as, color, flavor, and visually smooth surface [141]. Cucumber and orange peels were used for manufacturing of single-cell protein using *Saccharomyces cerevisiae* through a submerged fermentation. The yield of single-cell protein was 30.5% [115]. In an investigation, *Saccharomyces cerevisiae* was cultivated on sweet orange peel waste for the generation of single-cell protein and the yield was 26.26% [5].

Table 4.1 represents biological activities of some bioactive compounds that were isolated or produced from fruit wastes.

4.3 ISOLATION TECHNIQUES OF BIOACTIVE COMPOUNDS FROM FRUIT WASTES

Several technologies have been developed to isolate bioactive compounds from fruit wastes. Extraction method may depend on the targeted bioactive compound. Preparation of sample is one of the critical aspects to regulate the type and quantity of bioactive compound for extraction. Usually, the extraction techniques are categorized as conventional and emerging or novel [147].

4.3.1 CONVENTIONAL EXTRACTION METHODS

4.3.1.1 SOXHLET EXTRACTION

The traditional methods are regarded as conventional techniques as they have been applied since ages. The conventional extraction techniques include (1) Soxhlet extraction, (2) hydro-distillation and (3) maceration, etc. [77].

TABLE 4.1 Summary of Bioactive Compounds from Fruit Wastes

Bioactive Compound	Source	Bioactive Role	References
Amino acids	Kinnow-mandarin waste; peels of pineapple and papaya; seeds of watermelon.	Protein source; Protein enrichment of foods.	[17, 48]
Anthocyanidins (Cyanidin, Delphinidin, Pelargonidin)	Grape waste; Berry pomace; Blackcurrant fruit skin; Blackcurrant pomace; Sapota pomaces; Litchi pericarp and seeds; Apple pomace; Apple skin.	Food colorant, Antioxidant, Anti-Inflammatory, Antimicrobial	[17, 78, 167, 185]
Anthocyanins (*Cyanidin 3-galactoside; Delphinidin 3-rutinoside)*	Watermelon peels, Grape waste, Pomegranate pith, and carp; Blackcurrant fruit skin, Blackcurrant pomace.	Food colorant, Antioxidant; Anti-inflammatory; Antimicrobial.	[9, 17, 78, 185]
Bioactive lipids	Mango kernel; Pomegranate seed.	Essential fatty acids; Cardiovascular health; Skin health.	[16, 17, 103]
Carotenoids *(α-carotene; β-carotene; xanthophylls)*	Mango peel; Papaya peel; Banana peel.	Antioxidant; Vitamin pre-cursors; Food colorant.	[17, 64, 154, 163]
Dietary fiber	Watermelon rind; Passion fruit peel; Mango peel; Blackcurrant pomace.	Intestinal health promoter; Food texture improver.	[17, 32, 50, 157, 176]
Enzymes invertase	Mango kernel fermentation by *Fusarium solani*; Orange waste by *Streptomyces sp.*	Hydrolytic enzyme as processing aid.	[84, 99]
Ethyl butyrate	Apple pomace fermentation by *Ceratocystis fimbriata.*	Flavoring compound.	[90, 147]
Flavan-3-ols (catechin, epicatechin)	Seeds from winery wastes of *V. Vinifera*	Antioxidant; Anti-proliferative.	[150]
Flavonols (*Quercetin*)	Pomegranate peels, orange peels, mango peels	Antioxidant; Anti-microbial.	[17, 25, 159, 165]
Flavonones (*Naringenin*)	Citrus peels and seeds.	Antioxidant, Anti-inflammatory	[14, 17]
Glycosides	Apple peels; Banana stem	Antioxidant.	[17, 113]

TABLE 4.1 *(Continued)*

Bioactive Compound	Source	Bioactive Role	References
Pectins	Citrus fruit peels; Apple pomace; Banana peels; Mango peel.	Thickening and gelling food additive.	[49, 85, 128]
Phenolic acids (*Elagic Acid; Chlorogenic Acid; Gallic Acid; Ferulic Acid; Vanillic Acid; Caffeic Acid*)	Peels of mango and kernel; Guava seeds; Peels of orange; Passion fruit peel; Pomegranate seed; Citrus pomace and peel; Grape pomace.	Antioxidant; Anti-microbial; Anti-tumor.	[3, 8, 17, 50, 54, 83, 87, 140, 147, 180]
Saponins	Sapota seeds.	Antibacterial.	[17, 83]
Sterols (*β-Sitosterol; δ-Avenasterol; Campesterol; Stigmasterol*)	Mango seed kernels; Pomegranate seed;	Cholesterol-Lowering.	[2, 116, 132]
Stilbenes (*Resveratol*); Lignans and xyloglucans	Grape pomace; Pomegranate seed; Watermelon rinds; Mango peel; Blackcurrant pomace; Passion fruit peel.	Dietary fiber; Food additives; Fat substituents; Antioxidant.	[9, 17, 32, 34, 147]
Tannins Condensed tannins	Mango seed kernels; Mango peel; Grape stem;	Defense strategies of plants; Processing aid.	[12, 131, 139]
Terpenes (*limonin; d-limonene; ursolic acid*)	Citrus seed; Citrus pomace and Citrus peel.	Antimicrobial; anti-inflammatory; antioxidants; perfumery.	[136]
Vitamins (*α-Tocopherol; γ-Tocopherol*)	Mango seed kernel; Passion fruit seed; Pomegranate seed.	Growth and metabolism in the body	[101, 116, 175]

Soxhlet extraction is a solvent-based extraction process. Typically, soxhlet extraction is used, when the impurities are insoluble in that solvent and the desired compound has a limited solubility in a solvent [147].

A kinetic model of phenolic pigment production using soxhlet extraction was determined and those kinetic constants were implemented in large-scale system from orange peel waste, obtained from fruit juice production industry. The highest phenolic pigment yield was 57.3% (0.57 g phenolic pigment per g dry peel) from dried pulp. During extraction, temperature of 79°C, liquid/solid ratio (L/kg) of 40:1 and particle size smaller than 0.5 mm were maintained [81].

4.3.1.2 HYDRODISTILLATION

Hydro-distillation is used to extract oils and other bioactive compounds from plants. It can be applied before drying a plant sample. There are 3 kinds of hydro-distillation, such as, water distillation, water, and steam distillation [174]. The process of hydro-distillation starts with packing the plant sample in a still compartment. Then, boiling is performed with adequate quantity of water. Steam can be used as an alternative for extraction. The vapor mixture of water and oil is condensed and the condensed mixture is sent to a separator, where the bioactive compounds and oil are separated from the water [147, 156]. Hydro-distillation includes processes of hydro-diffusion, hydrolysis, and heat decomposition. Since, this process involves application of heat, it may not be suitable for heat-labile compounds [147].

In a study by Das et al. [38], fresh 1.5 kg peels of *Citrus reticulate* were hydro distilled for three hours, which yielded about 0.33% v/w of volatile oil [38].

4.3.1.3 MACERATION

Steps in the maceration process to extract bioactive compounds include: grinding of raw material, and followed by mixing with appropriate quantity of the solvent. Finally, the mixture is pressed and the solid residue is separated from solvent by filtration. The solvent carries the extracted bioactive compounds [147].

Total extracted polyphenol (TP) was 27.6 mg gallic acid equivalent g^{-1} obtained from chokeberry (*Aronia elanocarpa*) dried fruits by maceration extraction technique. This value was higher compared with ultrasonic-assisted extraction [37].

4.3.2 EMERGING TECHNOLOGIES

Conventional technologies have several disadvantages, such as, heat denaturation of heat labile bioactive compounds, low efficiency, longer extraction process, etc. Novel and emerging techniques overcome the limitations of traditional technologies [147].

4.3.2.1 MICROWAVE-ASSISTED EXTRACTION

MAE is a 'green' technique because it cuts down the use of organic solvents [10]. The electromagnetic field for microwaves ranges from 300 MHz to 300 GHz in the extraction process. It consists of 2 perpendicular fields, i.e., magnetic field and electric field. In this process, heating effect of microwaves on polar materials is considered [93]. Microwave energy is converted to heat energy through dipole rotation and ionic conduction [70]. Heat is generated due to the resistance of the medium through ionic conduction, while other side-ions get aligned in the field's direction and change directions randomly. This creates collision among molecules, which generates heat. Different steps for MAE for a sample from the matrix are: (a) splitting of solute molecules from sample matrix occurs due to increase of pressure and temperature; and (b) solvent diffusion into the sample matrix followed by release of solute into solvent from the sample [10].

MAE has been applied for pectin extraction from fruit rind waste of *Citrullus lanatus.* The highest pectin yield from waste *Citrullus lanatus* fruit rinds was 25.79% when microwave power 477 W was applied for 128 s. During MAV process, pH, and solid-liquid ratio were 1.52 and 1:20.3 g mL^{-1}, respectively [107].

4.3.2.2 SUPERCRITICAL FLUID-EMPLOYED EXTRACTION

Supercritical fluid extraction (SFE) is performed by applying temperature and pressure that transforms the gas in the supercritical fluid to a point, where the gas and liquid phases cannot be distinguished. The extraction is fast, selective without any need of further cleaning and can be performed with small samples [123]. It is a mass transfer operation, with convection occurring between the solid surface and fluid phase [157]. The steps in the process are: (a) solubilization of the compounds, which are in the solid matrix and subsequently, and separation in the supercritical solvent,

(b) the solvent passes through the packed bed and extracts solubilized compounds from the matrix, and (c) solvent then exits the extractor and by pressure reduction and temperature increase, it transforms to a solvent-free extract [158].

In comparison to liquid solvent employed in conventional extraction processes, supercritical fluids are characterized with lower viscosity and can spread easily in the solid matrix. Low surface tension allows faster infiltration of the solvent into the solid, resulting in improved extraction efficiency [132]. Supercritical extraction is primarily used to extract non-polar compounds, such as, carotenoids, and lipids. For the isolation of flavonoids, modifiers, and solvents (such as: ethanol, methanol, water, and acetone) are used [67].

In a study by Oliveira et al. [120], different extraction techniques were adopted to evaluate their performance, based on total phenolic content and antioxidant activity. Seeds and seed cake of Passion fruit were used for isolation of biomolecules. Out of which, SFE with CO_2 (SC-CO_2) at 40°C and 50°C, 150 bars and 250 bars of pressure and fluid velocity 0.5 $kgCO_2h^{-1}$ offered the superior results [120].

4.3.2.3 ULTRASOUND-ASSISTED EXTRACTION

UAE process depends on the cavitation, which causes the creation, expansion, and collapse of bubbles. Ultrasound comprises of sound waves (20 kHz to 100 MHz) [15]. The generated shear forces accelerate the mass transfer through distraction of cell wall of plants, leading to enhanced release of bioactive compounds [144]. Ultrasound is comparatively easy to use. It has more versatility, flexibility, and requires low initial investment compared to other extraction techniques [138].

Xu et al. reported the use of UAE of antioxidants from *Eucommia oliver* plant with distilled water as the solvent. Ultrasound enhanced the efficiency of extraction, providing superior yields and selectivity of antioxidants [182].

4.3.2.4 PULSED ELECTRIC FIELD-DEPENDENT EXTRACTION

In this technology, electrical field disrupts cell membrane of food by electroporation [134]. Due to disruptions of cell membrane in tissues, solutes are easily extracted from the cells. Energy input, electric field, applied temperature, properties of food material influence the effectiveness of the pulsed electric field-based isolation (PEFE) of bio-molecules [66]. An electric field of 500 V/cm and 1000 V/cm avoids temperature rise, with minute impact on heat-degradable tissue components [92].

Pulsed electric field-applied extraction was used for extraction of anthocyanin monoglucosides from grape by-products [36].

4.3.2.5 ENZYME-ASSISTED EXTRACTION

Enzymatic pretreatment is a novel and green method to extract compounds and amplify their yield of extraction [145]. Enzyme-aided extraction (EAE) depends on capacity of enzymes to break down the cell wall of plant tissue and release of bioactive compounds. There are 2 methods for enzyme-assisted extraction, such as: enzyme-assisted cold pressing (EACP) and enzyme-assisted aqueous extraction (EAAE) [89]. EAAE technique was originally developed for extraction of oils from plant seeds [152]. An increased content of phenolic compounds (25.90–39.72%), sugars (12–14 g L^{-1}) was recorded in enzyme-assisted extraction from grape pomace and citrus peel [135].

4.3.2.6 HIGH HYDROSTATIC PRESSURE

HHP disrupts hydrophobic bonds, salt bridges, causing protein denaturation. Further solvent can penetrate the cells and additional compounds can permeate through the cell membrane to the solvent [23].

Briones-Labarca et al. [23] compared the effectiveness of conventional-, ultrasound-, and HHP-applied extraction techniques. They reported that HHP-applied extraction technique was superior than the other techniques for extraction of bioactive molecules from papaya (*Vasconcellea pubescens*) [23]. George et al. reported that HHP leads to a greater infusion of the anthocyanins from apple in comparison to infusion at atmospheric pressure. The high-pressure treatment ensued cell permeabilization, which caused increased solid mass transfer [59].

4.3.2.7 COMBINATION TECHNIQUES

Rapid advances in research, development, and upgrading of extraction methods are paving great attention. To improve the yield of bio-molecules and efficiency of technology, UAE can be combined with other extraction methods (extraction heat-reflux, microwave, and supercritical CO_2).

Yang and Wei [183] reported that combined heat reflux extraction (conventional and solvent: ethanol) along with UAE with specific operating parameters (40 kHz frequency, power of 185 W and agitation) was efficient to extract bioactive compounds from *Rabdosia rubescens*. The authors established that amalgamation of various extraction techniques shall decrease the processing time and improve the extraction of bioactive molecules.

4.4 IDENTIFICATION AND CHARACTERIZATION OF DIFFERENT GROUPS OF BIOACTIVE COMPOUNDS

4.4.1 POLYPHENOLS

Compounds having aromatic ring with one or more hydroxyl groups in the structure develop simple phenolic molecule and their complex structure develops high-molecular weight polymer, known as polyphenol. These are secondary metabolites, obtained from pentose phosphate, shikimate, and phenylpropanoid pathways in plants [125].

Phenolic compounds can be classified as: flavonoids (anthocyanidins, anthocyanins, flavones, flavanones, flavonols, flavanonols, isoflavones, flavan-3-ols), tannnis (condensed and hydrolyzable), stilbenes, phenolic acids (benzoic and cinnamic acid derivatives), and lignans [147]. Phenolic compounds provide the color to different fruits and vegetables [125]. Many phenolic bioactivities compounds are accountable for antioxidant, anti-mutagenic, anti-carcinogenic, anti-inflammatory, and antimicrobial effects. They also contribute to cell death by arresting cell cycle, amending metabolism of carcinogen, ontogenesis expression, and cell adhesion, migration, proliferation, or differentiation and also blocking of signaling pathways [51].

Separation and quantification of the recovered bioactive compounds with potential antioxidant properties were carried out using high-pressure liquid chromatography (HPLC) and high proficient techniques, such as, liquid chromatography-mass spectrometry (LC-MS), spectrophotometric methods [51]. In general, phenolic acids were detected at 220–280 nm, flavones, and flavonols at 350–365 nm, anthocyanins at 460–560 nm [173]. Matrix-assisted-laser-desorption-ionization-time-of-flight (MALDI-TOF) techniques have also been used to characterize phenolic compounds in pomegranate peel [146].

Sánchez et al. employed liquid chromatography with tandem mass spectrometry (LC-MS-MS) to identify 60 phenolic compounds from apple pomace. Among these, 23 components were novel [149]. The advantages of MS/MS technique were exclusion of interferences and verification of the structures of different compounds present in an extract [29].

Deng et al. explored the phenolic contents, antioxidant capacities and their correlation, for water soluble and fat soluble extracts of 50 fruit-residues, including mango, apple, banana, blueberry, plantain, etc. The values of ferric reducing antioxidant power (FRAP), trolox equivalent antioxidant capacity (TEAC), and total phenolic content for peels and seeds were higher than those in pulp. This indicates that these fruit residues can be used at low-cost and are easily available sources of bioactive compounds [41]. Antimicrobial activity of the polyphenols can be determined by measuring the minimum inhibitory concentration (MIC) through conventional disc diffusion test against pathogenic bacteria and *in-vitro* anti-proliferative activity test on cancer cell lines through 3- (4,5-Dimethylthiazol-2-Yl)-2,5-diphenyl tetrazolium bromide (MTT) assay [51].

Narender et al. [117] observed that peel powder of few fruits (including watermelon, musk melon and citrus) offers antimicrobial activity against *Lactobacillus, Escherichia coli, Proteus vulgaris, Saccharomyces cerevisia,* and *Staphylococcus. aureus.* The authors suggested that the bactericidal and fungicidal actions of the peels of the fruits might be due to the presence of antioxidants present in them [184]. *In vitro* anti-proliferative activities by MTT assay of pulps, peels, seeds of 61 fruits (including banana, apple, grape fruit)were evaluated on cancer cell lines of A549 (human lung cancer cells), HepG2 (human hepatoma cells) MCF-7 (human breast cancer cells), and HT-29 (human colon cancer cells). The results documented that different fruits and different parts of each fruit exhibited diverse anti-proliferative capacities, mainly polyphenols. All 61 fruits showed noteworthy inhibitory effects on the four cancer cell lines and the inhibition was dose-dependent [51].

4.4.2 DIETARY FIBER

Dietary fiber can be classified into soluble dietary fibers (such as: pectin, gums) and insoluble dietary fibers (such as: hemicelluloses, cellulose, and lignin). Fruit waste is excellent source of both these fibers. Dietary fiber is associated with plant cell walls and tissues. Therefore, it is mostly located in peels, skins, pericarps, and stalks of the fruit [29].

In human, dietary fiber reduces intestinal transit time and increases the fecal bulk. They also induce the growth of colonic micro-flora, including probiotics, reducing blood cholesterol level and plummeting insulin responses [91]. Dietary fiber is also used in food processing for their positive effect on functional properties, such as: increased water holding capacity, oil holding capacity, gel formation and emulsification [20].

Dietary fiber in fruit wastes can be measured by the enzymatic-gravimetric-based method, such as, Association of Official Agricultural Chemists (AOAC) Prosky method (AOAC method 985.29) [1]. The procedure involves breakdown and removal of starch and protein by treating the sample with starch and protein degrading enzymes (α-amylase, protease, and amyloglucosidase), followed by alcohol precipitation, filtration, and weighing of dietary fiber. Protein and ash residue correction is also considered to avoid overestimation of dietary fiber [133].

Chen with co-workers [27, 28] investigated the water holding capacity of apple pomace fiber and their effects on some food products. The study concluded that addition of apple fiber diminished the loaf volume in bread, spreading in the cookies and increased the density of muffin.

4.4.3 ENZYMES

Both solid-state fermentation (SSF) and submerged fermentation have been employed for production of commercially important amylases using microorganisms mostly fungi using solid fruit waste residues. *Aspergillus niger*, *Bacillus subtilis* and *Rhizopus oryzae* are most commonly used microorganisms. Amylases are used in food industries for diverse products, including fruit juices, syrups, cakes, and in brewing, baking, preparation of digestive, etc., [90]. Examples include invertase production under optimized conditions (temperature 30°C, incubation time 4 days, inoculum size 3%, pH 5) by *Aspergillus flavus* fermenting fruit peel waste [172].

Invertase has lower crystallinity level than sucrose, which helps in keeping products soft and fresh for long time [85]. Reddy et al. [142] studied banana waste as fermentation substrate for *Aspergillus niger* to produce cellulase [167]. Okafor et al. [119] investigated production of pectinolytic enzymes using *Penicillium chrysogenum* and *Aspergillus niger. Penicillium chrysogenum* produced pectinase at a level of 220.3 IU mg protein^{-1} when pineapple peel was used as a substrate.

4.4.4 PROTEIN

Protein deficiency may lead to several abnormal situations for the development of body and biochemical functions. The non-edible residues of many fruits and vegetables are good alternative sources of proteins. Protein content in apple pomace, mosambi peel, mango peel, pineapple peel, banana peel and orange peel was 4.45 g, 5.4 g, 9.5 g, 8.7 g, 6.02 g and 5.97 g, respectively [152].

Surabhi [168] applied two separate protein extraction processes to isolate proteins from banana (cv. *Grand Naine*) fruit peel and pulp tissues. Maniyan et al. reported that according to Folin-Lowry's method, protein concentration in musk melon, passion fruit, sapota, mango, grape, and guava was 1.6 mg mL^{-1}, 1.33 mg mL^{-1}, 1.06 mg mL^{-1}, 1.06 mg mL^{-1}, 0.16 mg mL^{-1} and 0.19 mg mL^{-1}, respectively [104].

4.4.5 FLAVOR

Flavors and aromas can be extracted from fruit wastes. In SSF, many Flavoring compounds were extracted through microbial biotransformation process [186]. The pineapple flavor component "ethyl butyrate" was produced using fermentation by *Ceratocystis fimbriata* using an apple pomace as a substrate [147].

4.5 SUMMARY

This chapter highlights different bioactive compounds from fruit wastes. Extraction techniques have been described in a comprehensive way with suitable examples. The limitations of conventional extraction methods are: prolonged extraction time, necessity of biocompatible solvent with high cost and purity, low extraction selectivity, requirement of evaporation of the bulk amount of solvent, and chances of thermal and chemical decomposition of the bioactive compounds. To overcome these limitations, the emerging methodologies have been used. Instead of direct disposal of the fruit wastes into environment or its low-cost uses, such as production of biofertilizer and biogas, production of bioactive compounds from fruit wastes may be considered as a hallmark in the context of waste valorization. Furthermore, it is expected that production of biomolecules from fruit wastes through environmental-benign process will reduce the burden of environmental pollution, and economic dilemma in food and biopharmaceutical industries.

KEYWORDS

- **bioactive molecule**
- **enzyme-aided extraction**
- **fruit waste**
- **high hydrostatic pressure**
- **microwave-assisted extraction**
- **minimum inhibitory concentration**

REFERENCES

1. AOAC (Association of Official Analytical Chemists), (1990). 985.29: Total dietary fiber in food, enzymatic-gravimetric method. In: *Official Methods of Analysis of the Association of Official Analytical Chemists*: *Sec. 985.29, 1105* (Vol. 2). Association of Official Analytical Chemists. https://www.ncbi.nlm.nih.gov/pubmed/22816275 (accessed on 5 August 2020).
2. Abdalla, A. E. M., Darwish, S. M., Ayad, E. H. E., & Hamahmy, R. M., (2007). Egyptian mango by-product: Compositional quality of mango seed kernel. *Food Chemistry, 103,* 1134–1140.
3. Abdullah, M. B., (2008). Conversion of Pineapple juice waste into lactic acid in batch and batch fermentation systems. *Reaktor, 12*, 98–101.
4. Acun, S., & Gül, H., (2014). Effects of grape pomace and grape seed flour on cookie quality. *Quality Assurance and Safety of Crops and Foods, 6* (1), 81–88.
5. Adilah, Z. M., Jamilah, B., & Hanani, Z. N., (2018). Functional and antioxidant properties of protein-based films incorporated with mango kernel extract for active packaging. *Food Hydrocolloids, 74,* 207–218.
6. Ahmed, I. S., Tohami, S. M., Almagboul, Z. A., & Verpoorte, R., (2005). Characterization of anti-microbial compounds isolated from *Mangifera indica L* seed kernel. *University of Africa Journal of Sciences, 2*, 77–91.
7. Ajila, C. M., Leelavathi, K., & Rao, U. J. S., (2008). Improvement of dietary fiber content and antioxidant properties in soft dough biscuits with the incorporation of mango peel powder. *Journal of Cereal Science, 48* (2), 319–326.
8. Ajila, C. M., Aalami, M., Leelavathi, K., & Rao, U. J. S., (2010). Mango peels powder: A potential source of antioxidant and dietary fiber in macaroni preparations. *Innovative Food Science and Emerging Technologies, 11*, 219–224.
9. Al-Sayed, H. M. A., & Ahmed, A. R., (2013). Utilization of watermelon rinds and sharlyn melon peels as a natural source of dietary fiber and antioxidants in cake. *Annals of Agricultural Sciences, 58* (1), 83–95.
10. Alupului, A, C., Alinescu, I., & Lavric, V., (2012). Microwave extraction of active principles from medicinal plants. *UPB Sci. Bull Series B, 74*, 129–142.

11. Andrade, R. M. S., Ferreira, M. S. L., & Gonçalves, E. C. B. A., (2016). Development and characterization of edible films based on fruit and vegetable residues. *Journal of Food Science, 81* (2), E412–E418.
12. Arogba, S. S., (2000). Mango (*Mangifera indica*) kernel: Chromatographic analysis of the tannin, and stability study of the associated polyphenol oxidase activity. *Journal of Food Composition and Analysis, 13*, 149–156.
13. Arshad, Z. M., Amid, A., Yusof, F., Jaswir, I., Ahmad, K., & Loke, S. P., (2014). Bromelain: An overview of industrial application and purification strategies. *Applied Microbiology and Biotechnology, 98*, 7283–7297.
14. Attard, T. M., Watterson, B., Budarin, V. L., & Clark, J. H., (2014). Microwave-assisted extraction as an important technology for valorizing orange waste. *New Journal of Chemistry, 38* (6), 2278–2283.
15. Azmir, J., Zaidul, I. S. M., Rahman, M. M., & Sharif, K. M., (2013). Techniques for extraction of bioactive compounds from plant materials: A review. *Journal of Food Engineering, 117* (4), 426–436.
16. Babbar, N., Singh, H. O., & Kaur, S. S., (2015). Therapeutic and nutraceutical potential of bioactive compounds extracted from fruit residues. *Critical Reviews in Food Science and Nutrition, 55* (3), 319–337.
17. Banerjee, J., Singh, R., & Vijayaraghavan, R., (2017). Bioactives from fruit processing wastes: Green approaches to valuable chemicals. *Food Chemistry, 225,* 10–22.
18. (2016). *Bananas, more Waste than Product: Are They a Source of Bioenergy?* https://www.sciencedaily.com/releases/2016/05/160519082430.htm (accessed on 5 August 2020).
19. Bankar, A., Joshi, B., Kumar, A. R., & Zinjarde, S., (2010). Banana peel extract mediated novel route for the synthesis of silver nanoparticles. *Colloids and Surfaces A: Physicochemical and Engineering Aspects, 368* (1–3), 58–63.
20. Belitz, H. D., & Grosch, W., (1999). Aroma substances. In: *Food Chemistry* (pp. 319–377). Berlin, Heidelberg: Springer.
21. Bhushan, S., Kalia, K., Sharma, M., Singh, B., & Ahuja, P. S., (2008). Processing of apple pomace for bioactive molecules. *Critical Reviews in Biotechnology, 28*, 285–296.
22. Bocco, A., Cuvelier, M. E., Richard, H., & Berset, C., (1998). Antioxidant activity and phenolic composition of citrus peel and seed extracts. *Journal of Agricultural and Food Chemistry, 46,* 2123–2129.
23. Briones-Labarca, V., & Plaza-Morales, M., (2015). High hydrostatic pressure and ultrasound extractions of antioxidant compounds, sulforaphane, and fatty acids from Chilean papaya (*Vasconcellea pubescens*) seeds: Effects of extraction conditions and methods. *LWT-Food Science and Technology, 60* (1), 525–534.
24. Bhushan, S., & Gupta, M., (2013). Apple pomace: Source of dietary fiber and antioxidant for food fortification. In: *Handbook of Food Fortification and Health* (pp. 21–27). New York: Humana Press.
25. Cam, M., Icyer, N. C., & Erdogan, F., (2014). Pomegranate peel phenolics: Microencapsulation, storage stability and potential ingredient for functional food development. *LWT-Food Science and Technology, 55* (1), 117–123.
26. Cao, X., Wang, C., Pei, H., & Sun, B., (2009). Separation and identification of polyphenols in apple pomace by high speed counter-current chromatography and high performance liquid chromatography coupled with mass spectrometry. *Journal of Chromatography A, 1216,* 4268–4274.

27. Chen, H., Rubenthaler, G. L., Leung, H. K., & Baranowski, J. D., (1988). Chemical, physical, and baking properties of apple fiber compared with wheat and oat bran. *Cereal Chemistry, 65* (3), 244–247.
28. Chen, H., Rubenthaler, G. L., & Schanus, G., (1988). Effect of Apple fiber and cellulose on the physical properties of wheat flour. *Journal of Food Science, 53,* 304–309.
29. Chockchaisawasdee, S., & Stathopoulos, C. E., (2017). Extraction, isolation, and utilization of bioactive compounds from fruit juice industry waste. In: *Utilization of Bioactive Compounds from Agricultural and Food Production Waste* (pp. 272–313). Boca Raton, USA: CRC Press.
30. Choi, Y., Choi, J., Han, D., Kim, H., Lee, M., & Kim, H., (2015). Optimization of replacing pork back fat with grape seed oil and rice bran fiber for reduced-fat meat emulsion systems. *Meat Science, 84*, 212–218.
31. Choi, Y., Choi, J., Han, D., Kim, H., Lee, M., Jeong, J., Chung, H., & Kim, C., (2010). Effects of replacing pork back fat with vegetable oils and rice bran fiber on the quality of reduced-fat frankfurters. *Meat Science*, *84,* 557–563.
32. Choon, Y. C., Noranizan, M. A., & Russly, A. R., (2018). Current trends of tropical fruit waste utilization. *Critical Reviews in Food Science and Nutrition, 58* (3), 335, 336.
33. Choonut, A., Saejong, M., & Sangkharak, K., (2014). The production of ethanol and hydrogen from pineapple peel by *Saccharomyces cerevisiae* and *Enterobacter aerogenes*. *Energy Procedia, 52,* 242–249.
34. Cirqueira, M. G., Costa, S. S., & Viana, J. D., (2017). Phytochemical importance and utilization potential of grape residue from wine production. *African Journal of Biotechnology, 16* (5), 179–192.
35. Conte, A., Scrocco, C., Sinigaglia, M., & Del, N. M. A., (2007). Innovative active packaging systems to prolong the shelf life of mozzarella cheese. *Journal of Dairy Science, 90* (5), 2126–2131.
36. Corrales, M., Toepfl, S., Butz, P., Knorr, D., & Tauscher, B., (2008). Extraction of anthocyanins from grape by-products assisted by ultrasonics, high hydrostatic pressure or pulsed electric fields: A comparison. *Innovative Food Science and Emerging Technologies*, *9* (1), 85–91.
37. Ćujić, N., Šavikin, K., & Janković, T., (2016). Optimization of polyphenols extraction from dried chokeberry using maceration as traditional technique. *Food Chemistry*, *194*, 135–142.
38. Das, D. R., Sachan, A. K., Shuaib, M., & Imtiyaz, M., (2014). Chemical characterization of volatile oil components of citrus reticulata by GC-MS analysis. *World Journal of Pharmaceutical Sciences, 3*, 1197–1204.
39. Davidov-Pardo, G., & McClements, D. J., (2015). Nutraceutical delivery systems: Resveratrol encapsulation in grape seed oil nanoemulsions formed by spontaneous emulsification. *Food Chemistry*, *167,* 205–212.
40. Deliza, R., Rosenthal, A., & Abadio, F. B. D., (2005). Utilization of pineapple waste from juice processing industries: Benefits perceived by consumers. *J. Food Eng.*, *67,* 241–246.
41. Deng, G. F., Shen, C., Xu, X. R., & Kuang, R. D., (2012). Potential of fruit wastes as natural resources of bioactive compounds. *International Journal of Molecular Sciences, 13* (7), 8308–8323.
42. Dilas, S., Canadanovic-Brunet, J., & Cetkovic, G., (2009). By-products of fruits processing as a source of phytochemicals. *Chemical Industry and Chemical Engineering Quarterly, 15*, 191–202.

43. Dorta, E., Lobo, M. G., & Gonz´alez, M., (2012). Using drying treatments to stabilize mango peel and seed: Effect on antioxidant activity. *LWT-Food Science and Technology, 45,* 261–268.
44. Dorta, E., Lobo, M. G., & Gonzalez M., (2013). Optimization of factors affecting extraction of antioxidants from mango seed. *Food Bioprocess Technology, 6,* 1067–1081.
45. Dorta, E., Lobo, M. G., & Gonzalez, M., (2012). Reutilization of mango byproducts: Study of the effect of extraction solvent and temperature on their antioxidant properties. *Journal of Food Science, 77,* 80–88.
46. Du, W. X., Olsen, C. W., Bustillos, R. J., Friedman, M., & McHugh, T. H., (2011). Physical and antibacterial properties of edible films formulated with apple skin polyphenols. *Journal of Food Science, 76* (2), M149–M155.
47. Elfalleh, W., Hannachi, H., Tlili, N., Yahia, Y., & Nasri, N., (2012). Total phenolic contents and antioxidant activities of pomegranate peel, seed, leaf, and flower. *Journal of Medicinal Plants Research, 6,* 4724–4730.
48. El-Safy, F. S., Salem, R. H., & El-Ghany, M. E. A., (2012). Chemical and nutritional evaluation of different seed flours as novel sources of protein. *World Journal of Dairy and Food Sciences, 7* (1), 59–65.
49. Emaga, T. H., Ronkart, S. N., Robert, C., Wathelet, B., & Paquot, M., (2008). Characterization of pectins extracted from banana peels (Musa AAA) under different conditions using an experimental design. *Food Chemistry, 108* (2), 463–471.
50. Esparza-Martínez, F. J., Miranda-López, R., & Guzman-Maldonado, S. H., (2016). Effect of air-drying temperature on extractable and non-extractable phenolics and antioxidant capacity of lime wastes. *Industrial Crops and Products, 84,* 1–6.
51. Fang, L., Sha, L., & Hua-Bin, L., (2013). Antiproliferative activity of peels, pulps, and seeds of fruits. *Journal of Functional Foods, 5* (3), 1298–1309.
52. FAO (Food and Agriculture Organization of the United Nations), (2007). http://faostat.fao.org (accessed on 5 August 2020).
53. Ferreira, A. S., Nunes, C., Castro, A., Ferreira, P., & Coimbra, M. A., (2014). Influence of grape pomace extract incorporation on chitosan films properties. *Carbohydrate Polymers, 113,* 490–499.
54. Fischer, U. A., Carle, R., & Kammerer, D. R., (2011). Identification and quantification of phenolic compounds from pomegranate (*Punica Granatum* L.) peel, mesocarp, aril, and differently produced juices by HPLC-DAD-ESI/MSn. *Food Chemistry, 127,* 807–821.
55. Fontana, A. R., Antoniolli, A., & Bottini, R., (2013). Grape pomace as a sustainable source of bioactive compounds: Extraction, characterization, and biotechnological applications of phenolics. *Journal of Agricultural and Food Chemistry, 61* (38), 8987–9003.
56. Foo, L. Y., & Lu, Y., (1999). Isolation and identification of procyanidins in apple pomace. *Food Chemistry, 64,* 511–518.
57. Gallaher, D., & Schneeman, B. O., (2001). Dietary fiber. In: Bowman, B., & Russell, R. M., (eds.), *Present Knowledge in Nutrition* (pp. 700–710). Washington, DC: ILSI (Intl Life Sciences Inst).
58. García-Lomillo, J., & González, M. L., (2017). Applications of wine pomace in the food industry: Approaches and functions. *Comprehensive Reviews in Food Science and Food Safety Comprehensive Reviews, 16,* 3–22.
59. George, J. M., Selvan, T. S., & Rastogi, N. K., (2016). High-pressure-assisted infusion of bioactive compounds in apple slices. *Innovative Food Science and Emerging Technologies, 33,* 100–107.

60. González-Montelongo, R., Lobo, M. G., & González, M., (2010). Antioxidant activity in banana peel extracts: Testing extraction conditions and related bioactive compounds. *Food Chemistry, 119* (3), 1030–1039.
61. Gorinstein, S., Martın-Belloso, O., & Park, Y. S., (2001). Comparison of some biochemical characteristics of different citrus fruits. *Food Chemistry, 74*, 309–315.
62. Guerrero, R. F., Smith, P., & Bindon, K. A., (2013). Application of insoluble fibers in the fining of wine phenolics. *Journal of Agricultural and Food Chemistry, 61* (18), 4424–4432.
63. Gupta, K., & Joshi, V. K., (2000). Fermentative utilization of waste from food processing industry. In: *Postharvest Technology of Fruits and Vegetables: Handling, Processing, Fermentation, and Waste Management* (pp. 1171–1193). New Delhi: Indus Pub. Co.
64. Haque, S., Begum, P., Khatun, M., & Islam, S. N., (2015). Total carotenoid content in some mango (*Mangifera Indica*) varieties of Bangladesh. *International Journal of Pharmaceutical Sciences and Research, 40,* 4875–4878.
65. Hegazy, A. E., & Ibrahium, M. I., (2012). Antioxidant activities of orange peel extracts. *World Applied Sciences Journal, 18,* 684–688.
66. Heinz, V., Álvarez, I., Angersbach, A., & Knorr, D., (2001). Preservation of liquid foods by high intensity pulsed electric fields: Basic concepts for process design. *Trends in Food Science and Technology, 12* (3, 4), 103–111.
67. Herrero, M., Castro-Puyana, M., Mendiola, J. A., & Ibañez, E., (2013). Compressed fluids for the extraction of bioactive compounds. *Trends in Analytical Chemistry, 43*, 67–83.
68. Huang, Y. L., Chow, C. J., & Fang, Y. J., (2011). Preparation and physicochemical properties of fiber rich fraction from pineapple peels as a potential ingredient. *Journal of Food and Drug Analysis, 19* (3), 318–323.
69. Ismail, A. F. M., Salem, A. A. M., & Eassawy, M. M. T., (2016). Hepatoprotective effect of grape seed oil against carbon tetrachloride-induced oxidative stress in liver of Γ-irradiated rat. *Journal of Photochemistry and Photobiology, 160,* 1–10.
70. Jain, T., Jain, V., Pandey, R., Vyas, A., & Shukla, S. S., (2009). Microwave assisted extraction for phytoconstituents: An overview. *Asian Journal of Research in Chemistry, 2* (1), 19–25.
71. Joshi, R. V., (2007). Low calorie biscuits from banana peel pulp. *Journal of Solid Waste Technology and Management, 33* (3), 142–147.
72. Juneja, V. K., Bari, M. L., Inatsu, Y., Kawamoto, S., & Friedman, M., (2009). Thermal destruction of *Escherichia Coli* O157:H7 in sous-vide-cooked ground beef as affected by tea leaf and apple skin powders. *Journal of Food Protection, 72* (4), 860–865.
73. Kanazawa, K., & Sakakibara, H., (2000). High content of dopamine, a strong antioxidant, in Cavendish banana. *Journal of Agricultural and Food Chemistry, 48* (3), 844–848.
74. Ketnawa, S., Chaiwutb, P., & Rawdkuen, S., (2012). Pineapple wastes: A potential source for bromelain extraction. *Food Bioproducts Processing, 90*, 385–391.
75. Khan, J. A., & Hanee, S., (2011). Antibacterial properties of *Punica granatum* peels. *International Journal of Applied Biology and Pharmaceutical Technology, 2* (3), 23–27.
76. Khan, M., Khan, S. S., Ahmed, Z., & Tanveer, A., (2010). Production of single cell protein from *Saccharomyces cerevisiae* by utilizing fruit wastes. *Nanobiotechnica Universale (Nano Biotechnology Universal), 1* (2), 127–132.
77. Khoddami, A., Wilkes, M. A., & Roberts, T. H., (2013). Techniques for analysis of plant phenolic compounds. *Molecules, 18,* 2328–2375.

78. Khoo, H. E., Azlan, A., Tang, S. T., & Lim, S. M., (2017). Anthocyanidins and anthocyanins: Colored pigments as food, pharmaceutical ingredients, and the potential health benefits. *Food and Nutrition Research, 61* (1), 4. Article ID: 1361779.
79. Kittiphoom, S., & Sutasinee, S., (2013). Mango seed kernel oil and its physicochemical properties. *International Food Research Journal, 20,* 1145–1149.
80. Kodagoda, K. H. G. K., & Marapana, R. A. U. J., (2017). Development of nonalcoholic wines from the wastes of mauritius pineapple variety and its physicochemical properties. *Journal of Pharmacognosy and Phytochemistry, 6* (3), 492–497.
81. Kodal, S. P., & Aksu, Z., (2017). Phenolic pigment extraction from orange peels: Kinetic modeling. *Presented at the 15th International Conference on Environmental Science and Technology* (Vol. 31, pp. 798–803). Rhodes, Greece.
82. Kong, K. W., Ismail, A. R., Tyug, T. S., Prasad, K. N., & Ismail, A., (2010). Response surface optimization of extraction of phenolics and flavonoids from pink guava puree industry by-product. *International Journal of Food Science and Technology, 45,* 1739–1745.
83. Kothari, V., & Seshadri, S., (2010). *In Vitro* Antibacterial activity in seed extracts of manilkarazapota, *Anona squamosa* and *Tamarindus indica. Biological Research, 43* (2), 165–168.
84. Kumar, D., Yadav, K. K., Muthukumar, M., & Garg, N., (2013). Production and characterization of [alpha]-amylase from mango kernel by *Fusarium solani* NAIMCC-F-02956 using submerged fermentation. *Journal of Environmental Biology, 34* (6), 1053–1062.
85. Kumar, R., & Kesavapillai, B., (2012). Stimulation of extracellular invertase production from spent yeast when sugarcane press mud used as substrate through solid state fermentation. *Springer Plus, 1,* 81–88.
86. Lafka, T. I., Sinanoglou, V., & Lazos, E. S., (2007). On the extraction and antioxidant activity of phenolic compounds from winery wastes. *Food Chemistry, 3* (104), 1206–1214.
87. Lansky, E. P., & Newman, R. A., (2007). *Punica granatum* (Pomegranate) and its potential for prevention and treatment of inflammation and cancer. *Journal of Ethno Pharmacology, 109,* 177–206.
88. Larrauri, J. A., Rupérez, P., Borroto, B., & Saura-Calixto, F., (1996). Mango peels as a new tropical fiber: Preparation and characterization. *LWT-Food Science and Technology, 29* (8), 729–733.
89. Latif, S., & Anwar, F., (2009). Physicochemical studies of hemp (*Cannabis Sativa*) seed oil using enzyme-assisted cold-pressing. *European Journal of Lipid Science and Technology, 111,* 1042–1048.
90. Laufenberg, G., Schulze, N., & Waldron, K., (2009). Modular strategy for processing of fruit and vegetable wastes into value-added products. In: *Handbook of Waste Management and Co-Product Recovery in Food Processing* (pp. 286–353). New York: Woodhead Publishing Ltd.
91. Laurentin, A., Morrison, D., & Edwards, C., (2003). Dietary fiber in health and disease. *Nutrition Bulletin, 28,* 69–77.
92. Lebovka, N. I., Bazhal, M. I., & Vorobiev, E., (2002). Estimation of characteristic damage time of food materials in pulsed-electric fields. *Journal of Food Engineering, 54,* 337–346.
93. Letellier, M., & Budzinski, H., (1999). Microwave assisted extraction of organic compounds. *Analusis, 27,* 259–270.

94. Lin, C. S. K., Koutinas, A. A., & Stamatelatou, K., (2014). Current and future trends in food waste valorization for the production of chemicals, materials, and fuels: A global perspective. *Biofuels, Bioproducts, and Biorefining, 8* (5), 686–715.
95. Lota, M., Rocca, D., Serra, D., Tomi, F., Jacquemond, C., & Casanova, J., (2002). Volatile components of peel and leaf oils of lemon and lime species. *Journal of Agricultural and Food Chemistry, 50* (4), 796–805.
96. Lu, Y., & Foo, L. Y., (1997). Identification and quantification of major polyphenols in apple pomace. *Food Chemistry, 59,* 187–194.
97. Luby, J. J., (2003). Taxonomic classification and brief history. In: Ferree, D. C., & Warrington, I. J., (eds.), *Apples: Botany, Production, and Uses* (pp. 1–14). Cambridge, MA, USA; CABI Publishing.
98. Lun, O. K., Wai, T. B., & Ling, L. S., (2014). Pineapple cannery waste as a potential substrate for microbial biotranformation to produce vanillic acid and vanillin. *International Food Research Journal, 21* (3), 953–958.
99. Mahmoud, K., (2015). Statistical optimization of cultural conditions of an halophilic alpha-amylase production by halophilic *Streptomyces* sp. Grown on orange waste powder. *Biocatalysis and Agricultural Biotechnology, 4* (4), 685–693.
100. Maisuthisakul, P., & Gordon, M. H., (2009). Antioxidant and tyrosinase inhibitory activity of mango seed kernel by product. *Food Chemistry, 117,* 332–341.
101. Maisuthisakul, P., & Gordon, M. H., (2009). Antioxidant and tyrosinase inhibitory activity of mango seed kernel by-product. *Food chemistry, 117* (2), 332–341.
102. Malviya, S., Jha, A. A., & Hettiarachchy, N., (2014). Antioxidant and antibacterial potential of pomegranate peel extract. *JFST51, 51* (12), 4132–4137.
103. Mandawgade, S. D., & Patravale, V. B., (2008). Formulation and evaluation of exotic fat-based cosmeceuticals for skin repair. *Indian Journal of Pharmaceutical Sciences, 70* (4), 539–542.
104. Maniyan, A., John, R., & Mathew, A., (2015). Evaluation of fruit peels for some selected nutritional and anti-nutritional factors. *Emergent Life Science Research, 1,* 13–19.
105. Manthey, J. A., & Grohmann, K., (2001). Phenols in citrus peel byproducts. concentrations of hydroxycinnamates and polymethoxylated flavones in citrus peel molasses. *Journal of Agricultural and Food Chemistry, 49* (7), 3268–3273.
106. Manzanarez-López, F., Soto-Valdez, H., Auras, R., & Peralta, E., (2011). Release of [alpha]-tocopherol from poly-lactic acid films, and its effect on the oxidative stability of soybean oil. *Journal of Food Engineering, 104,* 508–517.
107. Maran, J. P., Sivakumar, V., Thirugnanasambandham, K., & Sridhar, R., (2014). Microwave assisted extraction of pectin from waste *Citrullus lanatus* fruit rinds. *Carbohydrate Polymers, 101,* 786–791.
108. Masibo, M., & He, Q., (2008). Major mango polyphenols and their potential significance to human health. *Comprehensive Reviews in Food Science and Food Safety, 7* (4), 309–319.
109. Masoodi, F. A., Sharma, B., & Chauhan, G. S., (2002). Use of apple pomace as a source of dietary fiber in cakes. *Plant Foods Human Nutrition, 57,* 121–128.
110. McCann, M. J., Gill, C. I. R., O'Brien, G., (2007). Anti-cancer properties of phenolics from apple waste on colon carcinogenesis *in Vitro. Food and Chemical Toxicology, 45* (7), 1224–1230.
111. Memon, J. R., Memon, S. Q., & Bhanger, I., (2009). Banana peel: A green and economical sorbent for the selective removal of Cr (VI) from industrial wastewater. *Colloids and Surfaces B, 70,* 232–237.

112. Memon, J. R., Memon, S. Q., Bhanger, I., & Khuhawar, M. Y., (2008). Banana peel: A green and economical sorbent for Cr (III) removal. *Pakistan Journal of Analytical and Environmental Chemistry*, *9* (1), 20–25.
113. Milner, J. A., & Romagnolo, D., (2010). Cancer biology and nutrigenomics. In: *Bioactive Compounds and Cancer* (pp. 25–43). Totowa: NJ: Humana Press.
114. Mohagheghi, M., Rezaei, K., Labbafi, M., & Ebrahimzadeh, M. S. M., (2011). Pomegranate seed oil as a functional ingredient in beverages. *European Journal of Lipid Science and Technology*, *113* (6), 730–736.
115. Mondal, A. K., Sengupta, S., Bhowal, J., & Bhattacharya, D. K., (2012). Utilization of fruit wastes in producing single cell protein. *International Journal of Science, Environment, and Technology*, *1* (5), 430–438.
116. Nadeem, M., Imran, M., & Khalique, A., (2016). Promising features of mango (*Mangifera Indica L.*) kernel oil: A review. *Journal of Food Science and Technology*, *53* (5), 2185–2195.
117. Narender, B. R., Rajakumari, M., Sukanya, B., & Harish, S., (2017). Antimicrobial activity on peels of different fruits and vegetables. *Journal of Pharmacy Research*, *7*, 1–7.
118. Nascimento, T. A., Calado, V., & Carvalho, C. W. P., (2012). Development and characterization of flexible film based on starch and passion fruit mesocarp flour with nanoparticles. *Food Research International*, *49* (1), 588–595.
119. Okafor, U. A., Okochi, V. I., & Chinedu, S. N., (2010). Pectinolytic activity of wild-type filamentous fungi fermented on agro-wastes. *African Journal of Microbiology Research*, *4*, 2729–2734.
120. Oliveira, D. A., Angonese, M., & Ferreira, S. R., (2013). Supercritical fluid extraction of passion fruit seeds and its processing residue (Cake). In: *III Iberoamerican Conference on Supercritical Fluids Cartagena De India's (Colombia)* (pp. 12–22).
121. Oreopoulou, V., & Tzia, C., (2007). Utilization of plant by-products for the recovery of proteins, dietary fibers, antioxidants, and colorants. In: *Utilization of By-Products and Treatment of Waste in the Food Industry* (pp. 209–232). Boston, MA: Springer.
122. Orijajogun, J. O., Batari, L. M., & Aguzue, O. C., (2014). Chemical Composition and phytochemical properties of mango (*Mangifera Indica.*) seed kernel. *International Journal of Advance Chemistry*, *2*, 185–187.
123. Oroian, M., & Escriche, I., (2015). Antioxidants: Characterization, natural sources, extraction, and analysis. *Food Research International*, *74*, 10–36.
124. (2017). *OIV (International Organization of Vine and Wine) Statistical Report.* http://www.oiv.int/public/medias/5479/oiv-en-bilan-2017 (accessed on 5 August 2020).
125. Ozcan, T., Akpinar-Bayizit, A., Yilmaz-Ersan, L., & Delikanli, B., (2014). Phenolics in human health. *International Journal of Chemical Engineering and Applications*, *5* (5), 393–397.
126. Padam, B. S., Tin, H. S., Chye, F. Y., & Abdullah, M. I., (2014). Banana by-products: An under-utilized renewable food biomass with great potential. *Journal of Food Science and Technology*, *51* (12), 3527–3545.
127. Palafox-Carlos, H., Ayala-Zavala, F., & González-Aguilar, G. A., (2010). The role of dietary fiber in the bioaccessibility and bioavailability of fruit and vegetable antioxidants. *Journal Food Science*, *76* (1), R6–R15.
128. Panchami, P. S., & Gunasekaran, S., (2017). Extraction and characterization of pectin from fruit waste. *International Journal of Current Microbiology and Applied Sciences*, *6* (8), 943–948.

129. Park, D., Lim, S. R., Yun, Y. S., & Park, J. M., (2008). Development of a new Cr (VI)-biosorbent from agricultural bio-waste. *Bioreources Technology, 99,* 8810–8818.
130. Park, S. I., & Zhao, Y., (2006). Development and characterization of edible films from cranberry pomace extracts. *Journal of Food Science, 71* (2), E95–E101.
131. Ping, L., Brosse, N., Sannigrahi, P., & Ragauskas, A., (2011). Evaluation of grape stalks as a bioresource. *Industrial Crops and Products, 33,* 200–204.
132. Pouliot, Y., Conway, V., & Leclerc, P. L., (2014). Separation and concentration technologies. In: *Food Processing: Principles and Applications* (2nd ed., pp. 33–60).
133. Prosky, L., Asp, N. G., & Furda, I., (1984). Determination of total dietary fiber in foods and food products: Collaborative study. *Journal of AOAC International, 68,* 677–679.
134. Puertolas, E., Lopez, N., Saldana, G., Alvarez, I., & Raso, J., (2010). Evaluation of phenolic extraction during fermentation of red grapes treated by a continuous pulsed electric fields process at pilot-plant scale. *Journal of Food Engineering, 98,* 120–125.
135. Puri, M., Sharma, D., & Barrow, C. J., (2012). Enzyme-assisted extraction of bioactives from plants. *Trends in Biotechnology, 30* (1), 37–44.
136. Putnik, P., BursaćKovačević, D., & RežekJambrak, A., (2017). Innovative green and novel strategies for the extraction of bioactive added value compounds from citrus wastes: A review. *Molecules, 22* (5), 680–686.
137. Rahimi, H. R., Arastoo, M., & Ostad, S. N., (2012). Comprehensive review of *Punica granatum* (Pomegranate) properties in toxicological, pharmacological, cellular, and molecular biology researches. *Iranian Journal of Pharmaceutical Research: IJPR, 11* (2), 385–389.
138. Rajha, H. N., Boussetta, N., & Louka, N., (2015). Effect of alternative physical pretreatments (pulsed electric field, high voltage electrical discharges and ultrasound) on the dead-end ultra-filtration of vine-shoot extracts. *Separation and Purification Technology, 146,* 243–251.
139. Ramakrishnan, K., & Krishnan, M. R. V., (1994). Tannin-classification, analysis, and applications. *Ancient Science of Life, 13* (3, 4), 232–236.
140. Ramirez-Lopez, L. M., & DeWitt, C. A., (2014). Analysis of phenolic compounds in commercial dried grape pomace by high performance liquid chromatography electro spray ionization mass spectrometry. *Food Science and Nutrition, 2* (5), 470–477.
141. Rana, S., Gupta, S., Rana, A., & Bhushan, S., (2015). Functional properties, phenolic constituents and antioxidant potential of industrial apple pomace for utilization as active food ingredient. *Food Science and Human Wellness, 4* (4), 180–187.
142. Reddy, G. V., Babu, P. R., Komaraiah, P., Roy, K. R. R. M., & Kothari, I. L., (2003). Utilization of banana waste for the production of lignolytic and cellulolytic enzymes by solid substrate fermentation using *P. ostreatus and P. sajor-caju. Process Biochemistry, 38,* 1457–1462.
143. Rosas-Domınguez, C. J. F., Vega-Vega, V., & Gonzalez-Aguilar, G. A., (2010). Antioxidant enrichment and antimicrobial protection of fresh-cut fruits using their own by products looking for integral exploitation. *Journal of Food Science, 75,* 175–181.
144. Roselló-Soto, E., Koubaa, M., & Moubarik, A., (2015). Emerging opportunities for the effective valorization of wastes and by-products generated during olive oil production process: Non-conventional methods for the recovery of high-added value compounds. *Trends in Food Science and Technology, 45* (2), 296–310.
145. Rosenthal, A., Pyle, D. L., & Niranjan, K., (1996). Aqueous and enzymatic processes for edible oil extraction. *Enzyme and Microbial Technology, 19,* 402–420.

146. Saad, H., Charrier-El, B. F., & Pizzi, A., (2012). Characterization of pomegranate peels tannin extractives. *Industrial Crops and Products*, *40*, 239–246.
147. Sagar, N. A., Pareek, S., Sharma, S., Yahia, E. M., & Lobo, M. G., (2018). Fruit and vegetable waste: Bioactive compounds, their extraction, and possible utilization. *Comprehensive Reviews in Food Science and Food Safety*, *17* (3), 512–531.
148. Sánchez-Alonso, I., Solas, M. T., & Borderías, A. J., (2007). Physical study of minced fish muscle with a white-grape by-product added as an ingredient. *Journal of Food Science*, *72* (2), 94–101.
149. Sánchez-Rabaneda, F., Jauregui, O., & Lamuela-Raventós, R. M., (2004). Qualitative analysis of phenolic compounds in apple pomace using liquid chromatography coupled to mass spectrometry in tandem Mode. *Rapid Communications in Mass Spectrometry*, *18* (5), 553–563.
150. Scola, G., Kappel, V. D., Moreira, J. C. F., Dal-Pizzol, F., & Salvador, M., (2011). antioxidant and anti-inflammatory activities of winery wastes seeds of *Vitis labrusca*. *Ciência Rural*, *41* (7), 1233–1238.
151. Selvamuthukumaran, M., & Shi, J., (2017). Recent Advances in extraction of antioxidants from plant by-products processing industries. *Food Quality and Safety*, *1*, 61–81.
152. Sharma, R., Oberoi, H. S., & Dhillon, G. S., (2016). Fruit and Vegetable processing waste: Renewable feedstocks for enzyme production. In: Dhillon, G. S., & Kaur, S., (eds.), *Agro-Industrial Wastes as Feedstock for Enzyme Production: Apply and Exploit the Emerging and Valuable Use Options of Waste Biomass* (pp. 23–59). London, UK: Academic Press Elsevier.
153. Shea, N., Arendt, E., & Gallagher, E., (2012). Dietary fiber and phytochemical characteristics of fruit and vegetable by-products and their recent applications as novel ingredients in food products. *Innovative Food Science and Emerging Technology*, *16*, 1–10.
154. Shen, Y. H., Yang, F. Y., Lu, B. G., Zhao, W. W., Jiang, T., Feng, L., & Ming, R., (2019). Exploring the differential mechanisms of carotenoid biosynthesis in the yellow peel and red flesh of papaya. *BMC Genomics*, *20* (1), 49–52.
155. Shinagawa, F. B., Santana, F. C., Torres, L. R. O., & Mancini-Filho, J., (2015). Grapeseed oil: A potential functional food. *Journal of Food Science and Technology*, *35*, 399–406.
156. Silva, L. V., & Nelson, D. L., (2005). Comparison of hydro distillation methods for the deodorization of turmeric. *Food Research International*, *38*, 1087–1096.
157. Silva, L. P. S., & Martínez, J., (2014). Mathematical modeling of mass transfer in supercritical fluid extraction of oleoresin from red pepper. *Journal of Food Engineering*, *133*, 30–39.
158. Silva, R. P. F. F., Rocha-Santos, T. A. P., & Duarte, A. C., (2016). Supercritical fluid extraction of bioactive compounds. *TRAC: Trends in Analytical Chemistry*, *76*, 40–51.
159. Singh, S., & Immanuel, G., (2014). Extraction of antioxidants from fruit peels and its utilization in paneer. *Journal of Food Processing and Technology*, *5* (7), 1–10.
160. Someya, S., Yoshiki, Y., & Okubo, K., (2002). Antioxidant compounds from bananas (*Musa Cavendish*). *Food Chemistry*, *79* (3), 351–354.
161. Sood, A., & Gupta, M., (2015). Extraction process optimization for bioactive compounds in pomegranate peel. *Food Bioscience*, *12*, 100–106.
162. Sossou, S. K., & Ameyaph, Y., (2009). Study of pineapple peelings processing into vinegar by biotechnology. *Pakistan Journal of Biological Science*, *12* (11), 859–865.

163. Subagio, A., Morita, N., & Sawada, S., (1996). Carotenoids and their fatty-acid esters in banana peel. *Journal of Nutritional Science and Vitaminology, 42* (6), 553–566.
164. Sudha, M. L., Baskaran, V., & Leelavathi, K., (2007). Apple pomace as a source of dietary fiber and polyphenols and its effect on the rheological characteristics and cake making. *Food Chemistry, 104,* 686–692.
165. Sudheeran, P., Feygenberg, O., Maurer, D., & Alkan, N., (2018). Improved Cold tolerance of mango fruit with enhanced anthocyanin and flavonoid contents. *Molecules, 23* (7), 1832.
166. Sun, J., Hu, X., Zhao, G., Wu, J., Wang, Z., Chen, F., & Liao, X., (2007). Characteristics of thin layer infrared drying of apple pomace with and without hot air pre-drying. *Food Science and Technology International, 13* (2), 91–97.
167. Sun-Waterhouse, D., & Luberriaga, C., (2013). Juices, fibers, and skin waste extracts from white, pink or red-fleshed apple genotypes as potential food ingredients: A comparative study. *Food Bioprocess Technology, 6,* 377–390.
168. Surabhi, G. K., (2016). Comparative method for protein extraction and proteome analysis by two-dimensional gel electrophoresis from banana fruit. *Horticultural Biotechnology Research, 2,* 8–13.
169. Tilay, A., Bule, M., Kishenkumar, J., & Annapure, U., (2008). Preparation of ferulic acid from agricultural wastes: Its improved extraction and purification. *Journal of Agriculture and Food Chemistry, 56,* 7664–7648.
170. Tow, W. W., Premier, R., Jing, H., & Ajlouni, S., (2011). Antioxidant and antiproliferation effects of extractable and nonextractable polyphenols isolated from apple waste using different extraction methods. *Journal of Food Science, 76* (7), T163–T172.
171. Tseng, A., & Zhao, Y. Y., (2013). Wine grape pomace as antioxidant dietary fiber for enhancing nutritional value and improving storability of yogurt and salad dressing. *Food Chemistry, 138* (1), 356–365.
172. Uma, C., & Gomathi, D., (2010). Production, purification, and characterization of invertase by *Aspergillus flavus* using fruit peel waste as substrate. *Advanced Biological Research, 4* (1), 31–36.
173. Valls, J., Millán, S., Martí, M. P., Borràs, E., & Arola, L., (2009). Advanced separation methods of food anthocyanins, isoflavones, and flavanols. *Journal of Chromatography A, 1216* (43), 7143–7172.
174. Vankar, P. S., (2004). Essential oils and fragrances from natural sources. *Resonance, 9,* 30–41.
175. Verardo, V., Garcia-Salas, P., & Baldi, E., (2014). Pomegranate Seeds as a source of nutraceutical oil naturally rich in bioactive lipids. *Food Research International, 65,* 445–452.
176. Vergara, V. N., & Granados, P. E., (2007). Fiber concentrate from mango fruit: Characterization, associated antioxidant capacity and application as a bakery product ingredient. *LWT-Food Science and Technology, 40,* 722–729.
177. Villas-Boas, S. G., & Esposito, E., (2003). Bioconversion of apple pomace into a nutritionally enriched substrate by *Candida utilis* and *Pleurotus ostreatus*. *World Journal of Microbiol Biotechnology, 19,* 461–467.
178. Wang, S., Chen, F., Wu, J., Wang, Z., Liao, X., & Hu, X., (2007). Optimization of pectin extraction assisted by microwave from apple pomace using response surface methodology. *Journal of Food Engineering, 78,* 693–700.
179. Wolfe, K., Wu, X., & Liu, R. H., (2003). Antioxidant activity of apple peels. *Journal of Agriculture and Food Chemistry, 51,* 609–614.

180. Wong, Y. S., & Sia, C. M., (2014). Influence of extraction conditions on antioxidant properties of passion fruit (*Passiflora edulis*) peel. *Acta Scientiarum Polonorum Technologia Alimentaria, 13* (3), 257–265.
181. (2018). *World Fruit Map-Global Trade Still Fruitful.* https://research.rabobank.com/far/en/sectors/regional-food-agri/world_fruit_map_2018.html (accessed on 5 August 2020).
182. Xu, J. K., Li, M. F., & Sun, R. C., (2015). Identifying the impact of ultrasound-assisted extraction on polysaccharides and natural antioxidants from *Eucommia ulmoides* oliver. *Process Biochemistry, 50* (3), 473–481.
183. Yang, Y. C., & Wei, M. C., (2015). Kinetic and characterization studies for three bioactive compounds extracted from *Rabdosia rubescens* using ultrasound. *Food and Bioproducts Processing, 94,* 101–113.
184. Younis, K., & Ahmad, S., (2015). Waste Utilization of apple pomace as a source of functional ingredient in buffalo meat sausage. *Cogent Food and Agriculture, 1* (1), 1–10.
185. Yu, J., Dandekar, D. V., Toledo, R. T., Singh, R. K., & Patil, B. S., (2007). Supercritical fluid extraction of limonoids and naringin from grapefruit seeds. *Food Chemistry, 105* (3), 1026–1031.
186. Zheng, Z., & Shetty, K., (1998). Solid-state production of beneficial fungi on apple processing wastes using glucosamine as the indicator of growth. *Journal of Agricultural and Food Chemistry, 46,* 783–787.
187. Zhu, F., Du, B., Zheng, L., & Li, J., (2015). Advance on the bioactivity and potential applications of dietary fiber from grape pomace. *Food Chemistry, 186,* 207–212.

CHAPTER 5

IDENTIFICATION OF BOTANICAL AND GEOGRAPHICAL ORIGINS OF HONEY-BASED ON POLYPHENOLS

ZSANETT BODOR, CSILLA BENEDEK, ZOLTAN KOVACS, and JOHN-LEWIS ZINIA ZAUKUU

ABSTRACT

Honey is produced by different species of bees (most commonly *Apis mellifera*) from nectar, from secretions of plants, or secretions of sap-sucking insects attacking the plants. Its chemical composition includes mainly different types of sugars and moisture, however, some other nutritionally active components, such as vitamins, minerals, enzymes, organic acids, phenolic components (phenolic acids, and flavonoids) are also present. The composition and thus health-promoting properties of honey depend mainly on its botanical source and geographical origin. Therefore, the determination of honey origin is a crucial task for food safety authorities, medical practitioners, and the scientific community. Numerous types of methods are employed to determine the origin of honey. Determination of floral markers (such as: phenolic compounds)is a useful technique. In literature, several markers have been suggested to assess the origin of honey, e.g., hesperetin in citrus honey, ellagic acid in heather honey, etc. Determination of antioxidant capacity, identification, and quantification of a variety of the phenolic components in honey has been reported to help in the identification of the botanical and geographical origins of honey.

5.1 INTRODUCTION

Honey has been used as a sweetener and a medical natural product since ancient times. Several archeological proofs are available for the existence of beekeeping practices and the use of honeys. Rock paintings from the Mesolithic cultures represent hunting activities of honey. The ancient Egyptians used honey as a medicine and a special product for embalming mummies. Biblical references are also available about the use of honey and honey was used as a medical product during the medieval times and in the 17th century [39].

The application of honey is very diverse around the globe: it is used in the traditional Chinese medicine, Ayurveda medicine, and also in apitherapy [14, 55]. Honey is a complex food matrix, containing more than 20 different types of sugar, water (around 20%), proteins, minerals, vitamins, organic acids, enzymes, and biologically active components [73]. As an exceptional natural product, honey contains a variety of phenolic compounds representing its key phytochemicals. This phytochemical acts as an important quality parameter and account for the color, sensory profile and antioxidant activity of honey. Total phenolic content of honey is a function of its botanical source and geographical origin, its amount varying from 46 to 753 $\mu g g^{-1}$. Honeys from different floral types are significantly different in their chemical composition and phenolic profile. Darker honeys are considered to have higher amounts of flavonoid components and less phenolic acid derivatives than the lighter ones.

Presence of polyphenols, especially flavonoids in honey, offers many health benefits. These also act as a relevant marker in identification of the botanical and eventually the geographical origin of honey. In order to explore the therapeutic value of honey, it is highly desirable to have a deep knowledge on the amount and composition of phenolic compounds in honey [49]. It is also known that the botanical and geographical origin of honey has an important role in its medicinal use. Manuka honeys for example are well known for their exceptional antimicrobial activity, linden honeys are usually used in cough, flu, and sinusitis, while chestnut honeys are considered to have effect on blood circulation, etc. [14].

This chapter gives an overview on main phytochemicals (phenolic acids and flavonoids)components in honey based on the different botanical origins. Chapter also discusses geographical origin, which has a direct impact on phenolic compounds, and thus the therapeutic value of honey.

5.2 CLASSIFICATION OF POLYPHENOLS

5.2.1 FLAVONOIDS

Honey contains about 6 $mg kg^{-1}$ of flavonoids, and this value is higher in pollen (0.5%) and even higher in propolis (up to 10%) [6]. Flavonoids are present mainly in the aglycone form in both propolis and honey, and have been identified as flavanones, flavonones, and flavanols. The flavonoid family is characterized by the generic presence of an 1-, 2- or 3-phenyl-1,4-benzopyrone. They are further classified in number of subfamilies, depending on the oxidation state of the carbon atoms and the substitution of the benzopyrone ring [22]. Variety of polyphenols in different honey sources were compiled and suggested as markers of origin by Gasic et al. [32]. Many other authors also suggest selected phenolics as useful indicators of honeys of different origins. Potential marker compounds in the literature are shown in Table 5.1 in bold and italics. Important flavonoid compounds identified in honey along with their botanical sources are indicated in Table 5.1. Common flavonoid components (such as: quercetin, pinocembrin, pinobanskin, crhysin, kaempferol, and apigenin) are present in different types of honeys. On the other hand, some flavonoids like naringenin and hesperetin are specific and were mainly detected in well-defined honey types (e.g., citrus). Some components like tricetin, acacetin, ellagic acid, catechin, and epicatechin were reported only in few cases.

TABLE 5.1 Major Flavonoids in Honey and Their Botanical Occurrence

Flavonoid Subclass and Structure	Flavonoid Type and Structure	Botanical Source and References
Flavones	Chrysin HO OH O	Acacia (*Robinia pseudoacacia*) [15, 50, 65] Chestnut *(Castanea sativa)* [15] Citrus *(Citrus spp.)* [25, 27, 28, 53, 63] Buckwheat (*Fagopyrum esculentum* Moench) [41] Eucalyptus (*Myrtaceae Eucalyptus* sp.) [63, 84] Fir (*Abies alba* Mill.) [15, 53] Gelam (*Melaleuca cajuputi)* [82] Heather *(Erica spp.)* [41, 76] Honeydew [15, 28, 54, 65] Linden *(Tilia spp.)* [15, 65] Lotus (*Fabaceae Lotus* sp.) [63] Manuka (*Leptospermum scoparium)* [4, 17, 18, 83] Pine *(Pinus L.)* [53] Rosemary (*Rosmarinus officinalis* L.) [28, 70] Sage (*Salvia officinalis* L.) [51] Spruce (*Picea abies* (L.) Karst) [15] Sunflower (*Helianthus annuus L.*) [6, 35, 65] Thyme (*Thymus* L.) [53]

TABLE 5.1 *(Continued)*

Flavonoid Subclass and Structure	Flavonoid Type and Structure	Botanical Source and References
	Luteolin 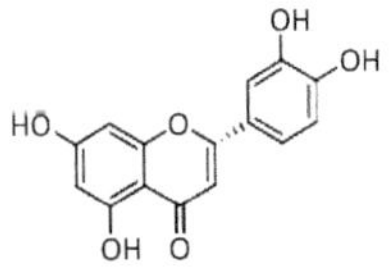	Acacia (*Robinia pseudoacacia*) [15, 18, 50, 65]
		Australian jelly bush *(Leptospermum polygalifolium)* [83]
		Azadirachta indica [25]
		Chestnut (*Castanea sativa*) [15]
		Citrus (*Citrus* spp.) [25, 27, 63]
		Eucalyptus (*Myrtaceae eucalyptus* sp., *Eucaliptus pilligaensis)* [25, 32, 36, 63]
		Fir (*Abies alba* Mill.) [15]
		Honeydew [15, 65]
		Lavender *(Lavandula spp.)* [6, 32]
		Linden (*Tilia* spp.) [15, 65]
		Litchi *(Litchi chinensis)* [86]
		Lotus (*Fabaceae Lotus* sp.) [36, 63]
		Manuka (*Leptospermum scoparium)* [17, 18, 83]
		Sage (*Salvia officinalis* L.) [51]
		Spruce (*Picea abies* (L.) Karst) [15]
		Sunflower (*Helianthus annuus L.*) [65]
		Thyme (*Thymus L.)* [18, 42]
		Tualang (*Koompassia excels*) [4, 18]
	Tricetin 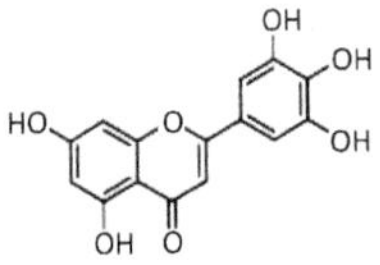	***Eucalyptus (****Myrtaceae Eucalyptus* sp., ***Eucaliptus camaldulensis***) [25, 32]
		Heather *(Erica spp.)* [6, 32]
	Apigenin 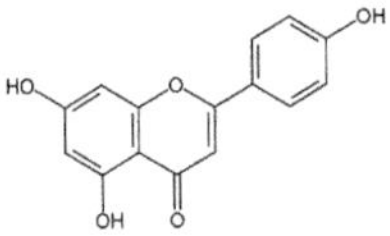	Acacia *(Robinia Pseudoacacia)* [15, 18, 24, 50, 56]
		Astralagus spp. [24]
		Chaste tree *(Vitexagnus-castus)* [24]
		Chestnut (*Castanea sativa*) [15, 24]
		Citrus (*Citrus* spp.) [63]
		Eucalyptus (*Myrtaceae Eucalyptus* sp.) [63]
		Fir (*Abies alba* Mill.) [15]
		Honeydew [15]

TABLE 5.1 *(Continued)*

Flavonoid Subclass and Structure	Flavonoid Type and Structure	Botanical Source and References
		Linden (*Tilia*spp) [15, 24]
		Lotus (*Fabaceae Lotus* sp.) [63]
		Manuka (*Leptospermum scoparium)* [4]
		Rhododendron *(Rhododendron spp.)* [24]
		Sage (*Salvia officinalis* L.) [51]
		Spruce (*Picea abies* (L.) Karst) [15]
		Thyme (*Thymus L.)* [42]
		Tualang (*Koompassia excels*) [4, 18]
Flavonols	Galangin	Acacia *(Robinia Pseudoacacia)* [15, 18, 50, 65]
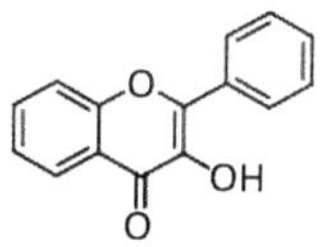	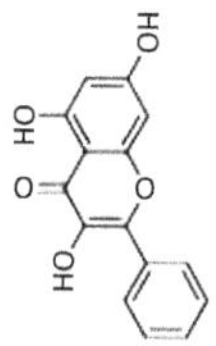	Buckwheat (*Fagopyrum esculentum* Moench) [41]
		Chestnut (*Castanea sativa*) [15]
		Citrus (*Citrus* spp.) [27, 28, 63]
		Eucalyptus (*Myrtaceae Eucalyptus* sp.) [63]
		Fir (*Abies alba* Mill.) [15]
		Heather *(Erica spp.)* [41]
		Honeydew [15, 28, 65]
		Linden (*Tilia* spp.) [15, 38, 65]
		Lotus (*Fabaceae Lotus* sp.) [63]
		Manuka (*Leptospermum scoparium)* [17, 18]
		Rosemary (*Rosmarinus officinalis* L.) [6, 28]
		Sage (*Salvia officinalis* L.) [51]
		Sunflower (*Helianthus annuus L.*) [6, 65]
	Kaempferol	Acacia (*Robinia pseudoacacia*) [15, 18, 24, 50, 56, 65]
	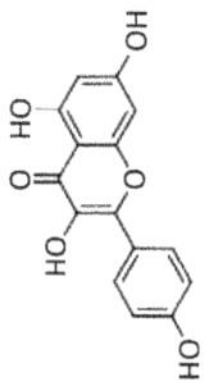	Buckwheat (*Fagopyrum esculentum* Moench) [41]
		Chestnut (*Castanea sativa*) [15]
		Citrus (*Citrus* spp.) [25, 27, 28, 53, 63]
		Diplotaxis tenuifolia [77]
		Eucalyptus (*Myrtaceae Eucalyptus* sp.) [25, 63]
		Fir (*Abies alba* Mill.) [15, 53]
		Ginger *(Zingiber officinale*) [25]
		Heather *(Erica spp.)* [41]
		Honeydew [15, 28, 65]
		Linden (*Tilia* spp.) [15, 57, 65]
		Litchi *(Litchi chinensis)* [86]

TABLE 5.1 *(Continued)*

Flavonoid Subclass and Structure	Flavonoid Type and Structure	Botanical Source and References
		Lotus (*Fabaceae Lotus* sp.) [63]
		Manuka (*Leptospermum scoparium)* [17, 83]
		Pine *(Pinus L.)* [24, 53]
		Rhododendron *(Rhododendron spp.)* [24]
		Rosemary (*Rosmarinus officinalis* L.) [28, 32, 70]
		Sage (*Salvia officinalis* L.) [51]
		Spruce (Picea abies (L.) Karst) [15]
		Strawberry tree *(Arbutus undedo L.)* [18]
		Sunflower (*Helianthus annuus L.*) [35, 65]
		Thyme (*Thymus L.)* [53]
		Tualang (*Koompassia excels*) [18]
	8-Methoxy-kaempferol 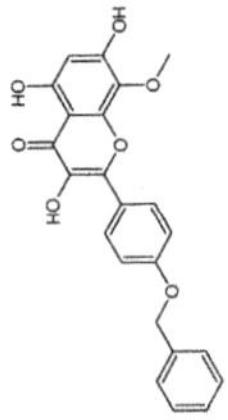	Manuka (*Leptospermum scoparium)* [17]
		Rosemary (*Rosmarinus officinalis* L.) [6, 32, 70]
	Myricetin 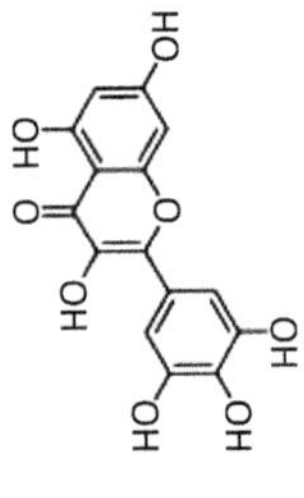	Acacia *(Robinia Pseudoacacia)* [18, 65]
		Azadirachta indica [25]
		Buckwheat (*Fagopyrum esculentum* Moench) [41, 76]
		Chestnut (*Castanea sativa*) [15]
		Citrus (*Citrus* spp.) [53]
		Eucalyptus (*Myrtaceae Eucalyptus* sp.) [25, 36, 63]
		Fir (*Abies alba* Mill.) [15, 53]
		Gelam (*Melaleuca cajuputi)* [43]
		Heather *(Erica spp.)* [18, 41, 76]
		Honeydew [15, 28, 65]
		Linden (*Tilia* spp.) [65]
		Lotus (*Fabaceae Lotus* sp.) [36]
		Pine *(Pinus L.)* [53]
		Spruce (Picea abies (L.) Karst) [15]
		Sunflower (*Helianthus annuus L.*) [65]
		Thyme (*Thymus L.)* [18, 42, 53]

TABLE 5.1 *(Continued)*

Flavonoid Subclass and Structure	Flavonoid Type and Structure	Botanical Source and References
	Myricetin-3-methylether 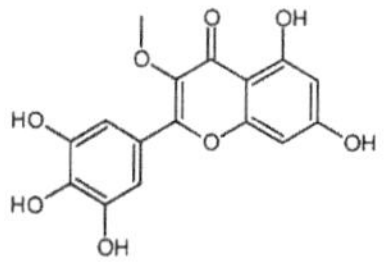	Australian jelly bush *(Leptospermum polygalifolium)* [83] ***Heather*** *(Erica spp.)* [6, 32] Manuka (*Leptospermum scoparium)* [83]
	Myricetin-3O-methylether 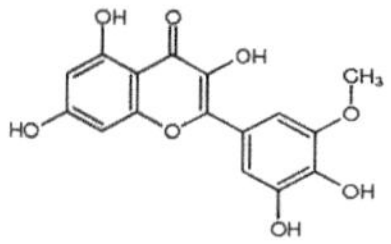	
	Quercetin 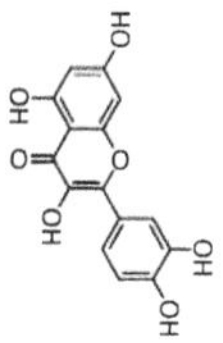	Acacia (*Robinia pseudoacacia*) [15, 18, 50, 65, 78] Australian jelly bush *(Leptospermum polygalifolium)* [83] *Azadirachta indica* [25] Chestnut (*Castanea sativa*) [15, 24] Citrus (*Citrus* spp.) [25, 27, 28, 53] Clover (*Triflolium spp.*) [18] *Diplotaxistenuifolia* [77] Eucalyptus (*Myrtaceae Eucalyptus* sp.) [25, 35, 36, 59, 63] Fir (*Abies alba* Mill.) [15, 53] Honeydew [15, 28, 64, 65] Gelam *(Melaleuca cajuputi)* [82] Ginger *(Zingiber officinale*) [25] Heather *(Erica spp.)* [24] Japanese grape *(Hovenia dulcis)* [59] Linden (*Tilia* spp.) [15, 57, 65] Lotus (*Fabaceae Lotus* sp.) [36, 63] Manuka (*Leptospermum scoparium)* [17, 18, 44, 83] Mastic (*Schinus terebinthifolius*) [59] Pine *(Pinus L.)* [24, 53] Quitoco (*Pluchea sagittalis*) [59] Rhododendron *(Rhododendron spp.)* [24]

TABLE 5.1 *(Continued)*

Flavonoid Subclass and Structure	Flavonoid Type and Structure	Botanical Source and References
		Rosemary (*Rosmarinus officinalis* L.) [28] Sage (*Salvia officinalis* L.) [51] Spruce (*Picea abies* (L.) Karst) [15] ***Sunflower*** *(Helianthus annuus L.)* [32, 35, 65, 70] Thyme (*Thymus L.)* [53] Tualang (*Koompassia excels*) [44]
	Isorhamnetin 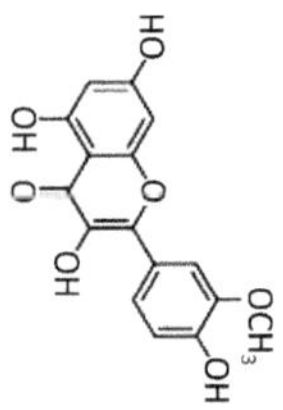 	Acacia *(Robinia Pseudoacacia)* [24, 65] *Azadirachta indica* [25] Citrus (*Citrus* spp.) [63] *Diplotaxis tenuifolia* [77] Eucalyptus (*Myrtaceae Eucalyptus* sp.) [25, 63] Honeydew [65] Linden (*Tilia* spp.) [65] Lotus (*Fabaceae Lotus* sp.) [63] Manuka (*Leptospermum scoparium)* [17, 83] Sunflower (*Helianthus annuus L.*) [65]
Isoflavones 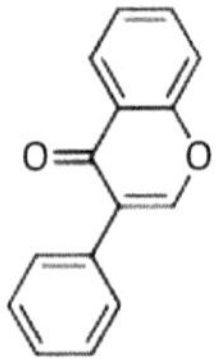 	Genistein 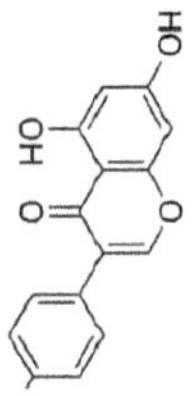 	Acacia *(Robinia Pseudoacacia)* [18] Clover (*Triflolium spp.*) [48] Cedrus [48]
Anthocyanidins 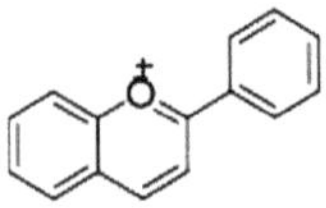 	—	*Acacia gerardii* [5] *Acacia tortilis* [5] Manuka (*Leptospermum scoparium)* [5] Tualang *(Koompassia excels*) [5]
Flavanones 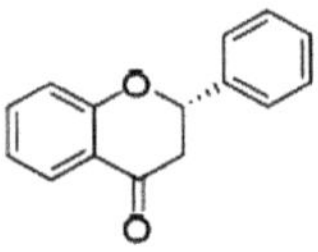 	***Hesperetin*** 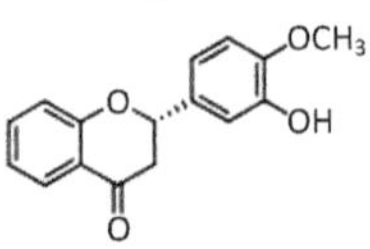 	***Citrus*** *(Citrus spp.)* [6, 25, 27, 28, 32, 63] Eucalyptus (*Myrtaceae Eucalyptus* sp.) [63] Gelam (*Melaleuca cajuputi)* [43, 82] Lotus (*Fabaceae Lotus* sp.) [63]

TABLE 5.1 *(Continued)*

Flavonoid Subclass and Structure	Flavonoid Type and Structure	Botanical Source and References
	Pinocembrin 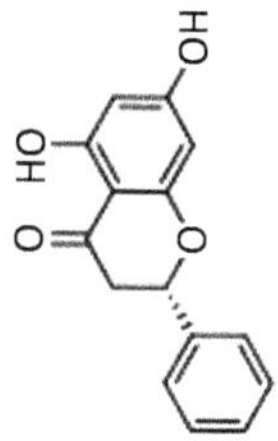	Acacia (*Robinia pseudoacacia*) [15, 18, 56, 65]
		Chestnut (*Castanea sativa*) [15]
		Citrus (*Citrus* spp.) [63]
		Eucalyptus (*Myrtaceae Eucalyptus* sp.) [25, 63, 85]
		Fir (*Abies alba* Mill.) [15]
		Honeydew [15, 28, 64, 65]
		Kamahi *(Weinmannia racemose)* [37]
		Leatherwood (*Eucryphia lucida*) [37]
		Linden (*Tilia* spp.) [15, 65]
		Lotus (*Fabaceae Lotus* sp.) [63]
		Manuka (*Leptospermum scoparium)* [4, 17, 18, 37]
		Rosemary (*Rosmarinus officinalis* L.) [6, 18, 28, 70]
		Spruce (Picea abies (L.) Karst) [15]
		Strawberry tree *(Arbutus undedo L.)* [18]
		Sunflower (*Helianthus annuus L.*) [6, 65]
	Naringenin 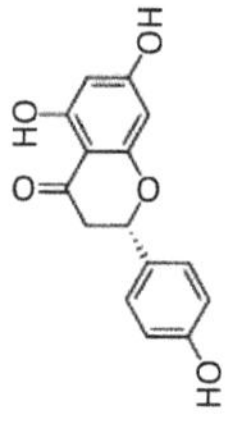	Citrus (*Citrus* spp.) [25, 27, 28]
		Fir (*Abies alba* Mill.) [15]
		Honeydew [15, 28, 68]
		Lavender *(Lavandula spp.)* [32]
		Lemon *(Citrus spp.)* [68]
		Linden (*Tilia* spp.) [15]
		Orange *(Citrus spp.)* [68]
		Rhododendron *(Rhododendron spp.)* [68]
		Rosemary (*Rosmarinus officinalis* L.) [28]
		Spruce (*Picea abies* (L.) Karst) [15]
		Tualang (*Koompassia excels*) [4]
	Acacetin (Apigenin 4'-methyl ether) 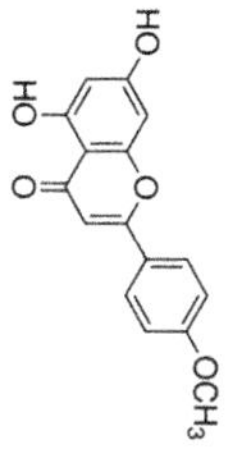	***Acacia (Romania)*** *(Robinia Pseudoacacia)* [56]

TABLE 5.1 *(Continued)*

Flavonoid Subclass and Structure	Flavonoid Type and Structure	Botanical Source and References
Flavanonols 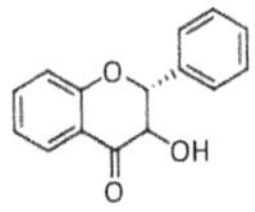	Pinobanksin 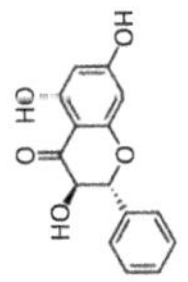	Acacia *(Robinia pseudoacacia)* [15, 18, 56]
		Chestnut (*Castanea sativa*) [15]
		Citrus (*Citrus* spp.) [63]
		Eucalyptus (European) (*Myrtaceae Eucalyptus* sp.) [63, 85]
		Fir (*Abies alba* Mill.) [15]
		Honeydew [15]
		Kamahi *(Weinmannia racemose)* [37]
		Leatherwood (*Eucryphia lucida*) [37]
		Linden (*Tilia* spp.) [15]
		Lotus (*Fabaceae Lotus* sp.) [63]
		Manuka (*Leptospermum scoparium)* [17, 37]
		Rosemary (*Rosmarinus officinalis* L.) [6, 18, 70]
		Spruce (Picea abies (L.) Karst) [15]
		Strawberry tree *(Arbutus undedo L.)* [18]
		Sunflower (*Helianthus annuus L.*) [6]
Flavan-3-ols 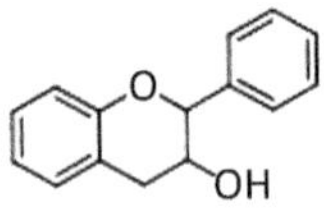	Catechin 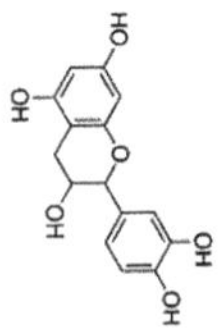	Heather *(Erica spp)* [38]
		Tualang (*Koompassia excels*) [4]
	Epicatechin 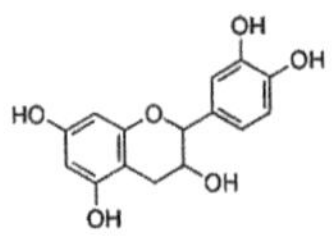	*Lavender (Lavandula spp.)* [24]
		Litchi *(Litchi chinensis)* [25]
		Oak *(Quercus)* [24]
Tannin derivatives	***Ellagic acid*** 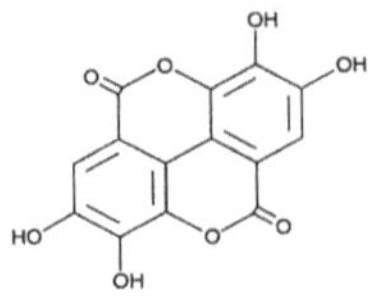	Australian jelly bush *(Leptospermum polygalifolium)* [83]
		Buckwheat (*Fagopyrum esculentum* Moench) [41]
		Gelam (*Melaleuca cajuputi)* [43]
		Heather *(Erica spp.)* [6, 32, 35, 41, 70]
		Manuka (*Leptospermum scoparium)* [83]

*Potential marker compounds proposed in the literature are in bold italics.

5.2.2 NON-FLAVONOID PHENOLIC COMPOUNDS IN HONEY

Some of the non-flavonoid polyphenols in honey are listed in Table 5.2. These are mainly phenolic acids in honeys (such as: syringic acid, gallic acid, p-coumaric acid, caffeic acid and. ferulic acid). On the other hand, isoferulic acid, m-coumaric acid, homoanisic acid and o-anisic acid were found only in few types of honey.

TABLE 5.2 Major Non-Flavonoid Phenolic Compounds in Honeys and their Occurrence According to Botanical Sources

Non-Flavonoid Polyphenol and its Structure	Phenolic Acid in Honey and its Structure	Botanical Source and References
Phenolic acids/ Benzoic acid derivatives 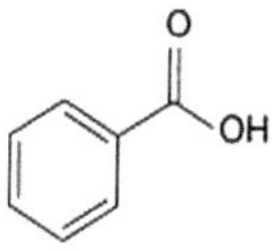	Benzoic acid 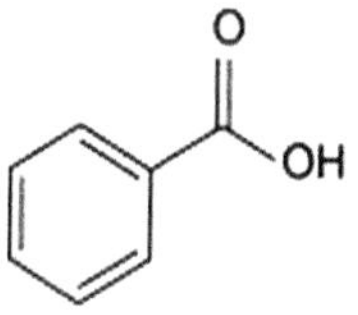	Citrus (*Citrus* spp.) [63] Eucalyptus (*Myrtaceae Eucalyptus* sp.) [63] Heather *(Erica spp.)* [26] Lotus (*Fabaceae Lotus* sp.) [63] Manuka *(Leptospermum scoparium)* [4] Tualang (*Koompassia excels*) [4]
	Syringic acid 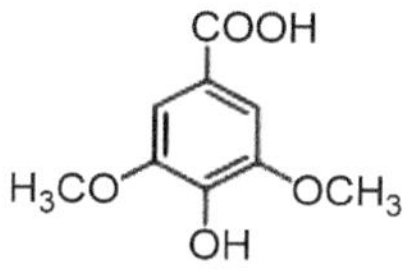	Acacia *(Robinia Pseudoacacia)* [24] *Azadirachta indica* [25] Buckwheat (*Fagopyrum esculentum* Moench) [41] Carob *(Ceratonia silique)* [48] Citrus (*Citrus* spp.) [53] Eucalyptus (*Myrtaceae Eucalyptus* sp.) [63] *Euphorbia milii* [48] Fir (*Abies alba* Mill.) [53] Ginger *(Zingiber officinale*) [25] Heather *(Erica spp.)* [41, 48, 57] *Lavender (Lavandula spp.)* [48] Leatherwood (*Eucryphia lucida*) [37] Lotus (*Fabaceae Lotus* sp.) [63] Manuka *(Leptospermum scoparium)* [4, 18] Oak *(Quercus)* [24] Pine *(Pinus L.)* [24, 48, 53] Thyme (*Thymus L.)* [18, 53] Tualang (*Koompassia excels*) [4, 18]

TABLE 5.2 *(Continued)*

Non-Flavonoid Polyphenol and its Structure	Phenolic Acid in Honey and its Structure	Botanical Source and References
	Methyl syringate* 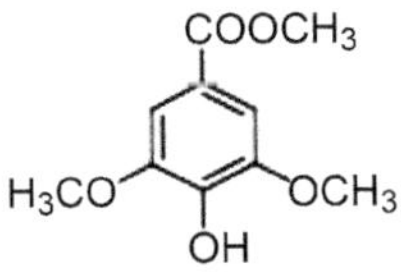 $COOCH_3$ / H_3CO / OCH_3 / OH	Acacia *(Robinia Pseudoacacia)* [40] ***Asphodel*** *(Asphodelus microcarpus)* [32] Canola (*Brassica napus*) [40] Kamahi *(Weinmannia racemose)* [37] Leatherwood (*Eucryphia lucida*) [37] Manuka *(Leptospermum scoparium)* [4, 37]
	4-Hydroxy-benzoic acid* 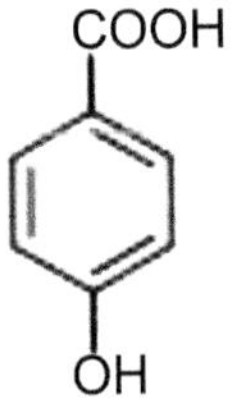 COOH / OH	Acacia *(Robinia Pseudoacacia)* [24, 56] *Astralagus spp.* [24] Buckwheat (*Fagopyrum esculentum* Moench) [6, 41] Chaste tree *(Vitexagnus-castus)* [24] ***Chestnut*** *(Castanea sativa)* [24, 26, 32] Citrus (*Citrus* spp.) [63] Clover (*Triflolium spp.*) [18, 24] Eucalyptus (*Myrtaceae Eucalyptus* sp.) [63] ***Heather*** *(Erica spp.)* [18, 24, 26, 41, 57] *Lavender (Lavandula spp.)* [24] Linden (*Tilia* spp.) [24, 57] Lotus (*Fabaceae Lotus* sp.) [63] Oak *(Quercus)* [24] Pine *(Pinus L.)* [24] Rhododendron *(Rhododendron spp.)* [24]
	3-Hydroxy-benzoic acid* 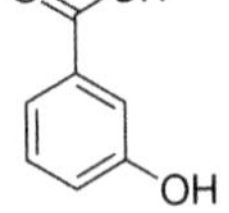 O / OH / OH	Buckwheat (*Fagopyrum esculentum Moench*) [6, 41] Carob *(Ceratonia silique)* [48] Heather *(Erica spp.)* [41] ***Linden*** *(Tilia spp.)* [26, 32]
	Methyl-4-hydroxy-benzoate 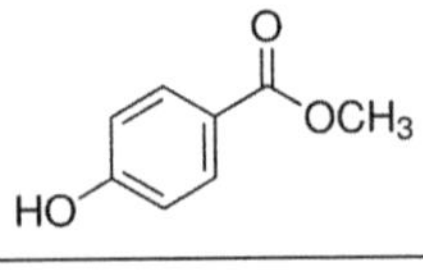 O / OCH_3 / HO	Canola *(Brassica napus)* [40] Orange *(Citrus spp.)* [40] Manuka *(Leptospermum scoparium)* [4]

TABLE 5.2 *(Continued)*

Non-Flavonoid Polyphenol and its Structure	Phenolic Acid in Honey and its Structure	Botanical Source and References
	Protocatechuic acid 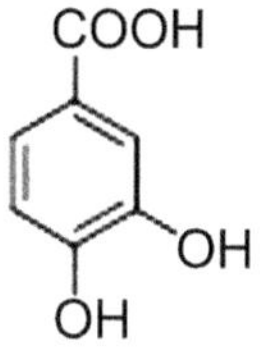	*Astralagus spp.* [24]
		Buckwheat (*Fagopyrum esculentum Moench*) [6]
		Chestnut *(Castanea sativa)* [18, 24, 26]
		Clover (*Triflolium spp.*) [24]
		Eucalyptus (*Myrtaceae Eucalyptus* sp.) [59, 63]
		Heather *(Erica spp.)* [18, 24]
		Linden (*Tilia* spp.) [24]
		Oak *(Quercus)* [24]
		Pine *(Pinus L.)* [18, 24]
		Rhododendron *(Rhododendron spp.)* [24]
		Thyme (*Thymus L.)* [48]
		Willow *(Salix L.)* [76]
	Gallic acid* 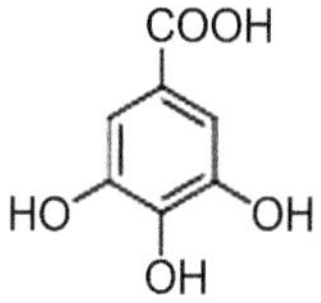	Acacia *(Robinia Pseudoacacia)* [65]
		Astralagus spp. [24]
		Australian jelly bush *(Leptospermum polygalifolium)* [83]
		Azadirachta indica [25]
		Chestnut *(Castanea sativa)* [24]
		Eucalyptus (*Myrtaceae Eucalyptus* sp.) [25, 59, 63]
		Gelam (*Melaleuca cajuputi)* [82]
		Ginger *(Zingiber officinale*) [25]
		Heather *(Erica spp.)* [18, 24, 26]
		Honeydew [65, 76]
		Japanese grape *(Hoveniadulcis)* [59]
		Kamahi *(Weinmannia racemose)* [37]
		Linden (*Tilia* spp.) [65]
		Litchi *(Litchi chinensis)* [25]
		Manuka (New Zealand) *(Leptospermum scoparium)* [18, 37, 83]
		Oak *(Quercus)* [24]
		Quitoco (*Pluchea Sagittalis*) [59]
		Rhododendron *(Rhododendron spp.)* [24]
		Sunflower (*Helianthus annuus L.*) [65]
		Tualang (*Koompassia excels*) [4]

TABLE 5.2 *(Continued)*

Non-Flavonoid Polyphenol and its Structure	Phenolic Acid in Honey and its Structure	Botanical Source and References
	Vanillic acid (COOH; OCH3; OH)	Acacia (*Robinia pseudoacacia*) [18, 56]
		Astralagus spp. [24]
		Buckwheat (*Fagopyrum esculentum* Moench) [41]
		Cedrus [48]
		Chestnut *(Castanea sativa)* [24, 26]
		Citrus (*Citrus* spp.) [25, 63]
		Eucalyptus (*Myrtaceae Eucalyptus* sp.) [48, 63]
		Ginger *(Zingiber officinale)* [25]
		Heather *(Erica spp.)* [18, 26, 41, 57]
		Litchi *(Litchi chinensis)* [25]
		Linden (*Tilia spp.)* [57]
		Lotus (*Fabaceae Lotus* sp.) [63]
		Pine *(Pinus L.)* [24, 48]
		Rhododendron *(Rhododendron spp.)* [24]
	Gentisic acid (HO; O; OH; OH)	Cedrus [48]
		Carob *(Ceratonia silique)* [48]
		Eucalyptus (*Myrtaceae Eucalyptus* sp.) [48]
		Pine *(Pinus L.)* [48]
		Thyme (*Thymus L.)* [48]
	Eudesmic acid* 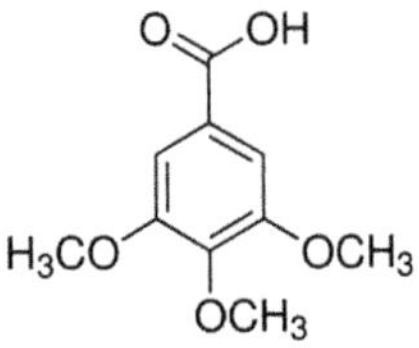 (O; OH; H_3CO; OCH_3; OCH_3)	***Manuka** (Leptospermum scoparium)* [32]
	o-Anisic acid* 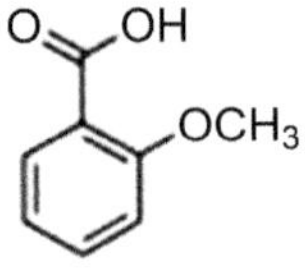 (O; OH; OCH_3)	***Manuka** (Leptospermum scoparium)* [32]

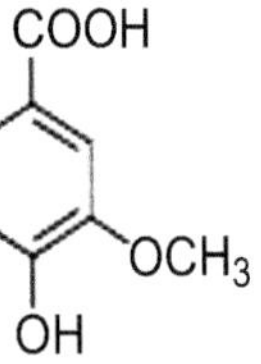

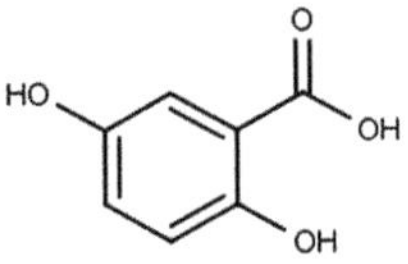

TABLE 5.2 *(Continued)*

Non-Flavonoid Polyphenol and its Structure	Phenolic Acid in Honey and its Structure	Botanical Source and References
Phenolic acids/4-Hydroxy-phenylacetic acid derivatives 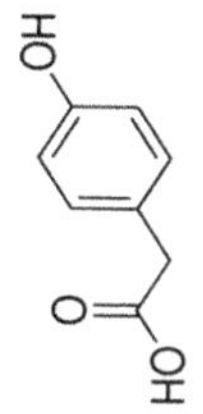	***4-Hydroxy-phenylacetic acid*** 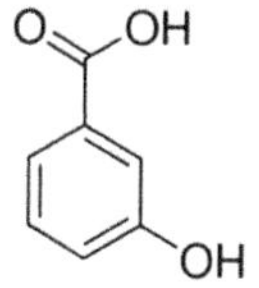	Buckwheat (*Fagopyrum esculentum Moench*) [6] ***Chestnut*** *(Castanea sativa)* [32] Eucalyptus (*Myrtaceae Eucalyptus* sp.) [63] Lotus (*Fabaceae Lotus* sp.) [63]
	Homoanisic acid* 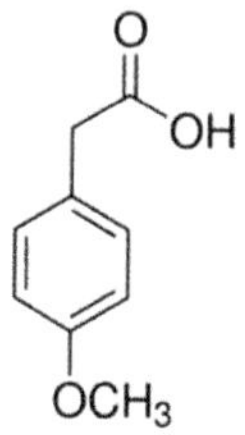	***Kanuka*** *(Kunzea ericoides)* [32]
	Homogentisic acid* 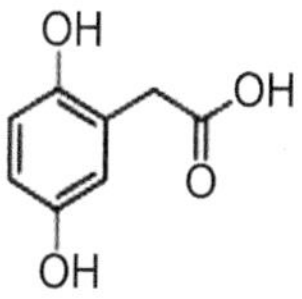	***Strawberry tree*** *(Arbutus undedo L.)* [32] Chaste tree *(Vitexagnus-castus)* [48] Thyme (*Thymus L.)* [48]
Phenolic acids/Cinnamic acid derivatives 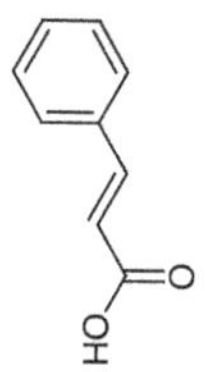	***Trans-Cinnamic acid**** 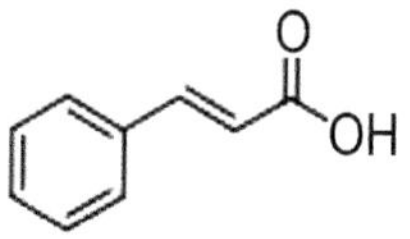	***Acacia*** *(Robinia pseudoacacia)* [32, 56] Clover (*Triflolium spp.*) [18] Heather *(Erica spp.)* [18, 26] Strawberry tree (*Arbutus unedo* L.) [18] Tualang (*Koompassia excels*) [4, 18]
	p-Coumaric acid* 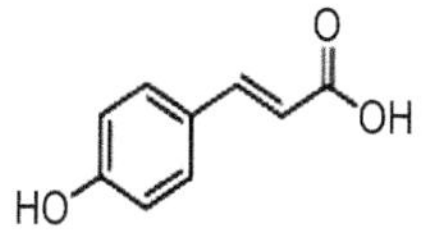	Acacia *(Robinia pseudoacacia)* [24, 56, 65] *Astralagus spp.* [24] *Azadirachta indica* [25] Buckwheat (*Fagopyrum esculentum* Moench) [6, 41, 76]

TABLE 5.2 *(Continued)*

Non-Flavonoid Polyphenol and its Structure	Phenolic Acid in Honey and its Structure	Botanical Source and References
		Capparis [25, 27, 28, 63] Chaste tree *(Vitexagnus-castus)* [24] ***Chestnut*** *(Castanea sativa)* [24, 26, 70] Citrus (*Citrus* spp.) [25, 27, 28, 63] ***Clover*** (*Trifolium spp.*) [25, 27, 28, 63] Eucalyptus (*Myrtaceae Eucalyptus* sp.) [25, 63] Heather *(Erica spp.)* [24, 26, 41, 76] Honeydew [15, 28, 65] Japanese grape *(Hoveniadulcis)* [59] *Lavender (Lavandula spp.)* [24] Linden (*Tilia* spp.) [24, 65] Lotus (*Fabaceae Lotus* sp.) [63] Manuka *(Leptoseprum scoparium* [17] Oak *(Quercus)* [24] Pine *(Pinus L.)* [24] Rhododendron *(Rhododendron spp.)* [24] Rosemary (*Rosmarinus officinalis* L.) [28] Sunflower (*Helianthus annuus L.*) [65] Tualang (*Koompassia excels*) [4] Wild carrot (*Daucus carota*) [35]
	m-Coumaric acid HO, OH, O	Acacia *(Robinia Pseudoacacia)* [26] Chestnut *(Castanea sativa)* [26]
	Caffeic acid HO, HO, OH, O	Acacia *(Robinia pseudoacacia)* [18, 24, 65] Australian jelly bush *(Leptospermum polygalifolium)* [30] *Azadirachta indica* [25] ***Capparis*** [48] Chaste tree *(Vitex agnus-castus)* [24] ***Chestnut*** *(Castanea sativa)* [24, 26, 70] Citrus (*Citrus* spp.) [25, 27, 28, 63] ***Clover*** (*Trifolium spp.*) [48]

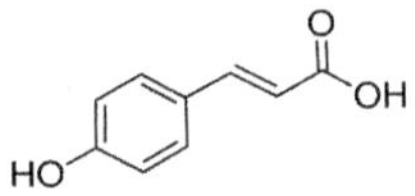

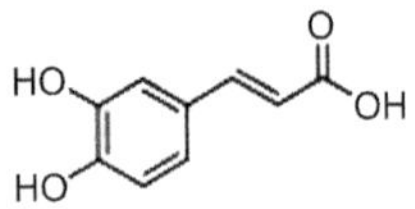

TABLE 5.2 *(Continued)*

Non-Flavonoid Polyphenol and its Structure	Phenolic Acid in Honey and its Structure	Botanical Source and References
		Eucalyptus (*Myrtaceae Eucalyptus* sp.) [25, 63]
		Gelam (*Melaleuca cajuputi)* [43]
		Ginger *(Zingiber officinale*) [25]
		Heather *(Erica spp.)* [24, 26, 41, 57, 76]
		Honeydew [15, 28, 65]
		Lavender (Lavandula spp.) [24]
		Linden (*Tilia* spp.) [24, 57, 65]
		Litchi *(Litchi chinensis)* [25]
		Lotus (*Fabaceae Lotus* sp.) [63]
		Manuka *(Leptoseprum scoparium)* [17, 18, 30]
		Oak *(Quercus)* [24]
		Pine *(Pinus L.)* [24]
		Rhododendron *(Rhododendron spp.)* [24, 48]
		Rosemary (*Rosmarinus officinalis* L.) [28]
		Sunflower (*Helianthus annuus L.*) [35, 65]
		Thyme (*Thymus L.)* [18]
		Tualang (*Koompassia excels*) [4, 18]
	Ferulic acid* H_3CO, HO, O, OH (structure)	Acacia *(Robinia pseudoacacia)* [18, 24, 56, 86]
		Astralagus spp. [24]
		Azadirachta indica [25]
		Buckwheat (*Fagopyrum esculentum* Moench) [6, 41]
		Canola *(Brassica napus)* [54]
		Chaste tree *(Vitex agnus-castus)* [24]
		Chestnut *(Castanea sativa)* [24, 26, 32, 70]
		Citrus (*Citrus* spp.) [25, 63]
		Eucalyptus (*Myrtaceae Eucalyptus* sp.) [25, 63]
		Ginger *(Zingiber officinale*) [25]
		Heather *(Erica spp.)* [18, 41]
		Honeydew [76]
		Jujube (*Ziziphusjujuba.* Mill.) [86]

TABLE 5.2 *(Continued)*

Non-Flavonoid Polyphenol and its Structure	Phenolic Acid in Honey and its Structure	Botanical Source and References
		Linden *(Tilia spp.)* [54] Litchi *(Litchi chinensis)* [25, 86] Lotus (*Fabaceae Lotus* sp.) [63] Manuka *(Leptoseprum scoparium)* [18] Oak *(Quercus)* [24] Pine *(Pinus L.)* [24] Raspberry *(Rubus)* [54] Rhododendron *(Rhododendron spp.)* [24, 48] Thyme (*Thymus L.)* [18] Willow *(Salix L.)* [76]
	Isoferulic acid* 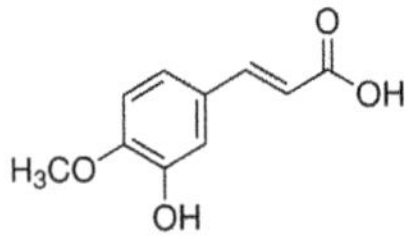	***Manuka*** *(Leptoseprum scoparium* [17]
	Rosmarinic acid* 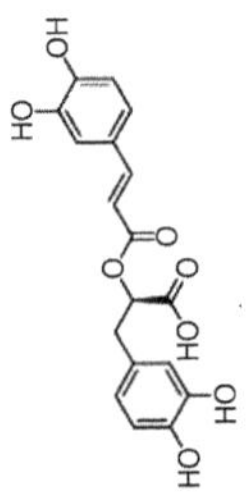	Buckwheat (*Fagopyrum esculentum Moench*) [41] Heather *(Erica spp.)* [41] Manuka *(Leptoseprum scoparium)* [76] ***Mint*** *(Mentha spp.)* [32]
	Chlorogenic acid* 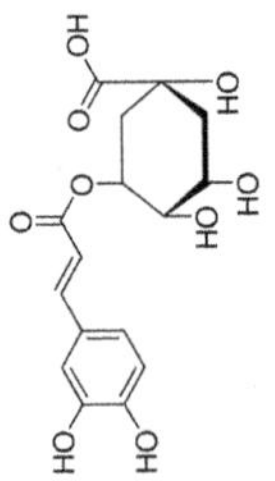	***Acacia*** *(Robinia pseudoacacia)* [18, 32] *Azadirachta indica* [25] Buckwheat (*Fagopyrum esculentum Moench*) [41] Gelam (*Melaleuca cajuputi)* [43, 82] Ginger *(Zingiber officinale*) [25] Heather *(Erica spp.)* [18, 41, 76] Honeydew [76] Thyme (*Thymus L.*) [18] Willow *(Salix L.)* [76]

TABLE 5.2 *(Continued)*

Non-Flavonoid Polyphenol and its Structure	Phenolic Acid in Honey and its Structure	Botanical Source and References
Metabolites of phenolic acids	***Phenylacetic acid**** 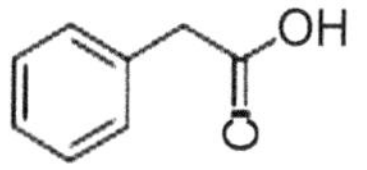	***Chestnut*** *(Castanea sativa)* [26] Heather *(Erica spp.)* [26] Linden *(Tilia spp.)* [26] ***Ling heather*** *(Erica spp., Calluna Vulgaris)* [26]
	3-Phenyllactic acid* 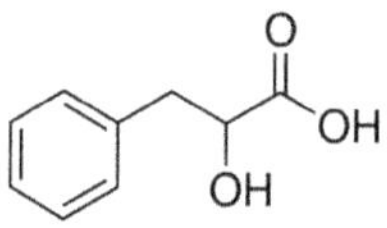	Chestnut *(Castanea sativa)* [26] ***Heather*** *(Erica spp.)* [26] ***Ling heather*** *(Erica spp., Calluna Vulgaris)* [26] Kamahi *(Weinmannia racemose)* [37] Leatherwood (*Eucryphia lucida*) [37] Manuka (*Leptospermum scoparium)* [4, 37] ***Milk thistle*** *(Silybummarianum)* [22, 32]
	Phenylpropanoic acid 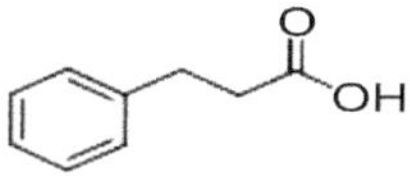	Canola *(Brassica napus*) [26] Kamahi *(Weinmannia racemose)* [37] Leatherwood (*Eucryphia lucida*) [37] Manuka (*Leptospermum scoparium)* [37]
	Mandelic acid 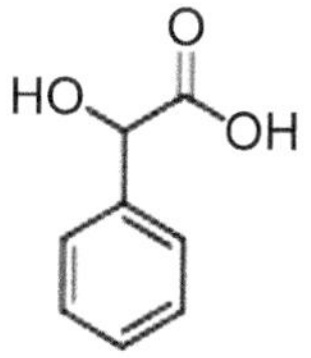	Heather *(Erica spp.)* [35]

*Potential marker compounds proposed in the literature are in bold italics.

5.3 ORIGIN AND IDENTIFICATION OF HONEY-BASED PHYTOCHEMICALS

5.3.1 CURRENT TRENDS IN DETERMINATION OF BOTANICAL AND GEOGRAPHICAL ORIGINS OF HONEY

Due to the variable composition of honey, identification of origin and detection of adulterations is a challenging task for food experts and authorities. There are no standard criteria in the European Union (EU) and worldwide, therefore it is a difficult task for identification of botanical and geographical origin of honeys. As quality of honey depends on several factors, such as, the nectar consumed by bees, natural structure of the nectar, climatic conditions and geographical origin, therefore the prediction of composition of honey also remains a challenging task [2]. There are several studies dealing with identification of origin based on physicochemical properties [10, 20, 45], spectral measurements [13, 23, 34], data acquired by electronic sensor instruments [46, 79, 80, 81], and sophisticated techniques like nuclear magnetic resonance (NMR), high performance liquid chromatography (HPLC), time-of-flight (TOF)-mass spectrometry (MS) [72], etc.

TPC according to techniques, such as: Folin-Ciocalteu (TPC), ferric reduction antioxidant power (FRAP), cupric ion reducing antioxidant capacity (CUPRAC), ABTS (2, 2'-azino-bis (3-ethylbenzothiazoline-6-sulfonic acid) radical scavenging. Total flavonoid content (TFC) are also studied for identification of origin of different monofloral honeys [1, 7, 11, 33, 47, 62, 67]. However, these methods cannot provide information on the type of the distinct molecules providing the antioxidant properties, therefore individual phytochemical components may serve as more appropriate marker molecules [16] for the identification of honeys from different botanical sources.

Sections 3.2, 3.3, and 3.4 give an account of different types of phenolic components and flavonoids that are present in monofloral honeys and can qualify as possible tracer molecules for authentication.

5.3.2 IDENTIFICATION OF BOTANICAL AND GEOGRAPHICAL ORIGINS BASED ON TOTAL ANTIOXIDANT CAPACITY

Studies report that the antioxidant capacity (AOC) and total phenolic and flavonoid contents in honeys are directly correlated.

5.3.2.1 METHODS FOR THE DETERMINATION OF ANTIOXIDANT PROPERTIES OF HONEY

Flavonoids and phenolic acids are considered as efficient plant-derived antioxidants. According to literature, there are many assays available to determine the TP and flavonoid contents in honey. The total AOC of honey can be measured through polyphenols content in honey. According to main mechanisms governing these reactions, the techniques can be classified into two main groups, such as: (1) reactions based on hydrogen atom transfer (HAT);

(2) single-electron transfer (SET); and (3) mixed mechanisms, involving both pathways.

5.3.2.1.1 Methods Based on Hydrogen Atom Transfer (HAT) Reactions

HAT based techniques are aimed to determine the ability of antioxidants to quench the free radicals by hydrogen donation. In these reactions, the bond dissociation enthalpy is an important parameter to evaluate the antioxidant action: lower enthalpies of the antioxidants facilitate the reaction. These methods are usually fast and need only seconds or minutes to get the results. The two main HAT-based methods are [69]: the oxygen radical absorbance capacity (ORAC) and the total-radical trapping antioxidant parameter (TRAP).

In ORAC method, the AOC is measured by measuring the intensity of a fluorescent signal from a probe that is quenched in the presence of free radicals. The antioxidants absorb the generated free radicals, reactive oxygen species, allowing thus the fluorescent signal to persist. The free radicals are generated from 2,2'-azobis (2-methylpropionamidine) dihydrochloride (AAPH) that produces a free radical (peroxyl).

The TRAP method monitors the capacity of the antioxidants in the sample to scavenge luminol-derived radicals, generated from AAPH decomposition. The reaction is followed by chemi-luminescence [69].

Other methods under HAT group are photochemi-luminescence (PCL), chemi-luminescence (CL) and total antioxidant scavenging capacity (TOSC) [69].

5.3.2.1.2 Methods Based on Single Electron Transfer (SET)

In SET reaction, the antioxidant delivers an electron to the free radical and itself becomes a radical cation. In this reaction, the ionization potential of the antioxidant is the most important factor affecting the antioxidant action. Lower ionization potentials make the electron abstraction easier [69]. SET measurements are generally more popular than HAT measurements. This also applies to the analysis of honey. Assays like ferric reducing antioxidant power (FRAP) and cupric ion reducing antioxidant power (CUPRAC) belong to this group [69].

FRAP reaction is based on the measurement of reduction of Fe^{+++} to Fe^{++}, which forms a colored complex with 2,4,6-tris (2-tripyridyl)-s-triazine

(TPTZ). This color change is followed by spectrophotometry at 593 nm. The method is not suitable for the detection of thiols and proteins [69].

CUPRAC method is similar to FRAP method. It is based on the reduction of Cu^{++} to Cu^{+}, which latter forms a colored complex with neocuproine. The CUPRAC can be used for both hydrophilic and lipophilic antioxidants. It has the advantage because this method works at a neutral pH similar to the physiological pH [69].

5.3.2.1.3 Methods Based on Mixed Set and Hat Mechanisms

2,2'-azino-bis (3-ethylbenzothiazoline-6-sulfonic acid) (ABTS) or trolox equivalent antioxidant capacity (TEAC) measurement belongs to this group, where the antioxidant ability of the test compound is determined based on the loss of color that is determined by their reaction with the $ABTS^{\bullet+}$ radical cation, produced prior to the assay. The most appropriate wavelengths for following the radical scavenging reaction are 415 nm and 734 nm.

A similar assay is 2,2-diphenyl-1-picrylhydrazyl (DPPH) method, which is based on the reducing capacity of the antioxidants towards DPPH radical causing change of color that can be measured at 515 nm. The method has the drawback related with steric hindrance of DPPH radical, which is not easily accessible for the antioxidants in the solution. The method is suitable for hydrophilic and lipophilic antioxidants.

5.3.2.1.4 Other Methods for Determination of Total Polyphenol

The formal content of total polyphenols (TPC) is determined by popular Folin-Ciocalteu method. This is based on the reduction of a mixture of phospotungstate-phospomolibdate by the antioxidants in the sample, resulting in the appearance of a blue color measured usually at 750 nm or 765 nm by spectrophotometry. The basic mechanism is a redox reaction. During this reaction, the phenolic group is oxidized and the metal ion is reduced. The method has the drawback of low specificity. It can detect many other reducing compounds, such as reducing sugars [75].

The TFC assay is often used to measure AOC of honey. This technique is based on the complication of flavonoids with aluminum chloride, which results in a change of color, detectable by spectrophotometry at 430 nm [74].

The total antioxidant status (TAS) and total oxidant status (TOS) measurements are also available. Usually, commercially available kits are used for the determination of these properties. Using TAS and TOS values, an oxidative stress index can be calculated. The spectrophotometrically detected change of color is dependent on the total amount of antioxidant molecules in the sample. The methods can be calibrated with H_2O_2 or trolox (6-hydroxy-2,5,7,8-tetramethylchroman-2-carboxylic acid).

As mentioned above, several *in vitro* assays are available for the detection of AOC in foods, however, all of them have unique advantages and disadvantages. Although new assays are continuously being developed, yet there is no "universal" method. Therefore, usually several techniques have been used to characterize the samples [31, 66].

5.3.2.2 APPLICATION OF AOC, TPC, AND TFC IN THE DISCRIMINATION OF BOTANICAL AND GEOGRAPHICAL ORIGINS OF HONEY

Although there has been no attempt to identify or quantify the individual compounds in these studies, yet the overall antioxidant properties of honeys are indicative in some cases for the botanical or geographical origins of honey.

Turkish researchers determined the TAS and TOS of honey samples from seven different regions in Turkey. They found significant differences between honeys from the different regions in terms of both parameters, due to the differences in the climatic conditions and the soil quality in the study zones [8].

Three different botanical types of honey (mainly: *Acacia tortilis* (Summer), *Ziziphus spina-crhisti* (Sidr) and multiflora honey) were investigated for their total polyphenol (TPC), total flavonoid (TFC) contents and their antioxidant capacity (DPPH) in Oman. The results revealed the differences between different types of honeys. Summer honey is richest in antioxidant compounds compared to other types. The researchers reported that their results were different from those reported from other countries, which can be the consequence of different climatic conditions [1].

Portuguese researcher investigated thyme, orange, strawberry, locust pod-shrub, rosemary, eucalyptus, and heather honeys. Honeys were classified based on their antioxidant capacities and concentration of minerals. Results showed that rosemary and orange honeys had lower TPC and AOC. AOC was increased in the following order: thyme honey>strawberry honey> locust pod-shrub honey> heather honey. Due to their low AOC, rosemary honey and orange honey were completely distinguished from the other types of honey. In addition, results showed that if AOC parameters are further combined with the mineral profile, this could represent a more efficient tool in the differentiation between honeys from various botanical types [7].

Bertoncelj and his co-workers [11] studied TPC, antioxidant activities of acacia, linden, chestnut, fir, spruce, forest, and multiflora honeys with FRAP and DPPH methods. For checking the significant differences among samples from different botanical origin, ANOVA test was applied, followed by Duncans's post-hoc test. ANOVA test showed that there were significant differences between different floral groups. Acacia honeys had significantly lower AOC and TPC than other botanical types. Linden honeys showed significantly higher values in TPC and DPPH in comparison to acacia honey, but significantly lower than other types. Forest and fir honeys had the significantly highest TPC and FRAP values. These results show that the phenolic content of honey is responsible for its antioxidant power, and these parameters can be suitable to obtain a satisfactory differentiation among samples of different botanical origin [11]. In this study, a linear discriminant analysis (LDA) model was used among antioxidant parameters, water content, electrical conductivity, pH, acidic parameters, color (L*a*b*) and optical rotation. Based on this statistical analysis, acacia, and multiflora honeys were 100% correctly classified; and linden, chestnut, and fir honeys also showed good separation and classification scores (>80%). Therefore, if routine physicochemical parameters are combined with the antioxidant properties and the results are chemo metrically evaluated, the botanical differentiation of honeys can be significantly improved [12].

Can and his co-workers [24] examined AOC of honeys from different floral sources (such as: chestnut, astralagus, heather, clover, lavender, lime, Jerusalem tea, common eryngo, chaste tree, rhododendron, oak, pine, acacia, and multiflora) by TPC, DPPH, FRAP, and TFC methods. Their results showed significant differences based on all parameters in relation to the botanical source. They concluded that honeydew, pine, and oak honeys were completely different from other types of honey in terms of each measured parameter. Heather honey was also outstanding due to high amount of phenolic compounds and elevated antioxidant activity [24].

In an analogous Romanian study, acacia, sunflower, forest, multiflora, linden, and sea buckthorn honeys were evaluated based on their pH, ash content, color, protein, free amino acid, TPC, and antioxidant capacities by ABTS and DPPH methods. Results indicated that TPC content varied significantly according to the botanical origin. Highest TPC level was obtained for forest honeys, while acacia honeys showed lowest values [12, 24]. DPPH and ABTS results showed similar trends, revealing similar differences related to the floral origin. Principal component analysis (PCA) model was built for measurement parameters revealing the different botanical groups [20].

Irish researchers reported concomitant measurements of TPC and physicochemical parameters of honeys (electrical conductivity, pH, and color). Simultaneous evaluation of these properties proved to play an important role in differentiating among botanical types of honey [47].

Group of Romanian researchers evaluated the TPC and TFC of acacia, linden, sunflower, and honeydew honeys. The results showed differences according to the botanical origin. In agreement with other reports, acacia honeys had lowest TPC and TFC, followed by linden and sunflower honeys, while honeydew honey had the highest values [3].

Also, Spanish researchers examined total concentration of polyphenols and flavonoids in chestnut, blackberry, heather, eucalyptus, multiflora, and honeydew honeys. Significant differences were found with ANOVA test followed with pair wise-comparison. Honeydew, heather, and chestnut honeys showed significantly higher TPC and TFC content than other types of honeys [29].

Brazilian researchers determined the antioxidant capacities by DPPH and FRAP methods and total phenolic content in nectar and honeydew honeys. Results confirmed the superior antioxidant qualities of honeydew honeys compared to blossom honeys. Therefore, these methods can be eligible for the differentiation between these two main types of honey [19].

Gül and Pehlivan [33] studied 23 different types of honeys for their antioxidant activities and total phenolic compounds by FRAP and DPPH methods. TPC was highest for chestnut, parsley, carob, and rhododendron honeys, while acacia and mint honeys showed lower values. All results revealed significant differences among the samples [33].

Maurya and co-workers [58] reviewed the antioxidant properties of honeys from different botanical and geographical origins. They reported that total phenolic content and antioxidant activities by FRAP, ORAC, DPPH, and ABTS methods were powerful tools to differentiate honeys from different botanical sources. The authors also concluded that there were differences in honeys from the same floral sources but different geographical origins. These differences might be attributed to climate and environmental factors, soil, and the surrounding flora [58].

Results of research carried out in Hungary were in agreement with these abovementioned studies regarding the relationships between antioxidant qualities and the botanical and geographical origins. In an investigation, antioxidant activities and TPC measured by CUPRAC, FRAP methods in 79 honeys were determined from various botanical and geographical sources. Acacia, canola, silkgrass, and sunflower honeys reached the lowest total polyphenol concentration and antioxidant activities, measured by FRAP method. With the same method, chestnut, pine, and honeydew honeys showed highest scores [13, 46].

In another study, classification models were built based on pH, electrical conductivity, color, ash content, and TPC of acacia, chestnut, linden, and multiflora honeys. Evaluation of the set of data by LDA provided 89.3% and 90.5% recognition and prediction abilities, respectively for the classification of honeys according to their botanical origin. Models were also built for samples from the same floral source but different geographical zones. Results showed that 100% correct classification was obtained in case of chestnut honey for different topographical sources; while the results for acacia honey were less conclusive [45]. Other researchers obtained lowest antioxidant levels for acacia and significantly higher values for chestnut honeys [3, 11, 29, 33].

5.3.3 BOTANICAL ORIGIN: IDENTIFICATION BASED ON POLYPHENOLS AS FLORAL MARKERS

The research studies dealing with origin authentication based on individual polyphenol as a marker molecule in different types honeys are presented in this section. Identification and quantification of these compounds require more complex, state-of-the-art techniques, typically chromatographic methods with UV or MS or MS^n detection.

According to Slovenian researchers, ten flavonoids (such as: myricetin, apigenin, luteolin, naringenin, pinobaksin, quercetin, pinocembrin, kaempferol, chrysin, galangin) and abscisic acid (plant hormone) and its derivatives were determined in acacia, linden, chestnut, fir, spruce, honeydew, and multiflora honeys by liquid chromatography, with diode array detection and electrospray ionization (ESI) mass spectrometry. A LDA was conducted for pinobaksin, galangin, and the two abscisic acid components. Results showed that acacia and linden honeys could be classified correctly, while other types showed misclassifications, overlapping between each other. The flavonoid profile of different honeys was not found to be suitable for the identification of botanical origin of the samples, as they were no identifiable marker molecules for any of the monofloral honeys [15].

Spanish researchers quantified some phenolic compounds and flavonoids (such as: myricetin, caffeic acid pinocembrin, hesperetin, naringenin, chrysin, kaempferol, quercetin, p-coumaric-acid, and galangin) by HPLC-diode array detector (DAD) method for four types of honeys. Hesperetin was detected as an exclusive marker molecule for citrus honeys. Principal component evaluation showed that kaempferol, chrysin, pinocembrin, naringenin, and caffeic acid content of rosemary honeys contributed to a separation of this honey from the other types. The relatively higher content of myricetin, quercetin, p-coumaric acid and galanginin honeydew honeys provided their separation from rosemary, citrus, and polyfloral honeys [28].

In another study, flavonoid, and phenolic components were quantified by HPLC and were used successfully for the distinction of citrus honeys (lemon and orange blossom honey) from each other using ANOVA test and PCA. The profile showed that significantly higher values of luteolin, caffeic acid, p-coumaric-acid, pinocembrin, naringenin, chrysin, kaempferol, quercetin, galangin contents in lemon blossom honeys, while no significant differences were found in hesperetin content [27].

Some inconsistency in the literature was found regarding the compounds based on markers. Hesperetin was a good marker of citrus honeys, however, hesperetin was also detected in eucalyptus, gelam, and lotus honeys [43, 63, 82]. Similarly, naringenin was suggested as a marker for lavender honey (Table 5.1) in fir, honeydew, linden, rhododendron, rosemary, spruce, and Tualang honeys [4, 15, 28, 32, 68].

An Algerian study concluded that flavonoids and phenolic acids can be used as floral markers to differentiate different types of honeys. The study assigned p-coumaric and caffeic acids as markers for *Trifolium* and *Capparis* honey [63], however, these compounds, especially caffeic acid, is also present in several other types of honey (Table 5.2).

According to Chan et al. [17], luteolin can serve as a marker for Manuka (*Leptoseprum scoparium honey*). According to other research studies, galangin, apigeninkaempferol, quercetin, isorhamnetin, and ellagic acids were also detected in Manuka honey (Table 5.1).

Australian researchers stated that flavonoids could be used for the identification of the origin of honeys from different botanical sources. Monofloral honeys have characteristic flavonoid profiles. Therefore, not only qualitative identification, but also quantitative discrimination is the answer for origin determination. Different types of monofloral honeys may contain same types of flavonoids, but their different concentrations can be a marker to differentiate varieties of honeys. They analyzed tea tree, crow ash, brush box, heath, and sunflower honeys. The flavonoid profile was determined by HPLC, and the mean and standard deviation was calculated for each honey [85].

Research group in India evaluated different types of monofloral and polyfloral honeys. Phenolic compounds with their exact amount were determined by liquid chromatography, with mass detection and ESI mass spectrometry technique and PCA was applied for the statistical evaluation. Based on the phenolic characteristics, one of the multifloral honey from Biligiri Rangana Hills (BRH) was completely distinguished from other types. Ellagic acid was considered as a marker molecule and this honey had the highest phenolic content. Another group of honeys was differentiated due to their intermediate phenolic content. In the intermediate case, gingerol was found as marker in ginger honey, homovanillic acid for neem honey and tricetin for eucalyptus honey. Lemon, litchi, and another polyfloral honeys were also separated as a group. Hesperetin and naringenin were marker molecules for lemon honey, while procyanidins were specific for litchi honey [25].

Oroian et al. [65] studied physicochemical properties (pH, electrical conductivity, L*a*b* color, total soluble dry matter, ash content, water activity, moisture content) and phenolic compounds (with HPLC method) in different types of honeys. PCA and LDA models were used to determine the botanical origin of honeys (acacia, linden, honeydew, polyfloral, and sunflower). Results showed that phenolic components may provide only a little chance for differentiation, while the combination of physicochemical parameters may be suitable for the identification of botanical origin of honeys [65].

In a research by Australian group, differentiation of monofloral and polyfloral honeys were based on the phenolic components quantified by ultra-high-performance liquid chromatography (UHPLC)-QTOF MS. Manuka, leatherwood, kamanhi honeys were separated from each other successfully using PCA analysis [37].

5.3.4 GEOGRAPHICAL ORIGIN: IDENTIFICATION BASED ON POLYPHENOLS AS MARKERS

In a study by Croatian researchers, flavonoid profile of acacia honey was investigated with HPLC method. Researchers stated that flavonoid components cannot be used as floral markers for acacia honey, as the quantity of these components had wide variation in the same type of honey, due to diverse geographical origin of the samples. Results showed that climatic conditions have a significant effect on the flavonoid content of honeys, because plants produce flavonoids against oxidative stress. Therefore, flavonoids are more likely to be used for geographical discrimination of monofloral honeys [50].

Argentinian researchers evaluated eucalyptus and lotus honeys from different geographical regions in Argentina. Flavonoid components were quantified by HPLC equipped with a UV detector. Their results showed that eucalyptus honey had higher quercetin and luteolin contents, and these together accounting for 88% of the TFC, while lotus honeys had lower quercetin and luteolin contents than eucalyptus and mixed (multifloral) honeys and a significantly higher myricetin content. The geographical origin was a determinant factor in the amount of different flavonoid compounds [36].

Researchers in Austria evaluated 49 honeys from 22 different geographical zones by UHPLC connected to quadrupole time-of-flight mass spectrometer (Q-TOF-MS) with ESI technique (negative electrospray ionization (ESI^-) and positive electrospray ionization (ESI^+)). In the ESI^- mode, Turkish, and Spanish honeys were discriminated from the others, while this separation was not obtained in case of positive ionization. In another model, honeys from New Zealand and Australia were also separated from each other. These honeys had also different botanical origins; and separation according to the botanical origin was also detected. Results showed that climatic conditions have a significant impact on the phenolic characteristics of honey [37].

Chinese researchers determined the phenolic profile of acacia honeys from six different geographical origins by HPLC equipped with MS detector. Geographical origin was a determination factor in the phenolic profile of honeys and the differences were detected using Duncan's test at $p<0.05$ significance level, according to the places of collection [71].

Greek investigators evaluated different honeys from various geographical regions in Greece. They analyzed different phenols and flavonoids in honey samples by HPLC method, only identifying different phytochemicals, but without quantifying them. Results showed that p-coumaric, syringic, vanillic acids, galangin, and chlorogenic acid in honeys from some geographical sources were not detected, while these compounds were found in honeys collected from other geographical sources. The results portray that geographical origin can be a decisive factor, especially in honey from the same botanical source [61].

Serbian researchers evaluated phenolic profile by UHPLC coupled with hybrid mass spectrometry for polyfloral honeys from five geographical origins. For the statistical evaluation of the quantities of different phenolic compounds, partial least squares discriminant analysis (PLS-DA) method, which showed significant separation of two geographical groups, while the other three groups could not be distinguished [52].

A unique polyphenol, acacetin, was identified as a compound present exclusively in acacia honey samples collected in Romania. For the quantification of acacetin and other polyphenols, LC-MS technique was used.

Other phenolic compounds (such as: p-hydroxybenzoic acid, vanillic acid, p-coumaric acid, cinnamic acid) and flavonoids (such as: apigenin, kaempferol, pinocembrin, and chrysin) were detected in the acacia honeys [56].

5.3.5 SUMMARY FOR MARKERS AND ORIGIN IDENTIFICATION

5.3.5.1 MARKER MOLECULES

Hesperetin is generally recognized as marker molecule for lemon and citrus honeys [9, 25, 27, 28, 63]. The p-coumaric acid and caffeic acids can be used as markers for *Trifolium* and *Capparis* honeys [63]. In another study, ellagic acid was considered marker of heather honey [6], kaempferol for rosemary honey [9], quercetin for sunflower honey, p-coumaric acid, ferulic acid, and caffeic acid for chestnut honey [9], homogentisic acid for strawberry tree honey [21], gallic acid for eucalyptus honey [63], and luteolin for manuka honey [17]. Methyl-syringate is considered as a marker for asphodel honey [22].

5.3.5.2 PERSPECTIVES OF ORIGIN: IDENTIFICATION BASED ON POLYPHENOL MARKERS

Phenolic acids and flavonoid components can be used for the identification of botanical [6, 9, 15, 20, 22, 25, 65, 85] and geographical [36, 37, 48, 57, 67] origins of honey. For the evaluation of quantitative data, usually, PCA and discriminant analysis (linear, canonical) are used [15, 27, 28, 36, 37], but significant differences can also be detected for the individual components [52, 71].

Based on the literature review, phenolic compounds can be used for the detection of the botanical origin of honeys from the same botanical type do not have high variability in the amount of bio-components. The qualitative determination of these components is probably not sufficient for the differentiation of honeys, as these might contain the same flavonoids and phenolic acids. Therefore, the quantitative determination has a high importance in the identification of origin.

With regards to geographical origin, phenolic profile of honeys from different zones can have differentiating values based on the fact that climatic conditions or soil could be a determining factor due to response of plants to oxidative stress.

Some molecules can be dominant in some types of honeys, thus qualifying them as potential marker molecules. Different phenolic acids and flavonoids are present in several types of honeys. Chrysin, luteolin, pinocembrin, quercetin, luteolin, apigenin, galangin, kaempferol, myricetin, naringenin, and pinobanksin were detected in more than 10 monofloral types. Also, syringic acid, 4-hydroxy-benzoic acid, vanillic acid, p-coumaric acid, gallic acid, caffeic acid, and ferulic acid were detected in more than 10 honeys from various botanical origins. Therefore, these are not real qualitative marker molecules, and their quantitative determination might be a unique tool to differentiate honeys from different botanical and geographical origins.

5.4 SUMMARY

Depending on their origin, honeys have remarkable antioxidant properties. For example, lighter honeys show generally lower antioxidant levels than the darker ones. Widely known spectrophotometric *in vitro* assays have been employed to determine global AOC, total phenolic or flavonoid content of honeys. It can be concluded that antioxidant assays are noteworthy tools in the identification and differentiation of honeys belonging to different botanical sources. Moreover, if these methods are combined with routine physico-chemical parameters, they can further improve the identification capacity of these techniques. Complex statistical analysis of the data (i.e., application of chemometric methods including classification, pattern recognition, clustering) are indispensable in the interpretation of such data sets. More complex analytical techniques have been employed during identification and quantification of individual phenolic compounds in honey. According to several studies, some phenolic acids or flavonoids could serve as marker molecules for many types of honey. The presence and the amount of these components are highly dependent on the botanical and geographical origins.

ACKNOWLEDGMENTS

Zoltan Kovacs acknowledges the support New National Excellence Program of the Ministry for Innovation and Technology (ÚNKP-19-4-SZIE-27) and the Bolyai János Scholarship from the Hungarian Academy of Sciences. Zsanett Bodor acknowledges the support New National Excellence Program of the Ministry for Innovation and Technology (ÚNKP-19-3-I-SZIE-71). Zsanett Bodor and John-Lewis Zinia Zaukuu acknowledge the support from the Doctoral School of Food Science, Szent István University. The project was supported by the European Union and co-financed by the European Social Fund (grant agreement no. EFOP-3.6.3-VEKOP-16-2017-00005).

KEYWORDS

- **antioxidant capacity**
- **botanical origin**
- **chemi-luminescence**
- **cupric ion reducing antioxidant capacity**
- **flavonoids**
- **polyphenols**

REFERENCES

1. Al-Farsi, M., Al-Amri, A., Al-Hadhrami, A., & Al-Belushi, S., (2018). Color, flavonoids, phenolics, and antioxidants of Omani honey. Heliyon, 4, 1–14.
2. Al-Nahari, A. A. M., Almasaudi, S. B., El-Ghany, E. S. M. A., Barbour, E., Al Jaouni, S. K., & Harakeh, S., (2015). Antimicrobial activities of Saudi honey against Pseudomonas aeruginosa. Saudi Journal of Biological Sciences, 22, 521–525.
3. Al, M. L., Daniel, D., Moise, A., Bobis, O., Laslo, L., & Bogdanov, S., (2009). Physico-chemical and bioactive properties of different floral origin honeys from Romania. Food Chemistry, 112, 863–867.
4. Ahmed, S., & Othman, N. H., (2013). Review of the medicinal effects of tualang honey and a comparison with manuka honey. The Malaysian Journal of Medical Sciences: MJMS, 20, 6–13.
5. Alqarni, A. S., Owayss, A. A., & Mahmoud, A. A., (2016). Physicochemical characteristics, total phenols and pigments of honeys in Saudi Arabia. Arabian Journal of Chemistry, 9, 114–120.
6. Anklam, E., (1998). Review of the analytical methods to determine the geographical and botanical origin of honey. Food Chemistry, 63, 549–562.
7. Alves, A., Ramos, A., Gonçalves, M. M., Bernardo, M., & Mendes, B., (2013). Antioxidant activity, quality parameters and mineral content of Portuguese monofloral honeys. Journal of Food Composition and Analysis, 30, 130–138.
8. Akyol, E., Selamoglu, Z., Dogan, H., Akgul, H., & Unalan, A., (2015). Determining the total antioxidant status and oxidative stress indexes of honey samples. Fresenius Environmental Bulletin, 24, 1204–1208.
9. Alvarez-Suarez, J. M., Tulipani, S., Romandini, S., Vidal, A., & Battino, M., (2009). Methodological aspects about determination of phenolic compounds and in vitro

evaluation of antioxidant capacity in the honey: A review. *Current Analytical Chemistry, 5*, 293–302.

10. Boussaid, A., Chouaibi, M., Rezig, L., Hellal, R., Donsì, F., Ferrari, G., & Hamdi, S., (2018). Physicochemical and bioactive properties of six honey samples from various floral origins from Tunisia. *Arabian Journal of Chemistry, 11*, 265–274.
11. Bertoncelj, J., Doberšek, U., Jamnik, M., & Golob, T., (2007). Evaluation of the phenolic content, antioxidant activity and color of Slovenian honey. *Food Chemistry, 105*, 822–828.
12. Bertoncelj, J., Golob, T., Kropf, U., & Korošec, M., (2011). Characterization of Slovenian honeys on the basis of sensory and physicochemical analysis with a chemometric approach. *International Journal of Food Science and Technology, 46*, 1661–1671.
13. Bodor, Z., Koncz, F. A., Rashed, M. S., Kaszab, T., Gillay, Z., Benedek, C., & Kovacs, Z., (2018). Application of near-infrared spectroscopy and classical analytical methods for the evaluation of Hungarian honey. *Progress in Agricultural Engineering Sciences, 14*, 1786–2335.
14. Bogdanov, S., (2013). Honey in medicine. *Deutsche Medizinische Wochenschrift, (German Medical Weekly), 138*, 2647–2652.
15. Bertoncelj, J., Polak, T., Kropf, U., Korošec, M., & Golob, T., (2011). LC-DAD-ESI/MS analysis of flavonoids and abscisic acid with chemometric approach for the classification of Slovenian honey. *Food Chemistry, 127*, 296–302.
16. Biluca, F. C., Schulz, M., Braghini, F., Vitali, L., Costa, A. C. O., Gonzaga, L. V., Micke, G. A., & Rodrigues, E., (2017). Phenolic compounds, antioxidant capacity and bioaccessibility of minerals of stingless bee honey (Meliponinae). *Journal of Food Composition and Analysis, 63*, 89–97.
17. Chan, C. W., Deadman, B. J., Manley-Harris, M., Wilkins, A. L., Alber, D. G., & Harry, E., (2013). Analysis of the flavonoid component of bioactive New Zealand mānuka (*Leptospermum scoparium*) honey and the isolation, characterization, and synthesis of an unusual pyrrole. *Food Chemistry, 141*, 1772–1781.
18. Cianciosi, D., Forbes-Hernández, T. Y., Afrin, S., Gasparrini, M., Reboredo-Rodriguez, P., Manna, P. P., Zhang, J., & Lamas, L. B., (2018). Phenolic compounds in honey and their associated health benefits: A review. *Molecules, 23*, 1–20.
19. Costa, A. C. O., Fett, R., Gonzaga, L. V., Seraglio, S. K. T., & Bergamo, G., (2018). Physicochemical characteristics of bracatinga honeydew honey and blossom honey produced in the state of Santa catarina: An approach to honey differentiation. *Food Research International, 116*, 745–754.
20. Cimpoiu, C., Hosu, A., Miclaus, V., & Puscas, A., (2013). Determination of the floral origin of some Romanian honeys on the basis of physical and biochemical properties. *Spectrochimica Acta-Part A: Molecular and Biomolecular Spectroscopy, 100*, 149–154.
21. Ciulu, M., Serra, R., Caredda, M., Salis, S., Floris, I., Pilo, M. I., Spano, N., & Panzanelli, A., (2018). Chemometric treatment of simple physical and chemical data for the discrimination of unifloral honeys. *Talanta, 190*, 382–390.
22. Ciulu, M., Spano, N., Pilo, M. I., & Sanna, G., (2016). Recent Advances in the analysis of phenolic compounds in unifloral honeys. *Molecules, 21*, 1–32.
23. Chen, L., Wang, J., Ye, Z., Zhao, J., Xue, X., Heyden, Y. V., & Sun, Q., (2012). Classification of Chinese honeys according to their floral origin with near infrared spectroscopy. *Food Chemistry, 135*, 338–342.

24. Can, Z., Yildiz, O., Sahin, H., Akyuz, T. E., Silici, S., & Kolayli, S., (2015). An investigation of Turkish honeys: Their physicochemical properties, antioxidant capacities and phenolic profiles. *Food Chemistry, 180*, 133–141.
25. Devi, A. & Jangir, J. (2018). Chemical characterization complemented with chemometrics for the botanical origin identification of unifloral and multifloral honeys from India. *Food Research International, 107*, 216–226.
26. Dimitrova, B., Gevrenova, R., & Anklam, E., (2007). Analysis of phenolic acids in honeys of different floral origin with solid-phase extraction and high-performance liquid chromatography. *Phytochemical Analysis, 18*, 24–32.
27. Escriche, I., Kadar, M., Juan-Borrás, M., & Domenech, E., (2011). Using flavonoids, phenolic compounds and headspace volatile profile for botanical authentication of lemon and orange honeys. *Food Research International, 44*, 1504–1513.
28. Escriche, I., Kadar, M., Juan-Borrás, M., & Domenech, E., (2014). Suitability of antioxidant capacity, flavonoids, and phenolic acids for floral authentication of honey. Impact of industrial thermal treatment. *Food Chemistry, 142*, 135–143.
29. Escuredo, O., Míguez, M., Fernández-González, M., & Seijo, M., (2013). Nutritional value and antioxidant activity of honeys produced in a European Atlantic area. *Food Chemistry, 138*, 851–856.
30. Ferreres, F., Datta, N., Martos, I., Tomás-Barberán, F. A., Singanusong, R., Yao, L., Datta, N., & Tomás-Barberán, F. A., (2003). Flavonoids, phenolic acids and abscisic acid in Australian and New Zealand *Leptospermum* honeys. *Food Chemistry, 81*, 159–168.
31. Fraga, C. G., Oteiza, P. I., & Galleano, M., (2014). *In vitro* measurements and interpretation of total antioxidant capacity. *Biochimica et Biophysica Acta-General Subjects, 1840*, 931–934.
32. Gašić, U. M., Milojković-Opsenica, D. M., & Tesic, Z. L., (2017). Polyphenols as possible markers of botanical origin of honey. *Journal of AOAC International, 100*, 852–861.
33. Gül, A., & Pehlivan, T., (2018). Antioxidant activities of some monofloral honey types produced across Turkey. *Saudi Journal of Biological Sciences, 25*, 1056–1065.
34. Gan, Z., Yang, Y., Li, J., Wen, X., Zhu, M., Jiang, Y., & Ni, Y., (2016). Using sensor and spectral analysis to classify botanical origin and determine adulteration of raw honey. *Journal of Food Engineering, 178*, 151–158.
35. Istasse, T., Jacquet, N., Berchem, T., Haubruge, E., Nguyen, B. K., & Richel, A., (2016). Extraction of honey polyphenols: Method development and evidence of Cis isomerization. *Analytical Chemistry Insights, 2016*, 49–57.
36. Iurlina, M. O., Saiz, A. I., Fritz, R., & Manrique, G. D., (2009). Major flavonoids of Argentinean honeys. Optimization of the extraction method and analysis of their content in relationship to the geographical source of honeys. *Food Chemistry, 115*, 1141–1149.
37. Jandrić, Z., Frew, R. D., Fernandez-Cedi, L. N., & Cannavan, A., (2017). An investigative study on discrimination of honey of various floral and geographical origins using UPLC-QToF MS and multivariate data analysis. *Food Control, 72*, 189–197.
38. Jibril, F. I., Hilmi, A. B. M., & Manivannan, L., (2019). Isolation and characterization of polyphenols in natural honey for the treatment of human diseases. *Bulletin of the National Research Center, 43*, 4.
39. Jones, R., (2009). Prologue: Honey and healing through the ages. *Journal of ApiProduct and ApiMedical Science, 1*, 2–5.

40. Joerg, E., & Sontag, G., (1993). Multichannel coulometric detection coupled with liquid chromatography for determination of phenolic esters in honey. *Journal of Chromatography A*, *635*, 137–142.
41. Jasicka-Misiak, I., Poliwoda, A., Dereń, M., & Kafarski, P., (2012). Phenolic compounds and abscisic acid as potential markers for the floral origin of two Polish unifloral honeys. *Food Chemistry*, *131*, 1149–1156.
42. Karabagias, I., & Karabournioti, S., (2018). Discrimination of clover and citrus honeys from Egypt according to floral type using easily assessable physicochemical parameters and discriminant analysis: An external validation of the chemometric approach. *Foods*, *7*, 70–76.
43. Kassim, M., Achoui, M., Mustafa, M. R., Mohd, M. A., & Yusoff, K. M., (2010). Ellagic acid, phenolic acids, and flavonoids in Malaysian honey extracts demonstrate *in vitro* anti-inflammatory activity. *Nutrition Research*, *30*, 650–659.
44. Khalil, M. I., Alam, N., Moniruzzaman, M., Sulaiman, S. A., & Gan, S. H., (2011). Phenolic acid composition and antioxidant properties of Malaysian honeys. *Journal of Food Science*, *76*, 921–928.
45. Kaszab, T., Bodor, Z., Kovacs, Z., & Benedek, C., (2018). Classification models of Hungarian honey samples based on analytical and physical characteristics. *Hungarian Agricultural Engineering*, *7410*, 22–28.
46. Koncz, F. A., Bodor, Z., Rashed, M. S., Kaszab, T., Gillay, B., Kovacs, Z., & Benedek, C., (2017). Floral and geographical origin identification of Hungarian honey with electronic tongue and classical analytical methods. *Analecta Technica Szegedinensia*, *11*, 1–9.
47. Kavanagh, S., Gunnoo, J., Marques, P. T., Stout, J. C., & White, B., (2019). Physicochemical properties and phenolic content of honey from different floral origins and from rural versus urban landscapes. *Food Chemistry*, *272*, 66–75.
48. Kıvrak, Ş., & Kıvrak, İ., (2017). Assessment of phenolic profile of Turkish honeys. *International Journal of Food Properties*, *20*, 864–876.
49. Kaškonienė, V., Maruška, A., Kornyšova, O., Charczun, N., Ligor, M., & Buszewski, B., (2009). Quantitative and qualitative determination of phenolic compounds in honey. *Cheminė Technologija (Chemical Technology)*, *3*, 1392–1231.
50. Kenjerić, D., Mandić, M. L., Primorac, L., Bubalo, D., & Perl, A., (2007). Flavonoid profile of *Robinia* honeys produced in Croatia. *Food Chemistry*, *102*, 683–690.
51. Kenjerić, D., Mandić, M. L., Primorac, L., & Čačić, F., (2008). Flavonoid pattern of sage (*Salvia officinalis* L.) unifloral honey. *Food Chemistry*, *110*, 187–192.
52. Kečkeš, S., Tesic, Z., Gašić, U., Dabić, D., Trifković, J., Natić, M., & Milojković-Opsenica, D., (2013). Phenolic profile and antioxidant activity of Serbian polyfloral honeys. *Food Chemistry*, *145*, 599–607.
53. Karabagias, I. K., Vavoura, M. V., Nikolaou, C., Badeka, A. V., Kontakos, S., & Kontominas, M. G., (2014). Floral authentication of Greek unifloral honeys based on the combination of phenolic compounds, physicochemical parameters and chemometrics. *Food Research International*, *62*, 753–760.
54. Lachman, J., Orsák, M., Hejtmánková, A., & Kovářová, E., (2010). Evaluation of antioxidant activity and total phenolics of selected Czech honeys. *LWT-Food Science and Technology*, *43*, 52–58.
55. Meo, S. A., Al-Asiri, S. A., Mahesar, A. L., & Ansari, M. J., (2017). Role of honey in modern medicine. *Saudi Journal of Biological Sciences*, *24*, 975–978.
56. Marghitas, L. A., Dezmieram, D. S., Pocol, C. B., Ilea, M., Bobis, O., & Gergen, I., (2010). The development of a biochemical profile of acacia honey by identifying

biochemical determinants of its quality. *Notulae Botanical Horti Agrobotanical Cluj-Napoca, 38*, 84–90.
57. Michalkiewicz, A., Biesaga, M., & Pyrzynska, K., (2008). Solid-phase extraction procedure for determination of phenolic acids and some flavonols in honey. *Journal of Chromatography A, 1187*, 18–24.
58. Maurya, S., Kushwaha, A. K., Singh, S., & Singh, G., (2014). An overview on antioxidative potential of honey from different flora and geographical origins. *Indian Journal of Natural Products and Resources, 5*, 9–19.
59. Nascimento, K. S., Gasparotto, S. J. A., Lauer, M. L. F., Serna, G. C. V., Pereira, D. M. I. L., Da Silva, A. E., Granato, D., & Sattler, A., (2018). Phenolic compounds, antioxidant capacity and physicochemical properties of Brazilian *Apis mellifera* honeys. *LWT-Food Science and Technology, 91*, 85–94.
60. Nousias, P., Karabagias, I. K., Kontakos, S., & Riganakos, K. A., (2017). Characterization and Differentiation of Greek Commercial thyme honeys according to geographical origin based on quality and some bioactivity parameters using chemometrics. *Journal of Food Processing and Preservation, 41*, 13–61.
61. Nousias, P., Karabagias, I. K., & Riganakos, K. A., (2018). *Deep Inside Polyphenols of Hellenic Thyme Honey, 6*, 1–4.
62. Noor, N., Sarfraz, R. A., Ali, S., & Shahid, M., (2014). Antitumor and antioxidant potential of some selected Pakistani honeys. *Food Chemistry, 143*, 362–366.
63. Ouchemoukh, S., Amessis-Ouchemoukh, N., Gómez-Romero, M., Aboud, F., Giuseppe, A., Fernández-Gutiérrez, A., & Segura-Carretero, A., (2017). Characterization of phenolic compounds in Algerian honeys by RP-HPLC coupled to electrospray time-of-flight mass spectrometry. *LWT-Food Science and Technology, 85*, 460–469.
64. Orian, M., Ropciuc, S., Buculei, A., Paduret, S., & Todosi, E., (2016). Phenolic profile of honeydew honey from the North-East part of Romania. Bulletin of University of Agricultural Sciences and Veterinary Medicine Cluj-Napoca. *Food Science and Technology, 73*, 106–110.
65. Oroian, M., & Ropciuc, S., (2017). Honey authentication based on physicochemical parameters and phenolic compounds. *Computers and Electronics in Agriculture, 138*, 148–156.
66. Patton, S. R., Clements, M. A., George, K., & Goggin, K., (2016). "I don't want them to feel different": A mixed methods study of parents' beliefs and dietary management strategies for their young children with type 1 *Diabetes Mellitus. Journal of the Academy of Nutrition and Dietetics, 116*, 272–282.
67. Pontis, J. A., Da Costa, L. A. M. A., Da Silva, S. J. R., & Flach, A., (2014). Color, phenolic, and flavonoid content, and antioxidant activity of honey from Roraima, Brazil. *Food Science and Technology, 34*, 69–73.
68. Petrus, K., Schwartz, H., & Sontag, G., (2011). Analysis of flavonoids in honey by HPLC coupled with coulometric electrode array detection and electro spray ionization mass spectrometry. *Analytical and Bioanalytical Chemistry, 400*, 2555–2563.
69. Prior, R. L., Wu, X., & Schaich, K., (2005). Standardized methods for the determination of antioxidant capacity and phenolics in foods and dietary supplements. *Journal of Agricultural and Food Chemistry, 53*, 4290–4302.
70. Pyrzynska, K., & Biesaga, M., (2009). Analysis of phenolic acids and flavonoids in honey. *TrAC-Trends in Analytical Chemistry, 28*, 893–902.

71. She, S., Chen, L., Song, H., Lin, G., Li, Y., Zhou, J., & Liu, C., (2019). Discrimination of geographical origins of Chinese acacia honey using complex 13C/12C, oligosaccharides, and polyphenols. *Food Chemistry, 272*, 580–585.
72. Spiteri, M., Dubin, E., Cotton, J., Poirel, M., Corman, B., Jamin, E., Lees, M., & Rutledge, D., (2016). Data fusion between high resolution 1H-NMR and mass spectrometry: A synergetic approach to honey botanical origin characterization. *Analytical and Bioanalytical Chemistry, 408*, 4389–4401.
73. Silva, P. M., Gauche, C., Gonzaga, L. V., Costa, A. C. O., & Fett, R., (2016). Honey: Chemical composition, stability, and authenticity. *Food Chemistry, 196*, 309–323.
74. Silva, L. A. L., Pezzini, B. R., & Soares, L., (2015). Spectrophotometric determination of the total flavonoid content in *Ocimum basilicum* L. (Lamiaceae) leaves. *Pharmacognosy Magazine, 11*, 96–101.
75. Singleton, V. L., & Rossi, J. A., (1965). Colorimetry of total phenolics with phosphomolybdic phosphotungstic acid reagents. *American Journal of Enology and Viticulture, 16*, 144–158.
76. Stanek, N., & Jasicka-Misiak, I., (2018). HPTLC phenolic profiles as useful tools for the authentication of honey. *Food Analytical Methods, 11*, 2979–2989.
77. Truchado, P., Tourn, E., Gallez, L. M., Moreno, D. A., Ferreres, F., & Tomás-Barberán, F. A., (2010). Identification of botanical biomarkers in Argentinean diplotaxis honeys: Flavonoids and glucosinolates. *Journal of Agricultural and Food Chemistry, 58*, 12678–12685.
78. Tu, J. Q., Zhang, Z. Y., Cui, C. X., Yang, M., Li, Y., & Zhang, Y. P., (2017). Fast separation and determination of flavonoids in honey samples by capillary zone electrophoresis. *Kemija u industriji (Chemistry in Industry), 66*, 129–134.
79. Wei, Z., & Wang, J., (2011). Classification of monofloral honeys by voltammetric electronic tongue with chemometrics method. *Electrochimica Acta, 56*, 4907–4915.
80. Wei, Z., & Wang, J., (2014). Tracing floral and geographical origins of honeys by potentiometric and voltammetric electronic tongue. *Computers and Electronics in Agriculture, 108*, 112–122.
81. Wei, Z., Wang, J., & Liao, W., (2009). Technique potential for classification of honey by electronic tongue. *Journal of Food Engineering, 94*, 260–266.
82. Wen, C. T. P., Hussein, S. Z., Abdullah, S., Karim, N. A., Makpol, S., & Mohd, Y. Y. A., (2012). Gelam and nenas honeys inhibit proliferation of HT 29 colon cancer cells by inducing DNA damage and apoptosis while suppressing inflammation. *Asian Pacific Journal of Cancer Prevention : APJCP, 13*, 1605–1610.
83. Yao, L., Datta, N., Tomás-Barberán, F. A., Ferreres, F., Martos, I., & Singanusong, R., (2003). Flavonoids, phenolic acids and abscisic acid in Australian and New Zealand *Leptospermum* honeys. *Food Chemistry, 81*, 159–168.
84. Yao, L., Jiang, Y., D'Arcy, B., Singanusong, R., Datta, N., Caffin, N., & Raymont, K., (2004). Quantitative high-performance liquid chromatography analyses of flavonoids in Australian eucalyptus honeys. *Journal of Agricultural and Food Chemistry, 52*, 210–214.
85. Yaoa, L., Jiang, Y., Singanusong, R., Datta, N., & Raymont, K., (2005). Phenolic acids in Australian Melaleuca, *Guioa*, *Lophostemon*, *Banksia*, and *Helianthus* honeys and their potential for floral authentication. *Food Research International, 38*, 651–658.
86. Zhou, X. J., Chen, J., & Shi, Y. P., (2015). Rapid and sensitive determination of polyphenols composition of unifloral honey samples with their antioxidant capacities. *Cogent Chemistry, 1*, 1–10.

CHAPTER 6

PREPARATION AND HEALTH BENEFITS OF RICE BEVERAGES FROM ETHNOMEDICINAL PLANTS: CASE STUDY IN NORTH-EAST OF INDIA

VEDANT VIKROM BORAH, MAHUA GUPTA CHOUDHURY, and PROBIN PHANJOM

ABSTRACT

In North-Eastern region of India, different indigenous tribes and communities use different methods for preparation of fermentation beverage from ethnomedicinal plants. The feedstock is prepared by mixing different parts of various species of plants and herbs, which may offer some health benefits. The different types of microbial consortia (starter culture) along with feedstock (which is known as starter culture cake) are used for preparation of fermented alcoholic beverages. Being part of the socio-cultural life, rice beer is effective against amoebiasis, acidity, vomiting; and it helps in cholesterol reduction, endocrine function. Furthermore, it alters and maintains the gut microflora. This chapter discusses different types of fermented beverages; preparation methodologies; qualitative parameters and their health benefits.

6.1 INTRODUCTION

The North-Eastern region of India is a part of Indo-Malayan biodiversity covering nearly 262,379 km^2 [56]. Based on composition of flora and its local climate [71], it encompasses a sizeable ethnic population of about 8 million with its eight sister states [86]. Due to high degree of precipitation, Arunachal Pradesh (3000 mm) and Sikkim (5000 mm) are considered part of Eastern Himalayas. North-Eastern region of India embraces eight states, e.g., Manipur, Nagaland, Assam, Meghalaya, Mizoram, Arunachal Pradesh, Sikkim, and Tripura (Figure 6.1) [68]. These eight states are inhabited by different indigenous tribes, who have unique and specific dialects, attires, rituals, and living habitats [86].

Ethnobotany is the study of plants and their traditional practices and customs focusing on availability, use, and medicinal properties of local plants [2, 21]. The different tribes have their own unique cultural heritage and are also rich with variety of plant habitats. Without modern facilities in these regions, the tribal communities are compelled to depend on natural resources for their primary healthcare and also for preparing different kinds of economically viable plant-based products [18].

Traditional medicine includes knowledge, skills, and practices that are based on theories, beliefs, and practical experience in indigenous different cultures; and is used in the maintenance of health and in the prevention, diagnosis or treatment of physical and mental illness [108, 109].

Traditional medicine is gaining importance in North-Eastern region of India for use of traditional plants, which are collected from natural ecosystems and ethnic landscapes [3, 15, 32, 51, 53]. The medicinal plants are well-known to act as fighting agents against various diseases. About 21 plant species have been used to treat various ailments by tribes in North-Eastern region of India [44, 85].

Meghalaya state is very rich in natural resources, such as, water bodies, forests, and minerals. Due to diverse climatic and topographic conditions, this state can sustain a vast floral diversity [37, 57], which has become an important segment for the indigenous tribes. The tribes of Meghalaya (Khasi, Jaintia, and Garo) encounter diversity of medicinal plants in their native area for use in daily life. Whereas, the Naga tribal community is the most bio-diverse because of varying altitudes, topography, soil types and weather conditions. The Nagaland tribes take good care of fauna and protect it with heart and soul. The tribes of Nagaland also help to avoid extinction of these plants because of indigenous rules [64].

Assam has rich biodiversity of herbal medicinal plants, which are used in a variety of ways in daily life for health care [25, 38]. In the southern part of North-Eastern region, Tripura has a rich heritage of medicinal plants, with knowledge ranging from common household therapy to specialized treatment by traditional healers [28]. The tribes in Arunachal Pradesh use ethnomedicinal plants for diverse treatments, depending on availability and presence of experts known as Miri Abu [36].

The Manipur state has an Imphal Valley that is occupied by 30 different tribes of Kuki and Nagas. This state has about 50% of the total biodiversity

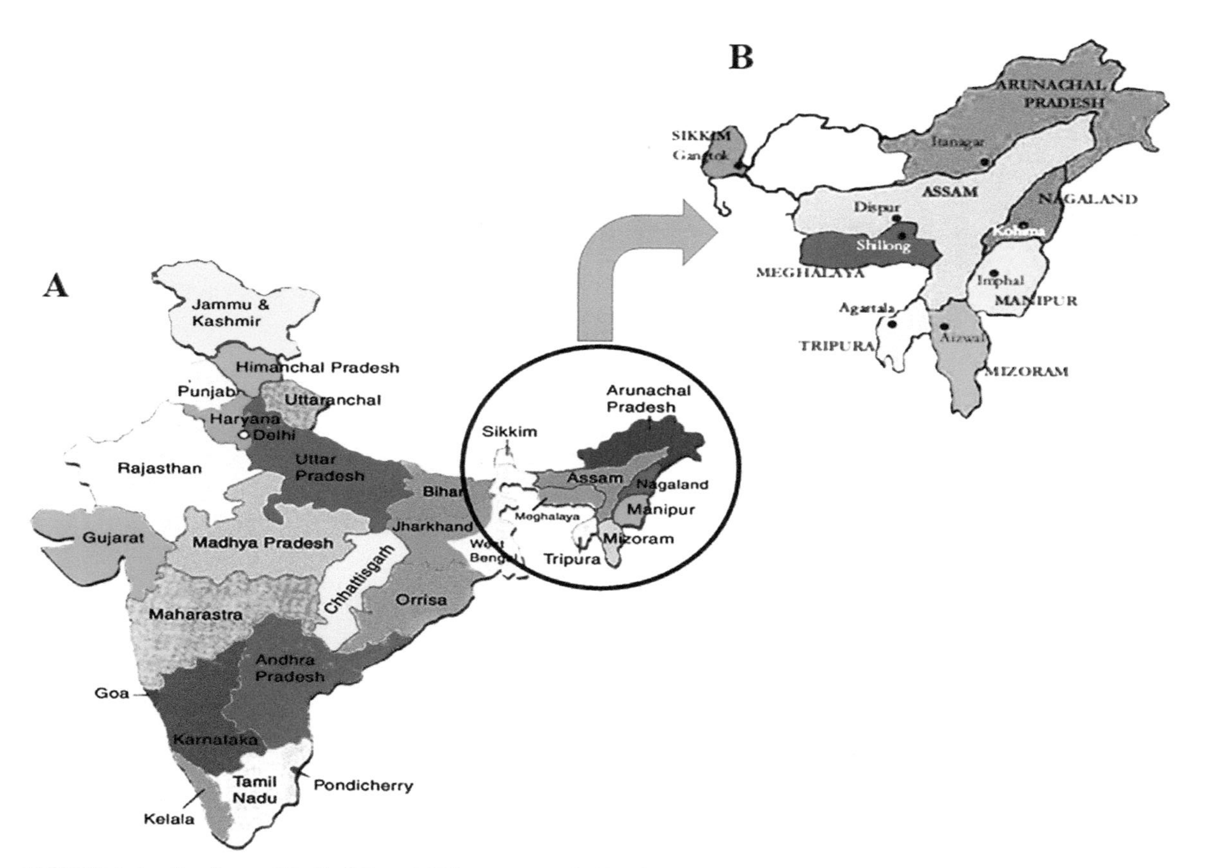

FIGURE 6.1 Territory of India (A) and different states in North-Eastern region of India (B).
Source: Self-developed with concepts from Ref. [68].

in India and hence contributes to 8% of the total geographical area of India. Eastern Himalayas in Sikkim possess wide range of unique endemic medicinal plants, containing myriad resources, such as, timber, fibers, medicinal, and edible plants [62]. Beside the use of plant habitat as folk medicine, ethnic, and tribal communities use several plant species for preparation of alcoholic beverages through fermentation route.

The word "fermentation" is drawn from the Latin word '*fervere.*' Fermentation refers to the use of yeast under anaerobic conditions for production of alcoholic from the extracts of fruit or malted grain. Starter culture during the production of alcoholic beverages yields in unique taste and aroma [84, 90]. It improves digestibility of products, produces organic acids, reduces allergenic activity, enrichment of nutritional constituents, and reduction in duration of cooking due to the presence of important phytochemicals [83].

Population outside of North-Eastern region of India has very limited idea on the diversity of fermented and non-fermented ethnic alcoholic beverages and food products [80]. North-Eastern Indian people have been consuming fermented foods and alcoholic beverages for more than 2500 years. The fermented foods and alcoholic beverages in North-East India are region-specific [84], with unique substrates (feedstock) and specific preparation methodologies. The tribal females engage in the preparation of fermented foods that are sold within North-East India [92]. Plant-based fermented foods in North-East India are classified into [93]:

- Fermented soybean products (hawaijar, aakhone, tungrymbai, peruyyan, bekantha, kinema) and non-soybean legume products (maseura, jalebi, hakua);
- Fermented vegetable products (inziang-dui, goyang, sinki, gundruk, khalpi, anishi, and inziang-sang);
- Bamboo shoot foods (soibum, mesu, siodon, soijim, ekung, iku, hikku, hirring, hitch, hitak, eup, hi, nogom, ipe, lung-siej, bastanga, tuaithur, tuairoi, miyamikhri);
- Fermented pulse foods and cereal products (chilra, tun-grymbai, bhatootu, kinema, and marchu);
- Fermented smoked fish products (ayaiba, ngari, tungtap, karati, bordia, lashim, mio, naakangba, and hentak);
- Preserved meat products (bagjinam, honohein grain, mogong grain);
- Beverages made from milk (lassi, kadi, gheu, chhurpi, nudu, and churpa);
- Mixed amylolytic starters made from non-food items (hamei, humao, nduhi, thiat, pham, khekhrii);

- Alcoholic beverages (ghanti, jann, daru, atingba, yu, jou, zutho, zhuchu, duizou, nchiangne, ruhi, madhu, dekuijao, apong, ponga, ennog, oh, kiad-lieh, judima, juharo, zu).

Various fermented alcoholic beverages are being used by different tribes of North-East India. Rice-beer is an integral part of many local tribes and is known in different names in different places [33, 47]. The ethnic communities in Mao Naga tribe of Nagaland prepare a domestic alcoholic product made from rice, i.e., zhuchu/zutho [55]. The Jaintia tribes in Meghalaya state prepare kiad from rice [45, 76]. Another alcoholic beverage, i.e., apong is used in Arunachal Pradesh state [102]. The Ahoms tribe of Assam prepares xajpani, which is a form of rice beer [75]. Judima, a beverage made from rice, is prepared by Dimasa tribe in Assam [17].

Kodo ko jaanr or chyang or chee is prepared and consumed by tribes of Bhutia, Monpa, Gorkha, Lepcha, and many ethnic groups. It is made from a specific finger-millet having a sweet-sour and acidic taste. This alcoholic beverage has an integral role in the religious beliefs and also in the dietary culture of these tribes [84, 91]. A staple fermented rice beverage prepared by the Gorkha is bhaati jaanr [72, 94].

Each tribe prepares its alcoholic beverage by a unique starter culture cake, which is made by mixing different parts of various plant species. The starter culture cake can be stored for longer duration, i.e., for several months. The process is carried out in earthen pots at 37°C and takes about 5–7 days to complete. The process of fermentation by different tribes is almost the same, except the difference in starter culture cake [97].

Following types plants are used for the preparation of rice beer by the tribes:

- *Albizia myriophylla,* is used by the Maiteis tribes in Manipur [89];
- *Amomum aromaticum* is used by the Jaintia tribe in Meghalaya [76];
- *Plumbago zeylanica, Buddleja asiatica, Vernonia cinerea,* and *Gingiber officinale* are used by Lepcha, Bhutia, and Nepale tribes in Sikkim [103];
- *Glycyrrhiza glabra*is used by the Dimasas tribe in Assam [17];
- *Ananas comosus, Artocarpus heterophyllus, Calotropis gigantea, Capsicum frutescens*, etc., are used by the Rabha tribe in Assam [30] and sprouted rice grains by Angamis tribe in the state of Nagaland [21, 75, 98].

This chapter focuses on different medicinal plants in North-East India and their way of using of these plants for the preparation of ethnic alcoholic beverages (rice beer). The chapter also includes discussions on nutritional benefits related with consumption of different types of rice beers that are prepared from different plants in North-East India.

6.2 TYPES OF ALCOHOLIC BEVERAGES IN NORTH EASTERN REGION OF INDIA

The consumption of different alcoholic beverages by different tribal populations of these regions, is associated with their socio-cultural life, such as: marriage, festivals, ceremonies, rituals, etc., [75] and some medicinal properties [37]. The different tribes have their own traditional method for preparation of rice-based alcoholic beverages with its own unique recipes, and these beverages are usually made from a culture cake of steamed rice mixed with herbs and spices. Usually, the fermentation process is performed for several days. After fermentation, the alcoholic beverages are mixed with water and used for consumption.

North-East India encompasses of eight sister states, such as (Figure 6.1): Arunachal Pradesh, Assam, Manipur, Meghalaya, Mizoram, Nagaland, Sikkim, and Tripura. Selected well-known indigenous fermented beverages in these 8 states are discussed in this section.

6.2.1 ASSAM: JOU

This fermentation-based alcoholic beverage is prepared from locally available rice by the Bodo tribe. Different types of substrates are used to produce beverages, which differ in terms of taste, appearance, and qualities. Amao is used as a feedstock in the preparation of Jou. Amao is prepared by soaking rice grains (known as ankur) and then grinding it along with plant materials using mortar and pestle. Then the mixture is molded into round flattened shaped cakes and inoculated with yeast from the old amao by sprinkling on it.

Jou is prepared by mixing the starter cake with the cooked rice and spreading it for 18–48 hours on a bamboo tray covered with leaves of bananas. Later, it is transferred to an earthen pot that has bamboo sieve for harvesting the alcohol. The mouth of the earthen pot is sealed tightly using a cloth. During summer, the fermentation incubation duration is about for 3 days compared to 7 days during winter. The rice beer obtained is distilled further to produce jou gwran (local alcoholic drink) [7].

6.2.2 ASSAM: CHU

Chu is a fermented rice-beer prepared by the Garo communities in Assam and Meghalaya. It is prepared by mixing cooked rice mixed wansi (starter cake) that is a good source of microorganisms, such as [58, 59]: *Lactobacillus plantarum, Saccharomyces cerevisiae, Wickerhamomyces anomalus* and *Rhodotorula mucilaginosa.*

For easy harvesting of the beer, the mixture is kept inside an airtight container with bamboo sieve. Using a piece of cloth, the mouth of the container is sealed and is kept under dark warm place for the fermentation process to occur. During winter, the fermentation rate is slower than the summer [61].

6.2.3 ASSAM: CHAKO

Chako is a fermented rice beer with a yellowish color and very strong taste; and it is prepared by the Rabha community. To prepare chako, water is first taken in an aluminum pot whose mouth is fitted with bamboo funnel that is closed using banana leaves. Through the funnel, washed rice is added and the whole arrangement is placed on an oven. The cooked rice is then spread on top of a jute bag for drying, which is then mixed with the starter culture cake (source of yeast in cooked rice: known as bakhor or phab), which is prepared using rice powder and herbs. After keeping for 30 minutes, it is then packed using jute bag that is kept in a shade or room to allow the action by microbes. The obtained mixture is then transferred to a motok (fermenter aluminum pot), which is covered with a banana leaf and is allowed to ferment. The fermentation time is generally 2 days during summer compared to 4 days in winter [30].

6.2.4 ASSAM: APONG

Apong is a fermented alcoholic beverage that is prepared by the Mishing community in Assam and local tribes of Arunachal Pradesh. The feedstock is known as ipoh, which contains different types of yeast, such as, *Saccharomyces cerevisiae, Hanseniaspora sp., Kloeckera sp., Pischia sp.* and *Candida sp.* to allow fermentation [102]. Ipoh is prepared by mixing dry powder of rice with paste of barks of locally available plants, such as:

- *Achyranthes aspera* L.;
- *Adhatoda Vasica (Nees);*
- *Ageratum conyzoides* L.;
- *Ananas comosus (L.) Merr.;*
- *Artocarpus heterophyllus;*
- *Asparagus racemosus Willd.;*
- *Capcicum annum L.;*
- *Centella asiatica L.;*
- *Cinnamomum bejolghata;*
- *Cinnamomum tamala Nees.;*
- *Clerodendrum viscosum L.;*
- *Costus speciosus (Koen. ex. Retz.) J. E. Smith;*
- *Drymeria cordata L.;*
- *Gomphostemma parviflora Wall.;*
- *Ipomea mauritiana Jacq.;*
- *Ipomoea aquatica Forsk;*
- *Kaempferia rotunda L.;*
- *Leucas plukenetii (Roth) Spreng.;*
- *Lygodium flexuosum;*
- *Melothrea heterophylla (Lour) Cogn;*
- *Microsorum punctatum (L.) Copel;*
- *Musa balbisiana Colla;*
- *Naravelia feylavica (D.C);*
- *Oldenlandia corymbosa Linn.;*
- *Oryza sativa L.;*
- *Phlogacanthus thyrsiformis (Hardw.) Mabb.;*
- *Piper* family: *Piper longum L.* and *Piper nigrum L.;*
- *Psidium guajava L.;*
- *Ptridium aquilinum Kuhn;*
- *Pueraria tuberose (Roxb. ex Willd.) DC;*

Subsequently, the mixture is allowed to dry over a fireplace or is left in a cool place for 3 to 4 days. To prepare apong, washed rice is first boiled using a large aluminum vessel with a wide bottom to make it soggy. Then, ipoh is mixed with the boiled rice in a specific ratio, which is then transferred to another vessel with lid on the top and a small amount of water is added to it. The mixture is incubated at room temperature for 3 to 5 days and the fermentation is allowed to produce alcohol. The alcohols are generally diluted with water before consumption [37].

6.2.5 ASSAM: XAJ-PANI

XAJ-PANI is a fermented rice beer that is prepared by Ahom community in Assam. This beer holds an important role in religious rites, festivals, rituals practices and is also used as a refreshing drink. The vekur pitha (starter culture) is prepared by mixing small amount of water, rice powder and powder of different herbs, e.g., *Cinamommum glanduliferum Meissn, Cissampelos Pereira, Lygodium flaxuosum Linn, Leucas aspera Spreng,* and *Scoparia dulsis Linn*.

Earlier prepared pitha (known as ghai pitha) as a source of yeast (starter culture) is added to this mixture. The mixture is then wrapped with banana leaves and is made into small cakes, which are kept for 4–5 days in an air tight condition for drying. Glutinous rice (known as bora rice belonging to sali variety) is mainly used for the preparation of beverages. The cooked rice is spread on an open plate and left for about an hour to allow drying before mixing with vekur pitha. The obtained mixture is then transferred to an earthen pot and is kept in the dark for 4–5 days in an airtight condition. The alcoholic beverage (known as xaj pani) is collected by filtration process. Xaj pani is alcoholic with strong aroma and sweet taste [75].

6.2.6 ASSAM: JUDIMA

Judima is a fermented rice beer that is prepared by Dimasa community in Assam. Humao (starter culture cake) is prepared from powder of 10–12 hours soaked brown rice at room temperature with dried powder of barks of a tree and the mixture is sun dried [37]. For the preparation of judima, cooked rice is mixed with humao and is spread on banana leaves for an overnight before transferring it to an earthen pot, which is not fully air tight. The fermentation takes about 3–4 days during summer compared to 6–7 days in winter [17].

6.2.7 ASSAM: ZU

It is a fermented rice beer that is prepared by Tiwa community in Assam. The feedstock (known as bakhor) is made by mixing rice powder with paste of several indigenous herbs; and the mixture is dried and small cakes are formed [27]. For the preparation of zu, the rice is boiled, cooled, and mixed with bakhor and is left for fermentation. After the incubation process, the fermented stock (zu) is collected, which is diluted with water before consumption [27].

6.2.8 ASSAM: SUJEN

It is an indigenous fermented rice beer, which is prepared by Deoris community in Assam. Perok kushi (starter culture) is prepared by mixing grounded paste of several herbs with rice powder. The mixture is then prepared into small balls that are sundried. For the preparation of rice beer, the rice is cooked, cooled, mixed with perok kushi and the mixture is then transferred to an earthen pot, which is covered with banana leaves to allow fermentation. After 4–5 days of incubation, the fermented stock (sujen) is diluted and filtered [24].

6.2.9 ARUNACHAL PRADESH: JUMIN

Several fermented beverages in Arunachal Pradesh are described under Sections 6.2.9–6.2.14. Jumin with taste like fruit-juice, sweet, salty, and sour is a fermented beverage that is prepared by Nocte tribe. The important ingredient for jumin preparation is the gelatinous rice (known as aahu dhan). Tapioca (*Manihot esculenta*) root is used in the preparation due to its presence of sugar contents. The feedstock (known as bichhi) is prepared by mixing flour of *Oryza sativa* or *Setaria italica* with Sala. Sala is prepared by mixing powder of plant parts of various medicinal herbs. The water is added to form the paste and then small cakes are formed. The mixture is kept in the dark for about 7 days to allow fermentation. To the newly prepared cakes, pieces of earlier made cake are mixed. It acts as a source of yeast before mixing with the cooked grains for the preparation of jumin. Then it is covered with a cloth or banana (*Musa paradisiaca* L.) leaves and kept in a pot in the dark room. The fermentation process can take about one day during summer compared two days during winter [9].

6.2.10 ARUNACHAL PRADESH: APONG

It is a white rice beer in Arunachal Pradesh. To prepare apong, the rice is cooked in a large clean vessel or pot at slow heat to get the right texture or soggy. Ipoh (starter culture) is then mixed with appropriate quantity of cooked rice in and transferred to a different utensil having small amount of water for fermentation, which takes place for about 3–5 days. Before consumption, the fermented stock is mixed with drinking water [102].

6.2.11 ARUNACHAL PRADESH: ENNOG

It is black rice beer, which is prepared using burned paddy husk and rice. Boiled rice is spread over the bamboo mat for cooling. The paddy husk is burned slowly in a tin-vessel and is then mixed well with boiled rice, which is allowed to cool down before mixing with ipoh (starter culture). It is then transferred to a conical bamboo basket, which is lined with wild leaves of *Phryium capitulum* and is tightly packed before storing in a cool dry place for the fermentation process, which takes about 3 days until a strong alcoholic aroma has been produced. Then the mixture is transferred to a different basket with wild leaves of *Phryium capitulum* for final storage, which is known as perop. To prepare the beer, the mixture (starter cake, boiled rice and paddy husk) is allowed to undergo fermentation for 10 days, after which, around 2 kg of the mixture is taken out into a cylindrical bamboo with a small opening at one end. Boiling water is poured slowly over the mixture to get the brewed product with high alcohol content. The first brew is called tok-til, which is always warm; and the second brew is called ennog, which is obtained after the first brew, having blackish color [102].

6.2.12 ARUNACHAL PRADESH: OPO

Opo is a blackish colored rice beverage that is prepared by Adi, Nyshing, and Mishmi communities mainly in Lohit-Changlang, east, and west Siang, upper, and lower Subansiri Districts of Arunachal Pradesh [24]. For preparation, the rice husk is first burned and is then mixed with cooked rice and is allowed to cool down before adding the starter culture cake. This fermented mixture is locally called pone, which is transferred to a wooden vessel for the fermentation. For extraction, a bamboo funnel filled with pone is used to which boiled water is poured slowly till the beer is extracted completely [87].

6.2.13 ARUNACHAL PRADESH: MADUA

Madua is a millet-based fermented alcoholic beverage and is prepared traditionally among all tribes in Arunachal Pradesh. The starter cake is prepared by mixing grounded rice powder with local herbs. For the preparation of madua, the finger millets are cooked for about 30 minutes, cooled, mixed with the starter culture cake and the mixture is kept in a perforated basket and is filled with wild leaves of *Phryium capitulum.* The mixture is incubated for about 4 days during summer and 7 days during winter to allow fermentation. After fermentation, warm water is added drop by drop and the filtrate is collected in a container. The color of the filtrate is golden yellow with a sweet taste sweet and a good alcoholic flavor [87].

6.2.14 ARUNACHAL PRADESH: THEMSING

Themsing is a fermented alcoholic millet beverage that is prepared only by Monpa community of Tawang district in Arunachal Pradesh. Themsing can be prepared from finger millet (locally known as kongpu) or from barley (locally known bong). It can also be prepared by mixing both of them. To prepare themsing, the millet is first cooked and cooled before mixing with the starter culture cake. Pinch of ash or charcoal is added before transferring the mixture (cooked millet with starter culture cake) into a wooden container. This mixture is covered with banana leaves to allow the fermentation. After fermentation, the mixture is transferred to a perforated bamboo basket and warm water is added slowly. The filtrate (themsing) has good aroma and is golden in color [87].

6.2.15 MANIPUR: YU OR ATINGBA

In Manipur, common fermented beverage among tribal communities is yu or atingba, which is prepared from glutinous rice. The starter culture cake (source of yeasts from cooked rice and herbs: hamei) is prepared by mixing bark powder of local tree and powder of raw rice. The associated yeast species are: *Saccharomyces cerevisiae, Candida tropicalis, Candida parapsilosis, Candida montana, Pichia anomala, Pichia guilliermondi, Pichia fabianii, Trichosporon sp., Torulaspora delbrueckii.* Hamei is mixed with cooked glutinous rice after cooling and is then kept in an earthen pot, which is then covered with hangla (*Alocasia sp.*) leaves for 3–4 days during summer and 6–7 days during winter to allow fermentation. Subsequently, submerged fermentation is performed in earthen pots for 2–3 days. The fermented alcoholic beverage is called atingba, whereas the distilled clear liquor (using traditional assembly) is called yu [47].

6.2.16 MEGHALAYA: KIAD

The kiadis prepared by Pnar tribes in Jaintia hills of Meghalaya. For preparation of kiad, kho-so (*Oryza sativa* L.: local red rice) is used. The starter culture (thiat) is prepared by mixing the rice powder mixed with powder of leaves of *Amomum aromaticum* Roxb) to prepare a sticky paste that is then used for making small round cakes of 4–5 cm in diameter and 0.8–1 cm thick. These starter cakes are sun-dried or dried over fire until it becomes hard and can be used as source of natural yeast for rice brewing.

For the preparation of kiad, kho-so is cooked in an earthen vessel and is then cooled by spreading over a round basket; and it is then mixed with starter culture, that. This mixture is put in a basket (shang) and fermentation is allowed for about 2–3 days. The fermented mixture (known as jyndem) is used for preparing locally yellowish white colored beer (known as sadhiar). Kiad is obtained by boiling of jyndem. Water is added for dilution depending on the taste [76].

6.2.17 NAGALAND: ZUTHO

In Nagaland, common fermented beverage among tribal communities is zutho, which is a traditional rice beer from gelatinous rice (*Oryza sativa* L.). Zutho preparation involves two steps: (1) polished grains of rice are soaked in water for 2 hours and the excess water is drained. It is then air dried and pounded into powder; (2) dehulled rice grains are allowed to germinate after soaking in water for about 2–3 days. The germinated rice grains are sun dried and are grounded to powder.

The powder of polished rice and germinated grains of rice are mixed in a ratio of 10:3 to make a paste using boiling water, which is then allowed to cool. The paste is incubated at room temperature for the fermentation process. Fermentation process takes about 4–5 days. The first fermented product in its pure form is called thoutshe; and is diluted with some water, called zutho [46].

6.2.18 SIKKIM: JNARD

In Sikim, common fermented beverage among tribal communities is jnard, which is made from finger millet (*Eleusine coracana*). The word jnard was derived from Mangaranti language among the Nepalese community. Jnard is also known by several names [95], such as, chiang in Tibet, chii among Rong tribe and toongba in Nepal.

For the preparation of jnard, the seeds of finger millet are mixed with a small quantity of wheat or corn grains. They are then boiled for about 30 minutes and spreaded on the leaves of banana, before mixing with the starter culture cake (source of yeasts in cooked wheat: known as murcha). The yeast species are *Saccharomycopsis fibuligera, Saccharomycopsis capsularis, and Saccharomyces bayanus, Pichia anomala* and *Pichia burtonii,* and *Candida glabrata* [103]. The mixture is kept for 24 hours in a pile heap before transferring to an earthen pot, which is covered with leaves and cow dung for the fermentation process. The grits are then placed in a bamboo vessel and subsequently water is added slowly. After 10 minutes, the beverage is ready to drink [95].

6.2.19 TRIPURA: LUNGI

In Tripura, common fermented beverage among tribal communities is lungi or gora, which is most popular traditional white rice beer that is prepared from rice (*Oryza sativa* L.). The starter cake for this preparation is called chuwan. The microbial species are *Sacchamycopsis fibuligera, Sacchomyces cerevisiae,* and *Lactobacillus fermentum.*

For the preparation of the beverage, rice is fully boiled in water for 20–30 minutes and then it is spreaded over a bamboo mat for cooling before mixing with chuwan, which is later packed tightly in an earthen container. Subsequently, it is covered with a lid or leaves for fermentation for around 3 days. After fermentation is over, the earthen container is filled with water up to the top and is kept for 6–8 hours, after which white transparent liquid floats on the top. This liquid is known as lungi or gora with a sweet taste. Traditional wine of Tripura is known as bangla, which is prepared after processing of lungi. Lungi is heated in an earthen pot, which has holes at the top. The holes are attached to a narrow pipe to collect the vapors in the form of droplets. These droplets are locally known as bangla or local wine [105].

6.3 USE OF HERBS AS STARTER: THERAPEUTIC PROPERTIES

The methods are same in the preparation of alcoholic fermented beverages by all tribes and communities in North-East India. However, the main ingredient for the preparation is the starter cake, which is a source of microorganisms along with some local herbs [23]. It is believed that herbs, which are mixed with the cake, have therapeutic properties. The exact ingredients and the quantities of herbs usually varies from one community to another due to availability of the plants, which affect the taste of the alcoholic beverage [72]. The exact ingredients for the preparation of the starter culture cake are also kept secret by the different tribes and communities. Less information is available on the local herbs and its medicinal properties. Therefore, more research is required for identification of the various herbs involved in the preparation of starter culture cake and understanding its important role in preparation of the fermented beverages.

Table 6.1 indicates some indigenous plant species, which are commonly used for the preparation of fermented beverages within different communities in North East India.

6.4 PHYSIOCHEMICAL AND BIOCHEMICAL PROPERTIES OF LOCAL RICE BEER

The physiochemical properties (smell, color, acidity, and alcohol content) and biochemical properties (concentrations of ethanol, carbohydrate, antioxidant, minerals, and microbial population) of local rice beers, produced among different communities in North-East of India are discussed in this section.

6.4.1 *TITRATABLE ACIDITY AND PH*

The titratable acidity and pH provide insights of the acidity of the food product and its quality. The pH provides the information about microbiota that is involved in the fermentation process and their ability to survive at a specific pH. Titratable acidity is contributed by various intrinsic organic acids in food product. These acids offer specific sensory property to the product [79].

The pH of rice beer is measured directly using a pH meter. Total acidity and volatile acidity are measured by titration-based methods, which are conducted against sodium hydroxide solution. Both the results are expressed as grams of tartaric acid and grams of acetic acid present per 100 mL of the sample, respectively. Phenolphthalein is used as an indicator [13].

In a comparative study, local rice beers (namely: judima, jumai, horo, and poro) showed pH range of 3.43 to 5.6, which is lower than pH of 6.2 for beer gin, 6.3 for vodka, 6.5 for rum and 6.6 for whiskey, respectively [5, 43]. Storage of local rice beer may alter its pH to some extent. For example, an

TABLE 6.1 Some Plant Species for Preparation of Starter Culture in North East India for Preparation of Fermented Beverages

Tribes/State	Fermented Beverage**	Starter Cake	Plants Used for Preparation of Starter Culture Cake			References
			Local Name	Scientific Name	Parts of Plants in Use	
Bodo/Assam	Jou	Amao	Mai	*Oryza sativa* L.	Grain; husk; straw	[7]
			Anaras	*Anana scomosus* L.	Leaves	
			Talir	*Musa paradisiaca* L.	Leaves	
			Kanthal	*Arthocarpus heterophyllus Lamk.*	Leaves	
			Bonfangrake	*Sco-paria dulcis* L.	Whole twigs	
			Lwkhwna	*Clerodendron infortunatum* L.	Leaves	
			Agwrsita	*Plumbago zeylanica* L.	Bark	
Garo/Assam and Meghalaya	Chu	Wanti	Achetra	*Plumbago zeylanica* L.	Leaves	[7]
				Capsicum annum L.	Fruits	
Rabha/Assam	Chako or jonga-mod	Bakhor, surachi or phap	Anaros (Pati)	*Ananas comosus (L.) Merr.*	Leaves	[30]
			Pan-chung (Rongdani)	*Artocarpus heterophyllus Lam.*	Leaves	
			Akhomhang (Rongdani)	*Calotropis gigantea (L.)* W. T. Aiton	Leaves	
			Jhaluk (Maitori)	*Capsicum frutescens L.*	Fruits	
			Holitita (Pati)	*Cleodendrum viscosum Vent*	Leaves	
			Bisdhinkia (Maitori)	*Dennstaedtia scabra (Wall.)* T. Moore	Fronds	

TABLE 6.1 *(Continued)*

Tribes/State	Fermented Beverage**	Starter Cake	Plants Used for Preparation of Starter Culture Cake			References
			Local Name	Scientific Name	Parts of Plants in Use	
			Kuchibun (Pati)	*Ochthochloa coracana Edgew*	Leaves	
			Agiachit (Rongdani)	*Plumbago indica*	Roots	
			Kurchi (Rongdani)	*Saccharum officinarum L.*	Leaves	
			Phap jibra (Rongdani)	*Scoparia dulcis L.*	Shoots	
Mishing/Assam	Apong	Ipoh	Bioni-hakuta	*Achyranthes aspera L.*	Leaves	[37]
			Patihonda	*Cinnamomum bejolghata*	Leaves	
			Titabahak	*Adhatoda Vasica (Nees)*	Leaves and Shoots	
			Gendelabon	*Ageratum conyzoides L*	Flower	
			Anaras	*Ananas comosus (L.) Merr.*	Tender leaf base	
			Kathalpat	*Artocarpus heterophyllus*	Matured leaf	
			Satmul	*Asparagus racemosus Willd.*	Tuberous root	
			Tezpat	*Cinnamomum tamala Nees.*	Leaves	
			Jalokia	*Capcicum annum L.*	Leaves	
			Bormanimuni	*Centella asiatica L.*	Whole plant	
			Dhapat tita	*Clerodendrum viscosum L.*	Leaves	
			Jomlakhuti	*Costus speciosus (Koen. ex. Retz.) J. E. Smith*	Leaves and Barks	
			Lai jabori	*Drymeria cordata L.*	Young leaves	
			Bhedaitita	*Gom Phostemma parvifloraWall.*	Tender leaves	

TABLE 6.1 *(Continued)*

Tribes/State	Fermented Beverage**	Starter Cake	Plants Used for Preparation of Starter Culture Cake			References
			Local Name	Scientific Name	Parts of Plants in Use	
			Bam kolmou	*Ipomoea aquatica Forsk*	Leaves	
			Bhui komora	*Ipomea mauritiana Jacq.*	Tubers	
			Bhumichampa	*Kaempferia rotunda L.*	Tubers	
			Durun	*Leucas plukenetii (Roth) Spreng.*	Leaves	
			Kopou dhekia	*Lygodium flexuosum*	Leaves	
			Belipoka	*Melothrea heterophylla (Lour) Cogn*	Leaves	
			Kapau dhekia	*Microsorum punctatum (L.) Copel*	Leaves	
			Bhimkol	*Musa balbisiana Colla*	Leaves	
			Goropsoi	*Naravelia feylavica (D. C)*	Leaves	
			Banjaluk	*Oldenlandia corymbosa Linn.*	Leaves	
			Dhan	*Oryza sativa L.*	Grains	
			Pipoli	*Piper longum L.*	Leaves	
			Jaluk	*Piper nigrum L.*	Seeds	
			Titaphool	*Phlogacanthus thyrsiformis (Hardw.)Mabb.*	Flower	
			Madhuriam	*Psidium guajava L.*	Leaves	
			Bhuin Komora	*Pueraria tuberose (Roxb. ex Willd.) DC*	Tuberous roots	

TABLE 6.1 *(Continued)*

Tribes/State	Fermented Beverage**	Starter Cake	Plants Used for Preparation of Starter Culture Cake			References
			Local Name	**Scientific Name**	**Parts of Plants in Use**	
			Bihlongoni	*Ptridium aquilinum Kuhn*	Leaves	
			Kuhiar	*Saccharum officinarum Linn*	Leaves	
			Bon chini	*Scoparia dulcis L.*	Leaves and Fluorescence	
			Selaginella	*Selaginella species (Lour) Cogn*	Leaves	
			Chirota tita	*Swernia Chirata (Buch-Hem)*	Leaves; Barks.	
Ahom/Assam	Xaj pani	Vekur pitha	Chepti-dhekia	*Lygodium flaxuosum Linn*	Leaves	[75]
			Durun khak	*Leucas aspera Spreng*	Leaves	
			Tubuki-lota	*Cissampelos Pereira*	Leaves	
			Bon-dhonia	*Scoparia dulsis Linn*	Leaves	
			Gondsoroi	*Cinamommum glanduliferum Meissn*	Leaves	
			Paan	*Piper betle Linn*	Leaves	
Tiwa/Assam			Nangol bhanga	*Clerodendron serratum (L.)*	Leaves	[27]
			Ghora-neem	*Melia azardirach (L.)*	Leaves	
			Kharua	*Streblus asper (Lour)*	Shoots	
			Halodhi	*Curcuma longa (L.)*	Shoots	
			Bor Bahaka	*Phlogocanthus thysiflorus (Roxb.)*		
			Boga Bahaka	*Adhatoda vasica (Nees.)*	Leaves	

TABLE 6.1 *(Continued)*

Tribes/State	Fermented Beverage**	Starter Cake	Plants Used for Preparation of Starter Culture Cake			References
			Local Name	Scientific Name	Parts of Plants in Use	
			Tita vekhuri	*Solanum indicum (L.)*	Leaves	
			Durun	*Leucus aspera (Spreng.)*	Leaves	
			Kothon	*Tabernaemontana coronaria (R. Br.)*	Flower	
			Bhatai	*Clerodendron infortunatum (Gaertn.)*	Flower	
			Kathal	*Artocarpus integrifolia (L.)*	Flower	
			Kuhiar	*Saccharum officinarum (L.)*	Leaves	
			Malbhog kol	*Musa velutina (Wendl.)*	Leaves	
			Anaras	*Ananas comosus*	Tender leaves	
			Missimi teeta	*Coptis teeta (Wall.)*	Tender leaves	
			Chenehi	*Scoparia dulsis (L.)*	Tender leaves	
			Bih Dhekia	*Polipodium sp*	Tender leaves	
			Parala lata	*Momordica sp*	Tender leaves	
			Horu bahaka	*Justicia betonica (Brum)*	Tender leaves	
Dimasa/Assam	Judima	Humao	—	*Glycyrrhiza glabra L.*	Bark	[17]
Deori/Assam			Bhatar duamali	*Jasminum sambac*	Leaves	[24]
			Thok thok	*Cinnamomum byolghata*	Leaves	
			Tesmuri	*Zanthoxylum hamiltonianum*	Leaves	
			Zing zing	*Lygodium flexuosum*	Leaves	

TABLE 6.1 *(Continued)*

Tribes/State	Fermented Beverage**	Starter Cake	Plants Used for Preparation of Starter Culture Cake			References
			Local Name	Scientific Name	Parts of Plants in Use	
			Zuuro	*Acanthus leucostychys*	Leaves	
			Bhilongoni	*Cyclosorus exlensa*	Leaves	
			Sotiona	*Alstonia scholaris*	Leaves	
			Dubusiring	*Alpinia malaccensis*	Leaves	
Nocte/Arunachal Pradesh	Jumin	Bichhi	Namninyng	*Ageratum conyzoides L.*	Leaves	[9]
			Torapat	*Alpinia nigra (Gaertn.) Burtt*	Rhizome	
			Sthul-ela	*Amomum subulatum Roxb.*	Rhizome	
			Ketekare	*Centella asiatica (L.) Urb*	Whole plant	
			Bappi	*Costus speciosus (J. Konig) C. Specht.*	Root	
			Chehsui yang	*Piper nigrum L.*	Seeds	
			Marcha	*Spilanthes oleracea L.*	Flower	
			Makat gutti	*Zanthoxylum armatum DC.*	Seeds and Bark	
			Makat gutti	*Zanthoxylum rhetsa DC.*	Seeds	
			Chehyui	*Zingiber officinale Roscoe*	Rhizome	
Meitei/Manipur	Yu or Atingba	Hamei	Yangli	*Albizia myriophylla*	Bark	[47]
Pnar/Meghalaya	Kiad	Thiat	Khaw-iang; Haw-iang	*Amomum aromaticum Roxb*	Leaves	[76]
Angami/Nagaland	Zutho	Akhri	Kemenya	*Oryza sativa L*	Grains	[46]
Nepalese and Tibetans/Sikkim	Jnard	Murcha	Aduwa; Bca sga	*Zingiber officinale*	Roots	[43]

TABLE 6.1 *(Continued)*

Tribes/State	Fermented Beverage**	Starter Cake	Plants Used for Preparation of Starter Culture Cake			References
			Local Name	Scientific Name	Parts of Plants in Use	
			Atis	*Aconitum napellus*	Plant	
Local tribes/ Tripura	Lungi or Gora	Chuwan	—	*Citrus sinensis (L) Osbeck*	Leaves	[39]
			—	*Litsea monopetala (Roxb) Pers.*	Leaves	
			—	*Markhami astiputala (Wall.) Seem.*	Leaves	
			—	*Combretum indicum (L.) DeFilipps*	Leaves	
			—	*Allophylus serratus Kurz.*	Leaves	
			—	*Aporusa diɔcia (Roxb.) Muell.*	Leaves	
			—	*Artocarpus heterophyllus Lam.*	Leaves	
			—	*Ananas comosus Mill.*	Leaves	

**Method of preparation of these fermented beverages is described in Sections 6.2.1 through 6.2.19 in this chapter.

increase in pH from 3.43 to 4.06 in poro apong [80] and to 4.29 in jou has been reported [4]. The lower value of pH ensures the inhibition of growth of most known pathogens, such as, *Clostridium perfringens*, *Vibrio cholerae*, *Campylobacter jejuni*, *Bacillus cereus*, and *Escherichia coli* [98]. However, members of *Enterobacteriaceae* can reduce the pH due to onset of the anaerobic fermentation process in preparation of local beer but acidophiles will eventually dominate with the progress of fermentation process [1, 74]. Ghosh et al. [13] collected rice beer from different communities of Tripura; and they reported the volatile fatty acid composition, i.e., 0.06–0.28 g in 100 mL of tartaric acid and 0.02–0.35 g in 100 mL of acetic acid.

The fresh rice beer samples were found to have total acidity value in the range of 0.225–0.637 (% tartaric acid), whereas the five days old rice beer samples showed the values in the range of 0.750–0.975 [2]. In poro apong, it was calculated to be 0.53 (% lactic acid) [40]. It has been reported that higher value of acidity in beer is found for long storage time [2, 21].

6.4.2 ETHANOL CONTENT

Fermentation is a process, where carbohydrates are converted to alcohol in presence of microbes. Hence quantity and alcohol contents primarily depend on the fermentative microbes. Therefore, studies reveal that presence of ethanol from initial day of the fermentation process is increased gradually until the completion of the process [4, 13]. The ethanol content has been reported to be dependent not only on microbial transformation process, but also on the quality of the rice and initial sugar [4]. It was reported that total ethanol content (volume by volume) in rice beer was between 5.30–22.05% in jou [4], 7.52–18.5% in poro apong [40] and 16% in judima [86], 12–13% in laboratory prepared rice beer samples [5], 6–10% in gora bwtwk [76], 26–35% in distilled form (chuwak) [13] and 30% in horo and jumai beers [43].

6.4.3 CARBOHYDRATE CONTENT IN RICE BEER

However, microbes are used for fermentation of carbohydrate, therefore its content in the finished product is reduced drastically due to the presence of higher polymers of glucose, such as, maltotetraose [12]. Storage and aging of beer also results in the decrease of total carbohydrate content due to the possible enzymatic degradation of the insoluble carbohydrate polymers. These soluble carbohydrates are then further converted to ethanol during the storage or aging process. This will result into decrease in the non-reducing sugar content from the first day of fermentation compared to the finished product.

Analysis of carbohydrates in the local rice beer reveals that amount of carbohydrate in the beer depends on the variety of rice and herbs. Laboratory analyses for qualitative estimation of carbohydrate in beer are: Benedict's test, Molisch's test and iodine test; while quantitative estimation of carbohydrate in beer are phenol-sulfuric acid assay [72], 3,5-dinitrosalicylic acid test [4]. Glucose is used as a standard for the quantitative analysis [4, 72].

Carbohydrate contents range is 0.4–0.8 $mgmL^{-1}$ in rice beer prepared by the Debbarma and Molsom tribe [32], 32.43 $mgmL^{-1}$ in judima [39], 48 $mgmL^{-1}$ in jou [4] and 46.62 $mgmL^{-1}$ in poro apong [40]. While, non-reducing sugar content ranges between 0.046–1.09 $mgmL^{-1}$ in rice beer prepared by Jamatia tribe and Koloi tribes, 0.355–0.784 $mgmL^{-1}$ in rice beer prepared by Debbarma and Molsom tribes, 3.47 $mgmL^{-1}$ in jou; and reducing sugar is 3.33 $mgmL^{-1}$ in poro apong. Carbohydrate content in local rice beer was higher than beer, gin, rum, etc., [43].

6.4.4 PROTEIN AND AMINO ACID CONTENTS IN RICE BEER

Biuret test is used as a standard method for the qualitative detection of protein in rice beer. Bovine serum albumin is used as a standard in Biuret test [13, 39, 40]. Reports suggest the presence of protein with concentration of 0.97 $mgmL^{-1}$ in judima [39], 1.05 $mgmL^{-1}$ in poro apong [40], 9.63–12.42 $mgmL^{-1}$ in rice beer prepared by tribes in Tripura [13], 3.2 $mgmL^{-1}$ in boro, 5.7 $mgmL^{-1}$ in poro, 4.2 $mgmL^{-1}$ in judima and 6.2 $mgmL^{-1}$ in jumai [43].

Kardong et al. [40] reported that the metabolic activity of the microbes in the fermentation process yields high level of free amino acids with value of 2.43 $mgmL^{-1}$ in poro apong (saimod) compared to 0.97–3.21 $mgmL^{-1}$ of amino acids in judima [35]. The amount of protein in the rice beer is proportional to the biomass of the fermentative microbes. More microbial growth causes more amount of protein in rice beer. Microbes in rice beer can change the nutritive value.

6.4.5 ANTIOXIDANT PROPERTY OF LOCAL RICE BEER

Antioxidants protect and minimize the damaging effects of free radicals and ROS [48]. Some of the commonly known antioxidants are: vitamin C, vitamin E and carotenoids. Besides, others include: vitamins, flavonoids, tannins, phenols, lignin, and minerals that are naturally present in plants.

Total antioxidant activity is estimated by DPPH radical scavenging assay. During the fermentation process up to 15 months, there is an increase in the antioxidant activity in the local rice beer, ranging from 0.86 mg mL^{-1} in fresh beer to 2.20 mg mL^{-1} in 15 months old beer in comparison to ascorbic acid equivalents. Local rice beer can have free radical scavenging activity between 39.9 to 75.7%. The polyphenol content in Gallic acid equivalent can be approximately 4.18 mg mL^{-1} Gallic acid equivalent. In jou, flavonoid content can range from 10 to 29 $mgmL^{-1}$ in quercetin equivalents in 15 months old beer [12]. Vitamin C content is between 13–39 mg in 100 mL [5].

Rice beers of North-Eastern India have an effective antioxidant activity, including total phenolic content and total flavonoids content [5, 12] and these are useful for human consumption. The phenolic and flavonoid compounds are different in rice beer from different communities and these impart a characteristic sensory property to the beer. The astringency (slight acidity and bitter smell and taste, which is a result of the presence of polyphenols in the rice beer) in the local rice beer is generated by polyphenols, such as, flavonoids during storage or aging. The sources for different polyphenols are: herbs, rice varieties and microbial metabolites that are produced during the fermentation and storage [12].

6.4.6 *MACRONUTRIENTS AND MICRONUTRIENTS IN RICE BEER*

Macronutrients and micronutrients function as cofactors of several cellular enzymes. Examples of these metals include sodium (Na), iron (Fe), copper (Cu), potassium (K), manganese (Mn), and zinc (Zn). Metal ions provide various structural, regulatory, and metabolic/catalytic activities in the cell (s). These metals involve the formation and development of bones, amino acid synthesis, cholesterol, and carbohydrate metabolisms; reduce toxicity; and participate in the oxido-reduction processes. Presence of Na, Cu, and Zn has been reported within the permissible limits in rice beer in different communities in Assam (namely: Ahom, Mising, and Deori) [14]. In comparison to barley and wheat beers, there was no significant change in levels of K, Fe, Ni, Mn, and Zn, while the levels of Co, Pb, and Cd were varied. The estimated intake of these metals in rice beer by an individual of that community was found to be within limits of the recommended dietary allowance, which indicated safe consumption of the local rice beer [14].

6.4.7 ROLE OF MICROBES IN THE PRODUCTION OF RICE BEER

A group of microorganisms (e.g.: filamentous fungi, yeast, and a group of lactic acid bacteria (LAB) categorized as probiotics) perform the fermentation process [69]. LAB-involved fermentation process provides lactic acid as the end product. LABs are acid tolerant (pH 5), non-motile, non-spore formers, Gram-positive cocci- and rod-shaped. Few examples of LABs are *Lacto-bacillus, Lactococcus, Leuconostoc, Pediococcus, Streptococcus, Corne-bacterium, Enterococcus,* and *Weissella.* They producehydrolase-enzymes, which can convert plant-derived polysaccharides, including pectin's, gums, and xylans. Besides this, they increase the levels of essential metabolites, including vitamins, aspartic acid, glutamic acid in the fermented product [29]. The major players in rice beer fermentation include: filamentous fungi and yeast, such as, *Saccharomyces, Candida, Saccharomycopsis, Aspergillus, Rhizopus,* etc. [7.60].

The secretion of extracellular amylolytic enzymes breaks the starch into monomeric forms of glucose and maltose [77]. The texture, aroma, and sensory qualities of rice beer are due to the metabolic activity of the yeast. Furthermore, these microorganisms produce gas, esters, alcohols, volatile acids and other compounds. Probiotics reduce the probability of growth of pathogens or destroy mycotoxins, endotoxins, and anti-nutrients. Further, they offer therapeutic values of the rice beer by enriching antioxidants and antimicrobials [60]. Several investigators isolated of microbes from rice beer [123], such as:

- Different species of *Kloeckera, Candida, Hanseniaspora,* and *Pischia* in apong in Assam and Arunachal Pradesh [97];
- Different types of LAB (such as: *Pediococcus pentosaceus, Lacto-bacillus plantarum* and *Lactobacillus brevis*) in hamei and marcha Manipur [5];
- *Hansenula anomala* and LAB in rice beer from different parts of North-Eastern region of India [94];
- *Pediococcus pentosaceus, Debaryomyces hansenii, Bacillus circu-lans, Bacillus firmus, Bacillus pumilus* and *Bacillus catarosporous* in judima [123];
- *Rhizopus sp.* in zutho, prepared in Nagaland [11]; and
- Yeasts (such as: *Saccharomyces cerevisiae*).

Saccharomyces cereviseae in all traditional fermented drinks [60], *Wickerhamomyces anomalus* inchubitchi and wanti in Garo hills [42] were also identified. A metagenomic approach in xaj-pitha of Assam revealed the presence of fungi *Mucor circinelloides*, *Rhizopus delemar* and *Aspergillus sp.*, budding yeast *Wickerhamomyces ciferrii, Meyerozyma guilliermondii,* and *Debaryomyces hansenii,* yeasts *Dekkera bruxellensis* and *Ogataea parapolymorpha* besides the dominance of LAB, like *Lactobacillus plantarum, Lactobacillus brevis, Lactobacillus pseudomesenteroides, Leuconostoc lactis, Lactococcus lactis, Weissella paramesenteroides,* and *Weissella cibaria* [11]. A study reported the isolation of *Enterococcus sp.* from fermented fish in Meghalaya [86], which produces bioactive compounds with antimicrobial properties [6].

6.5 NUTRITIONAL VALUE OF RICE AND HEALTH BENEFITS OF RICE BEER

Nutritional value of food depends on presence (both qualitative and quantitative) of various macro- and micro-nutrients, vitamins, free amino acids, phytochemicals (such as: flavonoids, polyphenols), protein, and carbohydrates, including prebiotics.

Rice beer offers versatile health benefits. The availability of rice beer is linked to the availability of the raw materials, the geographical barriers and environmental conditions [67]. Primarily, the beneficial effects of rice beer are seen in altering and maintaining the gut profile and introducing diversity in them. Being part of the socio-cultural life, rice beer is effective against ailments, such as, amoeboisis, acidity, vomiting, and it helps in cholesterol reduction and endocrine function. Rice beer comprises a vast array of beneficial microbes, including probiotics [110]. Biochemical properties reveal variation in the quality of rice beer among different indigenous communities in North-East India. The variation in the composition of rice beer depends on the methodology of preparation, ingredients used as a feedstock and starter culture. The organic acids present in rice beer contribute unique organoleptic properties [10]. Besides, it offers preservative activities [84]. Rice beer has therapeutic activity against conditions, such as: insomnia, headache, and body ache, reduction in inflammation of body parts, occurrence of diarrhea, cholera, and urinary problems, and expel worms [10, 26, 109]. Furthermore, some communities prepare diluted rice beer with water prior to consumption due to highly acidic nature of the rice beer.

Metabolites in local rice beer have beneficial effects on the immune system and have protective activities against several diseases [19]. The major health benefits from rice beer are offered by several biomolecules with antioxidant activities [100]. The phenolic compounds offer antioxidant activity by transferring hydrogen atoms to reactive oxygen species [63]. Reports suggested that phytochemicals in rice beer stimulate the immune system, regulate gene expression in cell proliferation and apoptosis [49]. Rice bran fermented with yeast (*Saccharomyces boulardii*) showed the potential to reduce the growth of human B-cell lymphomas [94]. Tocopherol and tocotrienols, lipid-soluble metabolites encourage anti-inflammatory cell signaling process [31, 94]. Rice beer can further suppress carcinogenesis in esophagus, stomach, liver, colon, and bladder [41].

Different components in rice beer (such as: polyphenols, phytic acid, trypsin inhibitor)can improve digestion [70]. Germinated rice when fermented with probiotics makes it a healthy food with natural fiber, inositol hexaphosphate and g-aminobutyric acid [63]. Researchers have established that LABs help in preventing diarrhea, favoring growth of natural microbiota in the gut, regularity in bowel movements and preventing lactose intolerance [37, 63, 88].

Flavonoids present in rice beer significantly reduce the risk of cardiovascular and other degenerative diseases [63]. Flavonoids increase antioxidant capacity (AOC) in blood. Intestinal cholesterol absorption is reduced by the phytosterols [96]. The alkaloids show pharmacological activities, such as, antimalarial, antiasthmatic, anticancer, and antibacterial properties [8].

Health benefits of few locally available rice beers are listed below:

- Apong is consumed as an energy booster drink, with added benefits of antimicrobial, antioxidant properties. It is also known to reduce age-related symptoms and maintaining kidney function [11, 34].
- Bhatti jaanr is consumed to regain physical strength in ailing persons and postnatal women. It is a mild alcoholic sweet tasting beverage [82].
- Haria has high antioxidant property and is consumed to protect from gastrointestinal ailments, like, vomiting, amoebiasis, acidity diarrhea, and dysentery [63]. Presence of maltotetrose, maltotriose, and maltose inhibit the growth of intestinal pathogens. The b-D-galactopyranose pentaacetate and b-D-mannopyranose pentaacetate stimulate the immune response and have anti-mutagenic activities [22].
- Hor achois used to cure dysentery and pharyngitis in rural areas [106].
- Jou is used as a refreshing drink and to keep the body relaxed. It also prevents jaundice and urinary disorders [11].
- Judima has bioactivities, such as: antioxidant, anti-allergy, and anti-inflammatory activity, antibacterial, antifungal, antispasmodic, hypotensive, hypolipidemic, hepatoprotective, neuro-protective, anti-diabetic, and antiaging [104].

- Kiad is believed to act as remedy for urinary bladder ailments and dysentery [100].
- Rice jann is consumed as a refreshing drink to boost the body's energy levels and protect against cold [65].
- Rokshiin is used as a beauty-care product [5].
- Yu is used to regulate irregular menstrual flow and infertility factors, besides being used to regulate obesity, loss of appetite and low nourishments of food [5].
- Zutho boosts the immune function, prevents loss of appetite, lowers blood insulin and cholesterol levels, prevent infection and helps in wound healing [11, 89].

Plant species and plant parts (such as: roots, stem, bark, and leaves) are used to prepare the starter culture and also to contribute towards the health-promoting effects in rice beer. The plants used to prepare the starter culture are also used as traditional medicine. Examples of such plants and their application are [4, 22]:

- *Allophylus serratus* helps for treatment of gastrointestinal disorder, inflammation, osteoporosis, and elephantiasis.
- *Ananas comosus* leaves are used for treatment of helminthic infections, dysuria, diarrhea, and rheumatism.
- *Artocarpus heterophyllus* are used for treatment of skin disease and has anthelminthic property.
- *Casearia aculeate* exhibits antibacterial activity.
- *Citrus sinensis* has antibacterial and antifungal activity, cardio protective, antidiabetic, anticancer, and anti-inflammatory properties.
- *Litsea monopetala* has antioxidant properties.
- *Markhamia stipulate* are used for the treatment of nervous disease.
- *Moringa olifera* is used for the treatment of jaundice and common cold.
- *Saccharum officinarum* relieves constipation.

6.6 SUMMARY

Local rice beer in North-East India plays important role in social and cultural activities, such as marriages, festivals, merrymaking, death ceremonies, community gathering, etc., and so on. The different tribes almost use same method for the beer preparation, except the variation in feedstock. Both feedstock and starter culture do not only impact the development of color, flavor, sweetness, etc., but also offer many medicinal properties. This chapter discusses different aspects of rice beer preparation in different tribal communities of North-East India.

KEYWORDS

- **fermented beverages**
- **lactic acid bacteria**
- **manganese**
- **medicinal plants**
- **potassium**
- **probiotics**

REFERENCES

1. Albalasmeh, A. A., Berhe, A. A., & Ghezzehei, T. A., (2013). Method for rapid determination of carbohydrate and total carbon concentrations using UV spectrophotometry. *Carbohydrate Polymers, 97* (2), 253–261.
2. Armstrong, D., & Cohen, J., (1999). Trends in infectious diseases. *JAMA, 281*, 61–66.
3. Anyinam, C., (1995). Ecology and ethnomedicine: Exploring links between current environmental crisis and indigenous medical practices. *Social Science and Medicine, 40*, 321–329.
4. Arjun, J., (2015). Comparative biochemical analysis of certain indigenous rice beverages of tribes of Assam with some foreign liquor. *Biotechnological Communication, 8* (2), 138–144.
5. Arjun, J., Verma, A. K., & Prasad, S. B., (2014). Method of preparation and biochemical analysis of local tribal wine judima: Indegenous alcohol used by dimasa tribes of North Cachhar Hills District of Assam, India. *International Food Research Journal, 21* (2), 463–470.
6. Bahiru, B., Mehari, T., & Ashenafi, M., (2006). Yeast and lactic acid flora of tej indigenous Ethiopian honey wine: Variations within and between production units. *Food Microbiology, 23* (3), 277–282.
7. Basumatary, T. K., Terangpi, R., Brahma, C., & Roy, P., (2014). Jou: Traditional drink of the boro tribe of Assam and North East India. *Journal of Scientific and Innovative Research, 3* (2), 239–243.
8. Bhaskar, B., & Khan, M. R., (2017). Effect of rice beer on gut bacteria. *Canadian Journal of Biotechnology, 1*, 120.
9. Bhatt, K. C., Malav, P. K., & Ahlawat, S. P., (2018). Jumin: Traditional beverage of nocte Tribe in Arunachal Pradesh: An ethnobotanical survey. *Genetic Resource and Crop Evolution, 65* (2), 671–677.

10. Bhuyan, B., & Baishya, K., (2013). Ethnomedicinal value of various plants used in the preparation of traditional rice beer by different tribes of Assam, India. *Drug Invention Today*, *5* (4), 335–341.
11. Bhuyan, D. J., & Barooah, M. S., (2014). Biochemical and nutritional analysis of rice beer of North East India. *Indian Journal of Traditional Knowledge*, *13* (1), 142–148.
12. Biswas, K., Upadhayay, S., Rapsang, G. F., & Joshi, S. R., (2017). Antibacterial and Synergistic activity against β-lactamase-producing nosocomial bacteria by bacteriocin of LAB isolated from lesser known traditionally fermented products of India. *HAYATI Journal of Biosciences*, *24* (2), 87–95.
13. Blandino, A., Al-Aseeri, M. E., Pandiella, S. S., Cantero, D., & Webb, C., (2003). Cereal-based fermented foods and beverages. *Food Research International*, *36* (6), 527–543.
14. Bora, S. S., Keot, J., Das, S., Sarma, K., & Barooah, M., (2016). Metagenomics Analysis of microbial communities associated with a traditional rice wine starter culture (Xaj-pitha) of Assam, India. *3 Biotech*, *6* (2), 1–13.
15. Bussmann, R. W., & Glenn, A., (2011). Fighting pain, Traditional peruvian remedies for the treatment of asthma, rheumatism, arthritis, and sore bones. *Indian Journal of Traditional Knowledge*, *10*, 397–412.
16. Chakrabarty, J., Sharma, G. D., & Tamang, J. P., (2014). Traditional technology and product characterization of some lesser-known ethnic fermented foods and beverages of North Cachar Hills District of Assam, India. *Indian Journal of Traditional Knowledge*, *13*, 706–715.
17. Chakrabarty, J., Sharma, G. D., & Tamang, J. P., (2009). Substrate utilization in traditional fermentation technology practiced by tribes of North Cachar Hills of Assam. *Assam University Journal of Science and Technology: Biological Sciences*, *4* (1), 66–72.
18. Choudhury, C., Devi, M. R., Bawari, M., & Sharma, G. D., (2011). Ethno-toxic plants of Cachar District in Southern Assam with special reference to their medicinal properties. *Biological and Environmental Sciences*, *7* (1), 89–95.
19. Coe, F. G., Parikh, D. M., Johnson, C. A., & Anderson, G. J., (2012). The good and the bad: Alkaloid screening and brine shrimp bioassays of aqueous extracts of 31 medicinal plants of Eastern Nicaragua. *Pharmaceutical biology*, *50* (3), 384–392.
20. Dahiya, D. K., Renuka, P. M., & Shandilya, U. K., (2017). Gut Microbiota modulation and its relationship with obesity using prebiotic fibers and probiotics: A review. *Frontiers in Microbiology*, *8*, 1–17.
21. Dam, P. K., Yadav, S. P., Ramnath, T., & Tyagi, B. K., (1998). Constraints in conservation of medicinal plants in the climatologically changing Thar Dessert. *Paper Presented at National Environment Science Academy, VIII Annual Congress* (p. 5). Gulberga.
22. Das, A. J., Khawas, P., Miyaji, T., & Deka, S. C., (2014). HPLC and GC-MS analyses of organic acids, carbohydrates, amino acids and volatile aromatic compounds in some varieties of rice beer from Northeast India. *Journal of Institute of Brewing*, *120*, 244–252.
23. Das, A. J., & Deka, S. C., (2012). Mini review fermented foods and beverages of the North-Eastern India. *International Food Research Journal*, *19* (2), 377–392.
24. Das, A. J., Deka, S. C., & Miyaji, T., (2012). Methodology of rice beer preparation and various plant materials used in starter culture preparation by some tribal communities of North-Eastern India: A survey. *International Food Research Journal*, *19* (1), 101–107.
25. Das, A. K., Sharma, G. D., & Dutta, B. K., (2014). Study of plant biodiversity and its conservation in Hailakandi District, Assam, Part-1: Flora. *Journal of Economic and Taxonomic Botany*, *28* (1), 213–228.

26. Das, A., Raychaudhuri, U., & Chakraborty, R., (2012). Cereal-Based functional food of Indian subcontinent: A Review. *Journal of Food Science and Technology, 49* (6), 665–672.
27. Das, A., (2016). Medicinal plants used traditionally for the preparation of rice beer by the tiwa tribe of Morigaon District of Assam, India. *International Journal of Current Research, 8* (11), 40940–40943.
28. Deb, D., Sarkar, A., Barma, B. D., Datta, B., & Majumdar, K., (2013). Wild Edible plants and their utilization in traditional recipes of Tripura, North-Eastern India. *Advances in Biological Regulation, 7*, 203–211.
29. Deka, A. K., Handique, P., & Deka, D. C., (2018). Antioxidant-activity and physico-chemical indices of the rice beer used by the bodo community in North-Eastern India. *Journal of the American Society of Brewing Chemists, 76* (2), 112–116.
30. Deka, D. C., & Sharma, G. C., (2010). Traditionally used herbs in the preparation of rice-beer by the rabha tribe of Goalpara District, Assam. *Indian Journal of Traditional Knowledge, 9* (3), 459–462.
31. Fitzgerald, M. A., McCouch, S. R., & Hall, R. D., (2009). Just a grain of rice: The quest for quality. *Trends in Plant Science, 14* (3), 133–139.
32. Gesler, W. M., (1992). Therapeutic landscapes: Medical issues in light of the new cultural geography. *Social Science and Medicine, 34* (7), 735–746.
33. Ghosh, C., & Das, A. P., (2004). Preparation of rice beer by the tribal inhabitants of tea gardens in Terai of West Bengal. *Indian Journal of Traditional Knowledge, 3* (4), 374–382.
34. Ghosh, K., Ray, M., Adak, A., & Dey, P., (2015). Microbial, saccharifying, and antioxidant properties of Indian rice based fermented beverage. *Food Chemistry, 168*, 196–202.
35. Ghosh, S., Rahaman, L., & Kaipeng, D. L., (2016). Community-wise evaluation of rice beer prepared by ethnic tribes of Tripura. *Journal of Ethnic Foods, 3* (4), 251–256.
36. Gibji, N., Ringu, N., & Dai, N. O., (2012). Medicinal knowledge among the adi tribes of lower Dibang Valley district of Arunachal Pradesh, India. *International Research Journal of Pharmacy, 3* (6), 223–229.
37. Gogoi, B., Dutta, M., & Mondal, P., (2013). Various ethnomedicinal plants used in the preparation of apong: Traditional beverage used by mising tribe of upper Assam. *Journal of Applied Pharmaceutical Science, 3* (4), 85–88.
38. Gogoi, P., (2017). Ethnobotanical Study of certain medicinal plants used by local people in Lakhimpur-Assam. *International Journal of Chem. Tech Research, 10* (9), 7–13.
39. Grumezescu, A., & Holban, A. M., (2018). *Production and Management of Beverages: Volume 1. The Science of Beverages* (p. 504). Sawston, Cambridge: Woodhead Publishing.
40. Handique, P., Kalita, D. A., & Deka, D. C., (2017). Metal profile of traditional alcoholic beverages prepared by the ethnic communities of Assam, India. *Journal of the Institute of Brewing, 123* (2), 284–288.
41. Henderson, A. J., Ollila, C. A., & Kumar, A., (2012). Chemopreventive properties of dietary rice bran: Current status and future prospects. *Advances in Nutrition, 3* (5), 643–653.
42. Holzapfel, W., (2002). Use of starter cultures in fermentation on a household scale. *Food Control, 8* (5), 241–258.
43. Hund, A. J., & Wren, J. A., (2018). *The Himalayas: An Encyclopedia of Geography, History, and Culture* (p. 349). ABC-CLIO, Science. online; https://www.abc-clio.com/ABC-CLIOGreenwood/product.aspx?pc=A4837C (accessed on 5 August 2020).

44. Jadhav, R. R., (2016). Ethnobotanical survey of Kadegaon, India. *Journal of Medicinal Plant Studies, 4*, 11–14.
45. Jaiswal, V., (2010). Ethnobotany of jaintia tribal community of Meghalaya in North-East India. *Indian Journal of Traditional Knowledge, 9*, 38–44.
46. Jamir, B., & Deb, C. R., (2014). Some fermented foods and beverages of Nagaland, India. *International Journal of Food and Fermentation Technology, 4* (2), 87–92.
47. Jeyaram, K., Singh, W. M., Capece, A., & Romano, P., (2008). Molecular identification of yeast species associated with hamei: Traditional starter used for rice wine production in Manipur, India. *International Journal of Food Microbiology, 124*, 115–125.
48. Kähkönen, M. P., Hopia, A. I., & Vuorela, H. J., (1999). Antioxidant Activity of plant extracts containing phenolic compounds. *Journal of Agricultural and Food Chemistry, 47* (10), 3954–3962.
49. Kardong, D., Deori, K., & Sood, K., (2012). Evaluation of nutritional and biochemical aspects of apong (saimod): Home-made alcoholic rice beverage of mising tribe of Assam, India. *Indian Journal of Traditional Knowledge, 11* (3), 499–504.
50. Kumar, M. B., Hati, S., Brahma, J., Patel, M., & Das, S., (2018). Identification and characterization of yeast strains associated with the fermented rice beverages of Garo Hills, Meghalaya, India. *International Journal of Current Microbiology and Applied Sciences, 7* (2), 3079–3090.
51. Kumar, M., Sheikh, M. A., & Bussmann, R. W., (2011). Ethnomedicinal and ecological status of plants in Garhwal Himalaya, India. *Journal of Ethnobiology and Ethnomedicine, 7* (1), 32–36.
52. Lobo, V., Patil, A., Phatak, A., & Chandra, N., (2010). Free radicals, antioxidants, and functional foods: Impact on human health. *Pharmacognosy Reviews, 4* (8), 118–126.
53. Majumder, J., Bhattacharjee, P. P., Datta, B. K., & Agarwala, B. K., (2014). Ethnomedicine Plants used by bengali communities in Tripura-India. *Journal of Forestry Research, 25*, 713–716.
54. Mangang, K. C. S., Das, A. J., & Deka, S. C., (2017). Comparative shelf-life study of two different rice beers prepared using wild-type and established microbial starters. *Journal of the Institute of Brewing, 123* (4), 579–586.
55. Mao, A. A., (1998). Ethnobotanical Observation of Rice Beer "Zhuchu" Preparation by the Mao Naga Tribe from Manipur (India). *Bulletin of the Botanical Survey of India, 40*, 53–57.
56. Mao, A., Hynniewta, T., & Sanjappa, M., (2009). Plant Wealth of north-eastern india with reference to ethnobotany. *Indian Journal of Traditional Knowledge, 8*, 96–103.
57. Meghalaya, Shillong–India; (2018). *Environmental Issues and Management of Natural Resources: Community Participation and Government Intervention in Meghalaya* (pp. 216–244). Meghalaya Human Development Report; Meghalaya, India: Planning Department, Government of Meghalaya.
58. Mishra, B. K., Hati, S., Brahma, J., Patel, M., & Das, S., (2018). Identification and characterization of yeast strains associated with the fermented rice beverages of Garo Hills, Meghalaya, India. *International Journal of Current Microbiology and Applied Sciences, 7* (2), 3079–3090.
59. Mishra, B. K., Hati, S., Das, S., & Patel, K., (2017). Biodiversity of *Lactobacillus* cultures associated with traditional fermented foods in West Garo Hills, Meghalaya, India. *International Journal of Current Microbiology and Applied Sciences, 6* (2), 1090–1102.

60. Mohan, V., Spiegelman, D., & Sudha, V., (2014). Effect of brown rice and white rice with legumes on blood glucose and insulin responses in overweight Asian Indians: A randomized controlled trial. *Diabetes Technology and Therapeutics, 16* (5), 317–325.
61. Narzary, Y., Brahma, J., Brahma, C., & Das, S., (2016). Study on the indigenous fermented foods and beverages of Kokrajhar, Assam, India. *Journals of Ethnic Foods, 2016*, 1–9.
62. Negi, C. S., & Palyal, V. S., (2007). Traditional Uses of animal and animal products in medicine and rituals by shoka tribes of Pithoragarh in Uttaranchal. *Ethnomedicine, 1* (1), 47–54.
63. Ogue-Bon, E., Khoo, C., & Hoyles, L., (2011). *In Vitro* Fermentation of rice bran combined with *Lactobacillus acidophilus* 14 150B or *Bifidobacterium longum* 05 by the canine fecal microbiota. *FEMS Microbiology Ecology, 75*, 365–376.
64. Pfoze, N. L., Kehie, M., Kayang, H., & Mao, A. A., (2014). Estimation of ethnobotanical plants in Nagaland of North East India. *Journal of Medicinal Plants Studies, 2* (3), 92–104.
65. Phutthaphadoong, S., Yamada, Y., Hirata, A., & Tomita, H., (2009). Chemopreventive effects of fermented brown rice and rice bran against 4- (methylnitrosamino)-1-(3-pyridyl)-1-butanone induced lung tumorigenesis in female A/J Mice. *Oncology Reports, 21*, 321–327.
66. Piewngam, P., Zheng, Y., & Nguyen, T. H., (2018). Pathogen elimination by probiotic *Bacillus* via signaling interference. *Nature, 562* (7728), 532–537.
67. Praveena, R. J., & Estherlydia, D., (2014). Comparative Study of phytochemical screening and antioxidant capacity of vinegar made from peel and fruit of pineapple (*Ananas comosus* L.). *International Journal of Pharma and Bio Sciences, 5*, 394–403.
68. Rao, R. R., (1994). *Biodiversity in India: Floristics Aspects* (p. 371). Dehradun-India: Bishan Singh and Mahindra Publishers.
69. Rapsang, G. F., & Joshi, S. R., (2012). Bacterial diversity associated with tungtap, an ethnic traditionally fermented fish product of Meghalaya. *Indian Journal of Traditional Knowledge, 11* (1), 134–138.
70. Ray, M., Ghosh, K., Singh, S., & Chandra, M. K., (2016). Folk to functional: An explorative overview of rice-based fermented foods and beverages in India. *Journal of Ethnic Foods, 3* (1), 5–18.
71. Rizvi, J., (1983). *Ladakh: Crossroads of High Asia* (p. 213). New Delhi: Oxford University Press.
72. Rodgers, W. A., & Panwar, H. S., (1998). *Planning Wildlife Protected Area Network in India* (Vol. 1, p. 267). Dehra Dun, India: Wildlife Trust of India.
73. Roy, B., Kala, C. P., Farooquee, N. A., & Majila, B. S., (2004). Indigenous fermented food and beverages: Potential for economic development of the high altitude societies in Uttaranchal. *Journal of Human Ecology, 15* (1), 45–49. online; https://doi.org/10.1080/09709274.2004.11905665 (accessed on 5 August 2020).
74. Ryan, E. P., Heuberger, A. L., Weir, T. L., Barnett, B., Broeckling, C. D., & Prenni, J. E., (2011). Rice bran fermented with *Saccharomyces boulardii* Generates novel metabolite profiles with bioactivity. *Journal of Agricultural and Food Chemistry, 59* (5), 1862–1870.
75. Saikia, B., Tag, H., & Das, A. K., (2007). Ethnobotany of foods and beverages among the rural farmers of tai-ahom of North Lakhimpur District, Assam. *Journal of Traditional Knowledge, 6*, 126–132.
76. Samati, H., & Begum, S. S., (2007). Kiad: Popular local liquor of pnar tribe in Jaintia Hills District, Meghalaya. *Indian Journal of Traditional Knowledge, 6* (1), 133–135.

77. Sanders, M. E., (2009). How do we know when something called "probiotic" is really a probiotic?: A guideline for consumers and health care professionals. *Functional Food Reviews, 1* (1), 3–12.
78. Sapan, C. V., Lundblad, R. L., & Price, N. C., (1999). Colorimetric Protein assay techniques. *Biotechnology and Applied Biochemistry, 108*, 99–108.
79. Sarkar, P. K., & Tamang, J. P., (1995). Changes in the microbial profile and proximate composition during natural and controlled fermentations of soybeans to produce kinema. *Food Microbiology, 12* (C), 317–325.
80. Sathe, G. B., & Mandal, S., (2016). Fermented products of India and its implication: Review. *Asian Journal of Dairy and Food Research, 35*, 1–9.
81. Satish, K, R., Kanmani, P., & Yuvaraj, N., (2013). Traditional Indian fermented foods: A rich source of lactic acid bacteria. *International Journal of Food Sciences and Nutrition, 64* (4), 415–428.
82. Savitri, S., & Bhalla, T. C., (2007). Traditional foods and beverages of Himachal Pradesh. *Indian Journal of Traditional Knowledge, 6* (1), 17–24.
83. Sekar, S., & Kandavel, D., (2002). Patenting microorganisms: Towards creating a policy framework. *Journal of Intellectual Properties Right, 7*, 211–221.
84. Sekar, S., & Mariappan, S., (2007). Use of traditional fermented products by Indian rural folks and IPR. *Indian Journal of Traditional Knowledge, 6*, 111–120.
85. Shaikh, A. M., Shrivastava, B., Apte, K. G., & Navale, S. D., (2016). Medicinal plants as potential source of anticancer agents: A review. *Journal of Pharmacognosy and Phytochemistry, 5*, 291–295.
86. Sharma, M., Sharma, C. L., & Debbarma, J., (2014). Ethnobotanical studies of some plants used by Tripuri of Tripura, NE India with special reference to magico religious beliefs. *International Journal of Plant Animal and Environmental Science, 4*, 518–528.
87. Shrivastava, K., Greeshma, A. G., & Srivastava, B., (2012). Biotechnology in tradition: Process technology of alcoholic beverages practiced by different tribes of Arunachal Pradesh in North East India. *Indian Journal of Traditional Knowledge, 11* (1), 81–89.
88. Sindhu, S. C., & Khetarpaul, N., (2002). Effect of probiotic fermentation on antinutrients and *in Vitro* protein and starch digestibility of indigenously developed RWGT food mixture. *Nutrition and Health, 16*, 173–181.
89. Singh, P. K., & Singh, K. I., (2006). Traditional alcoholic beverage (Yu) of meitei communities of Manipur. *Indian Journal of Traditional Knowledge, 5* (2), 184–190.
90. Stanbury, O. F., (1999). Fermentation technology. In: Stanbury, P. F., Whitaker, A., & Hal, S. J., (eds.), *Principles of Fermentation Technology* (pp. 1–24). London: Oxford-Butterworth Heinemann.
91. Tamang, J. P., Thapa, S., Tamang, N., & Rai, B., (1996). Indigenous fermented food beverages of Darjeeling Hills and Sikkim: Process and product characterization. *Journal of Hill Research, 9*, 401–411.
92. Tamang, J. P., Chettri, R., & Sharma, R. M., (2009). Indigenous knowledge on North-Eastern women on production of ethnic fermented soybean foods. *Indian Journal of Traditional Knowledge, 8*, 122–126.
93. Tamang, J. P., Tamang, N., Thapa, S., & Dewan, S., (2012). Microorganism and nutritive value of ethnic fermented foods and alcoholic beverages of North-Eastern India. *Indian Journal of Traditional Knowledge, 11*, 7–25.

94. Tamang, J. P., & Thapa, S., (2006). Fermentation dynamics during production of bhaati jaanr: Traditional fermented rice beverage of the Eastern Himalayas. *Food Biotechnology, 20*, 251–261.
95. Tamang, J. P., & Sarkar, P. K., (1988). Traditional Fermented foods and beverages of Darjeeling and Sikkim: Review. *Journal of Science of Food and Agriculture, 44*, 375–385.
96. Tannock, G. W., (2004). Special fondness for *Lactobacilli. Mini Reviews, 70* (6), 3189–3194.
97. Tanti, B., Gurung, L., Sarma, H. K., & Buragohain, A. K., (2010). Ethnobotany of starter culture used in alcohol fermentation by a few ethnic tribes of Northeast India. *Indian Journal of Traditional Knowledge, 9* (3), 463–466.
98. Teramoto, Y., Yoshida, S., & Ueda, S., (2002). Characteristics of a rice beer (Zutho) and a yeast isolated from the fermented product in Nagaland, India. *World Journal of Microbiology and Biotechnology, 18*, 813–816.
99. Teron, R., (2006). Hor: Traditional alcoholic beverage of karbi tribe in Assam. *Natural Product Radiance, 5*, 377–381.
100. Thokchom, S., & Joshi, S. R., (2012). Antibiotic resistance and probiotic properties of dominant lactic micro flora from tungrymbai: An ethnic fermented soybean food in India. *Journal of Microbiology, 50* (3), 535–539.
101. Thompson, M. D., (2009). Biomedical agriculture: A Systematic approach to food crop improvement for chronic disease prevention. *Advances in Agronomy, 102*, 1–54.
102. Tiwari, S. C., & Mahanta, D., (2007). Ethnological observations fermented food products of certain tribes of Arunachal Pradesh. *Indian Journal of Traditional Knowledge, 6*, 106–110.
103. Tsuyoshi, N., Fudou, R., & Yamanaka, S., (2005). Identification of Yeast strains isolated from Marcha in Sikkim: A microbial starter for amylolytic fermentation. *International Journal of Food Microbiology, 99*, 135–146.
104. Tyl, C., & Sadler, G. D., (2017). pH and titratable acidity. In: Nielsen, S., (ed.), *Food Analysis* (pp. 389–406). New York: Food Science Text Series, Springer.
105. Uchoi, D., Roy, D., Majumdar, R. K., & Debbarma, P., (2015). Diversified traditional cured food products of certain indigenous tribes of Tripura, India. *Indian Journal of Traditional Knowledge, 14* (3), 440–446.
106. Vriesekoop, F., Krahl, M., Hucker, B., & Menz, G., (2013). Bacteria in brewing: The good, the bad and the ugly. *Journal of Institute of Brewing, 118* (4), 335–345.
107. WHO (2013). *Traditional Medicine.*
108. WHO, (2002). *WHO Policy Perspectives on Medicines Report 002 on Traditional Medicine: Growing Needs and Potentials* (p. 6). Switzerland: World Health Organization.
109. Woyengo, T. A., Ramprasath, V. R., & Jones, P. J. H., (2009). Anticancer effects of phytosterols. *European Journal of Clinical Nutrition, 63*, 813–820.
110. Yang, L., Yan, Q. H., Ma, J. Y., & Wang, Q., (2013). High Performance liquid chromatographic determination of phenolic compounds in propolis. *Tropical Journal of Pharmaceutical Research, 12* (5), 771–776.

PART III

Biological Activities of Plant-Based Phytochemicals

CHAPTER 7

NATURAL PHYTOBIOACTIVES: LET'S EAT SMART!

MANISHA GUHA, NILANJAN DAS, and XAVIER SAVARIMUTHU

ABSTRACT

Plant-based bioactive compounds have drawn attention of all communities around the world due to their unique biochemical activities and health benefits. Research studies have confirmed the safeguarding effects of certain plant-based diets on cardiovascular diseases, obesity, cancer, diabetes, etc. Nutrigenomics and food biotechnology should encompass biofortification of foods with phytobioactives apart from just conventional nutrients to boost well-being of humans. Plants produce a wide range of bioactive compounds as primary and secondary metabolites. This chapter provides a comprehensive review of health beneficial roles of certain secondary metabolic compounds (such as: terpenes, alkaloids, and phenolics) from plants, and nutrigenomics. Furthermore in this chapter, different technologies for extraction of phytobioactives and their mode of delivery are discussed.

7.1 INTRODUCTION

The scenario of negligence of the necessity of a healthy life-style provides excellent opportunities for chronic diseases to thrive, such as: cardiovascular diseases (CVDs), diabetes, oxidative stress-related disorders, cancer, etc. According to recent reports by the World Health Organization (WHO), "*CVDs cause the death of about 17.9 million people worldwide, which is about 31% of all global deaths* [26]." The rate of occurrence of CVD and its related deaths has increased over the past few decades. WHO reports cancer to be "the second most dominant cause of death worldwide with about 9.6 million deaths annually [25]." WHO's global report on diabetes indicates that the number of persons suffering from diabetes has shot up from 108 million in 1980 to 422 million in 2014. In 2016, about 1.6 million deaths were caused by diabetes directly [38].

Obesity cases have nearly tripled across the world resulting in a situation, where most people live in countries with obesity claiming more lives than mal-nourishment [96]. In such a pathetic scenario, a healthy diet and physical activity are what we should turn to at the earliest.

Healthy food contains many components to benefit our bodies in multifarious ways. Such components are present in different categories of foods. Not all required food items are available in all geographical locations with equal abundance. Therefore, scientists have turned towards supplementing staple food items with the required nutrients, and also making the required supplements commercially available by large-scale industrial synthesis [78].

Numerous bio-fortified foods in the market, and the huge number of fermentation industries worldwide bear testimony to the success of research in this field. However, most of the engineered components belong to either the macro- or the micro-nutrient category. Of late, a new category named 'bioactive' is gaining increasing attention in the food and nutrition research studies. Bioactive food components are beneficial for health, although not essential. Bioactives encompass all non-essential bio-molecules present in food, which display the ability to modulate metabolic processes, contributing to better health. They are often found in myriad forms, such as: glycosylated, hydroxylated, esterified, etc.

Most of the available bioactive compounds are of plant origin. They are termed as "phytobioactives." They help not only in metabolism through acting as internal signaling components, but also by modifying genes. The interaction of dietary factors with the genome is important in the development and progression of diseases [42].

With rich culture and traditional knowledge, Indians knew that certain active ingredients of different plant items are beneficiary to health. "*Charaka Samhita*" describes an ancient and traditional system of Indian medication completely based on plant extracts and herbs. Five chapters of this book are completely dedicated to "*Aharatatva* or dietetics," which states that a healthy diet gives way to good health, whereas an unhealthy diet gives way to maladies.

"The tastes are six, such as: sweet, sour, saline, pungent, bitter, and astringent.
Properly used, they nourish our body.
Improperly used (excess or deficient), they lead to the provocation of the *Dosha*.
The *Dosha* are three: *Vayu*, *Pitta*, and, *Kapha*.
When they are in their normal state, they are beneficial to our body.
However, when they become disorganized,
they afflict the body with diseases of diverse kinds."
—Charaka Samhita, 3.1.3–3.1.4 [68, 105, 121, 127].

Similar traditional knowledge also existed in China, Africa, and a few other countries [61, 73]. Probably, their exact role in health, and modes of action were not known. With scientific revolution over the years, recent research studies in this field are bringing forth the secrets of bioactives in ameliorating health [85, 112]. Increasing consumer awareness and shift to a healthy lifestyle and diet also hold good market prospects commercial products supplemented with functional ingredients. Figure 7.1 indicates parts of the world, where traditional knowledge of plants exists and is being used by tribal communities.

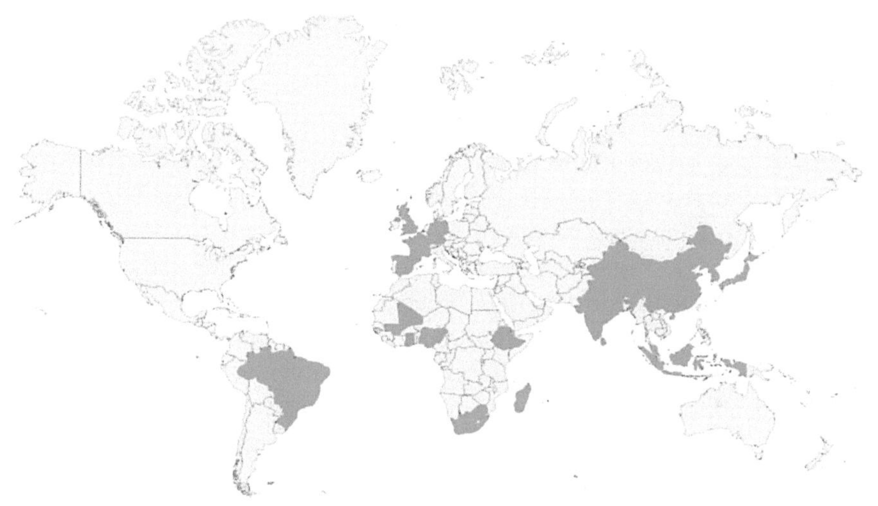

FIGURE 7.1 Distribution of traditional knowledge of plant components in the countries around the world. Countries marked with green color practice plant-based traditional knowledge.
Source: Map developed from: "https://www.armcharts.com/visited_countries/" with data adapted from Ameh et al. [8].

This chapter discusses types of phytobioactive compounds, various methods of their extraction and their roles in human health, with an objective to generate awareness against certain widespread chronic pathological conditions, and how we can contribute to reducing their prevalence by changing our diet and lifestyle.

7.2 TYPES OF PHYTOBIOACTIVES

Large number of bio-actives are present not only in plant and animal products but also in microbes. However, the majority are obtained from plants. The bio-actives obtained from plants are termed as phytobioactives. They vary widely both in structure and function. Accordingly, they belong to several groups, such as: flavonoids, carotenoids, phytosterols, tocopherols, carnitine, phytoestrogens, choline, creatine, coenzyme Q, anthocyanins, dithiolthiones, glucosinolates, prebiotics, polysaccharides, taurine, etc., [80].

The bioactives can be classified either based on their chemical structure or biosynthetic pathway [13]. Plant-based bioactive compounds can be classified broadly into three major categories based on structure, such as: terpenes and terpenoids, alkaloids, and phenolics.

7.2.1 TERPENE

Terpenes are hydrocarbons composed of isoprene units (basic structural element of terpenes). They are synthesized from intermediates of glycolysis or acetyl CoA; and are produced by the mevalonate and non-mevalonate (MEP) pathways. Terpenoids are derived from terpenes by some rearrangement in the structure or oxidation process. Taxol from bark and needles of *Taxus baccata* and forskolin from roots of *Coleus forskohlii* are diterpene phytobioactives having anti-cancer (against ovarian and breast cancer [90]) and anti-hypertension, anti-glaucoma properties [130], respectively. In Figure 7.2, *Taxus baccata* and *Coleus forskohlii* are presented, whereas Figure 7.3 indicates the structures of terpenoids.

FIGURE 7.2 *Taxus baccata* plant (A) [Modified from the Poison Garden website [122]]; bark of taxus (B) [Modified from the tree ebb [123]]; needles of taxus (C) [modified from the woodland Trust [138]]; *Coleus forskohlii* plant (D) [modified from ultimate herbal health [31]]; and roots of *Coleus forskohlii* (E) [modified from Kamohara et al. [65]].

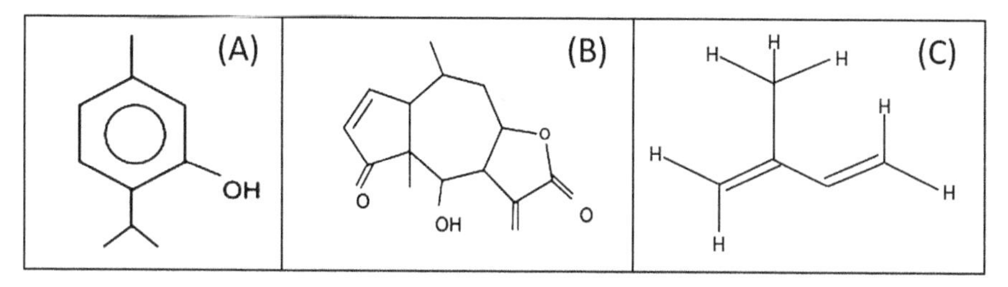

FIGURE 7.3 General structure of monoterpene (A) [Modified from [13]]; general structure of sesquiterpene (B) [Modified from [13]]; and structure of Isoprene unit (C) [modified from [5]].

7.2.2 ALKALOIDS

Alkaloids are naturally occurring organic substances with nitrogen in the heterocyclic ring (s). Alkaloids are mostly alkaline with the nitrogen atom protonated. Alkaloids are produced by the shikimic acid pathway. The first alkaloid to be isolated was morphine from the flowers of *Papaver somniferum* (the opium poppy). Quinine (from the bark of cinchona trees) and ephedrine (from *Ephedra*) are alkaloids with anti-malarial [1] and anti-asthmatic activities, respectively [87]. Examples of plants producing alkaloids are shown in Figure 7.4, while Figure 7.5 indicates the alkaloid structures.

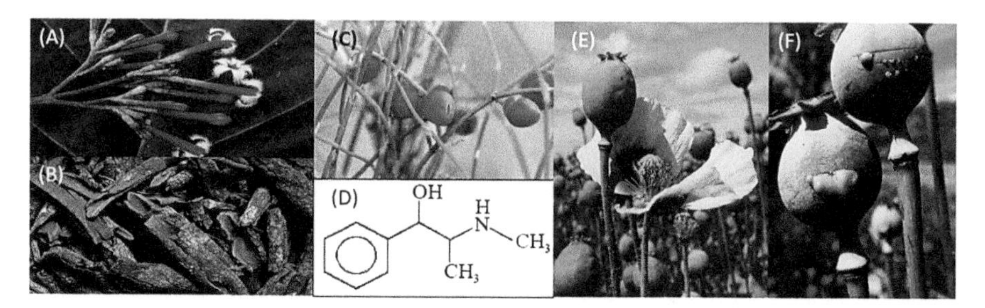

FIGURE 7.4 Cinchona plant (A) [modified from an open-access source Wikipedia [132]]; bark of cinchona (B) [modified from Britannica [22]]; ephedra (C) [modified from the NIH-national center for complementary and integrative health [91]]; structure of ephedrine (D) [modified from epharmacognosy.com [43]]; *Papaver somniferum* plant (E) [modified from herbs natural-medicinal herbs [58]]; and latex of papaver (F) [modified from Rosin Cerate [108]].

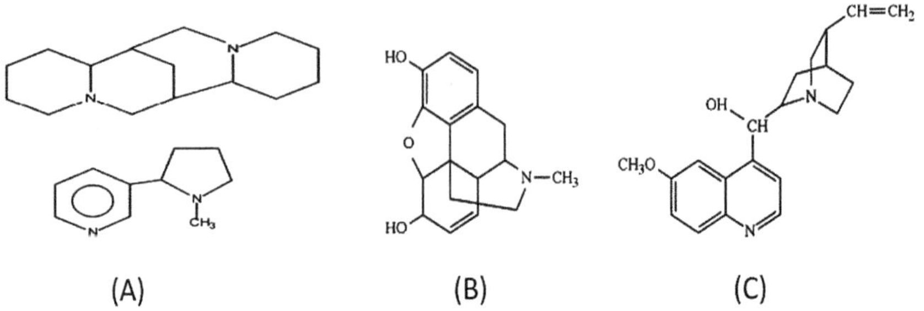

HO
O
HO
N—CH3
CH=CH2
OH
CH
N
CH3O
N
CH3
(A)
(B)
(C)

FIGURE 7.5 General structure of alkaloids (A) [modified from [13]]; structure of morphine (B) and quinine (C) [modified from Britannica [23]].

7.2.3 PHENOLICS

Phenolics are compounds with aromatic benzene rings with one or more hydroxyl groups. They are produced by plants in response to stress; and are produced through the shikimic or malonic acid pathway. Flavonoids, isoflavonoids, and anthocyanins are subcategories of phenolics [53]. These are the largest group of phytochemicals. Around 8000 dietary phenolics have been discovered, out of which over 4000 are flavonoids [124]. Plant phenolics are ubiquitous and serve as essential human dietary components [15]. Gallic acid equivalents in grapes act as anti-oxidants [32]. Figure 7.6 shows general structure of the phenolics. Figure 7.7 indicates synthesis pathways of major plant bioactive compounds.

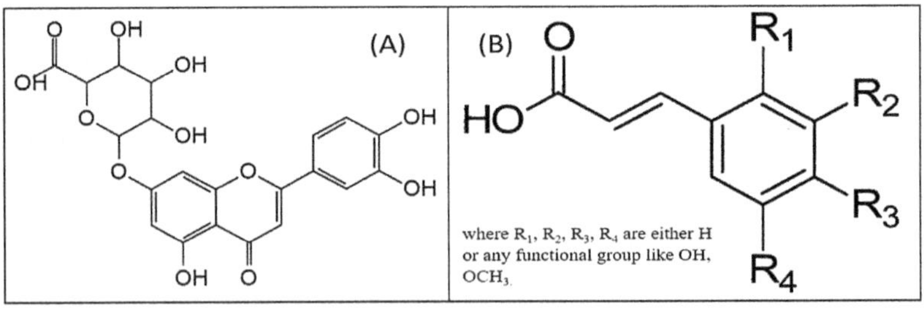

(A)
(B)
O
OH
OH
OH
OH
OH
OH
O
O
OH
O
HO
R1
R2
R3
R4
where R1, R2, R3, R4 are either H or any functional group like OH, OCH3.

FIGURE 7.6 General structure of phenolic compounds (A, B).
Source: Figure A and Figure B were modified from Refs. [13, 53], respectively.

Four major pathways for synthesis of phytobioactives are [119]:

- **Malonic Acid Pathway (for Phenolics):** Carbohydrate metabolism leads to the production of acetyl-CoA, which enters the malonic acid pathway to produce phenolics. However, the shikimic acid pathway is major biosynthesis pathway for phenolics.

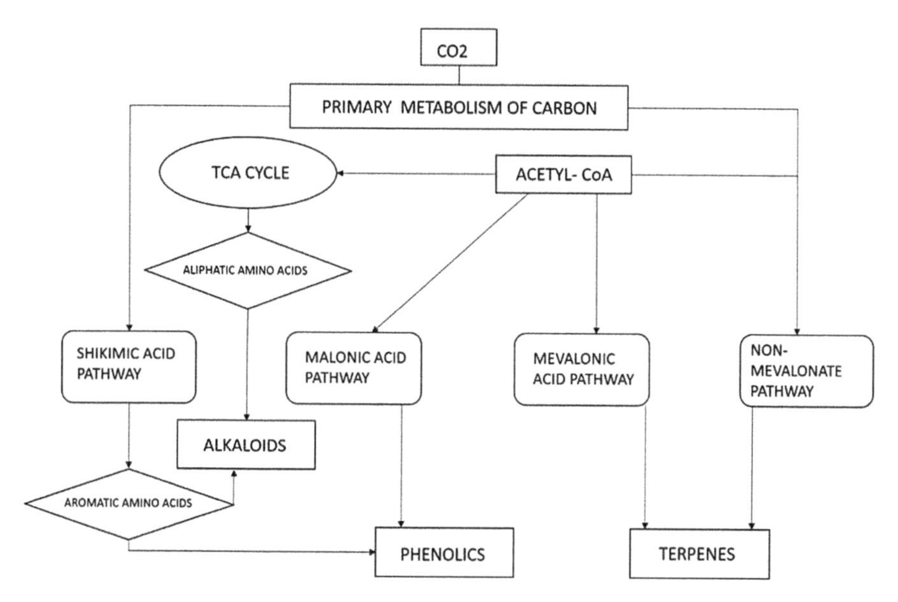

FIGURE 7.7 Flow chart for biosynthetic pathways of major plant bioactives.
Source: Modified from Ref. [13].

- **Mevalonate Pathway (for Terpenes):** A six-carbon intermediate, mevalonic acid, is formed by joining three acetyl-CoA molecules. This, in turn, is converted into isopentenyl diphosphate. This compound along with its isomer dimethylallyl diphosphate forms the building blocks of terpenoids.
- **Non-Mevalonate Pathway (for Terpenes):** Apart from mevalonic acid pathway, two carbon atoms from pyruvate combine with glyceraldehyde-3-phosphate (intermediates of glycolysis) to form an intermediate compound, which yields isopentenyl diphosphate via methylerythritol phosphate (MEP) pathway. The further synthesis of terpenes from isopentenyl diphosphate is similar to that in the mevalonate pathway.
- **Shikimic Acid Pathway (for Alkaloids and Phenolics):** It is a metabolic pathway for the biosynthesis of aromatic amino acids. This involves seven steps and is not found in animals. Carbohydrate precursors derived from the pentose phosphate pathway and glycolysis are converted to three aromatic amino acids, such as, phenylalanine,

tyrosine, and tryptophan. Phenolics and alkaloids are produced from these three amino acids.

7.3 EXTRACTION OF BIOACTIVE COMPOUNDS

Bioactive compounds from crude resources can be separated and identified only after their selective extraction from their natural sources. The extraction of bioactives from plant cells and tissues can be done by either the conventional or the non-conventional extraction procedures.

7.3.1 CONVENTIONAL METHODS

The classical or conventional extraction techniques rely on hydrophilicity and hydrophobicity of the solvent. Furthermore, the application of heat helps to enhance the efficiency of the solvent used. However, heat should be applied only when the compound under study is not heat-labile [45]. Three types of conventional extraction processes are mentioned in this section.

7.3.1.1 SOXHLET EXTRACTION

Soxhlet extraction makes use of a Soxhlet extractor (Figure 7.8), where a small quantity of dry sample is put in the thimble. Then the thimble is lodged in a distillation flask containing the solvent. Having reached an overflowing level, the solution in the thimble-holder is aspirated by a siphon, which discharges the solution into the distillation flask. Solute is retained in the distillation flask, while the solvent is returned to the plant solid bed. The process is repeated until complete [13].

7.3.1.2 MACERATION

To obtain bio-actives on a small scale, maceration involves fine grinding of plant substances to increase the surface area for proper mixing with an appropriate solvent (called menstruum) in a closed vessel. Having strained the liquid off, the left out solid residue (called marc) is pressed to recover the remaining liquid, which is later purified by filtration. Maceration requires occasional shaking for better extraction [13]. Figure 7.9 shows the steps for extraction by the maceration process.

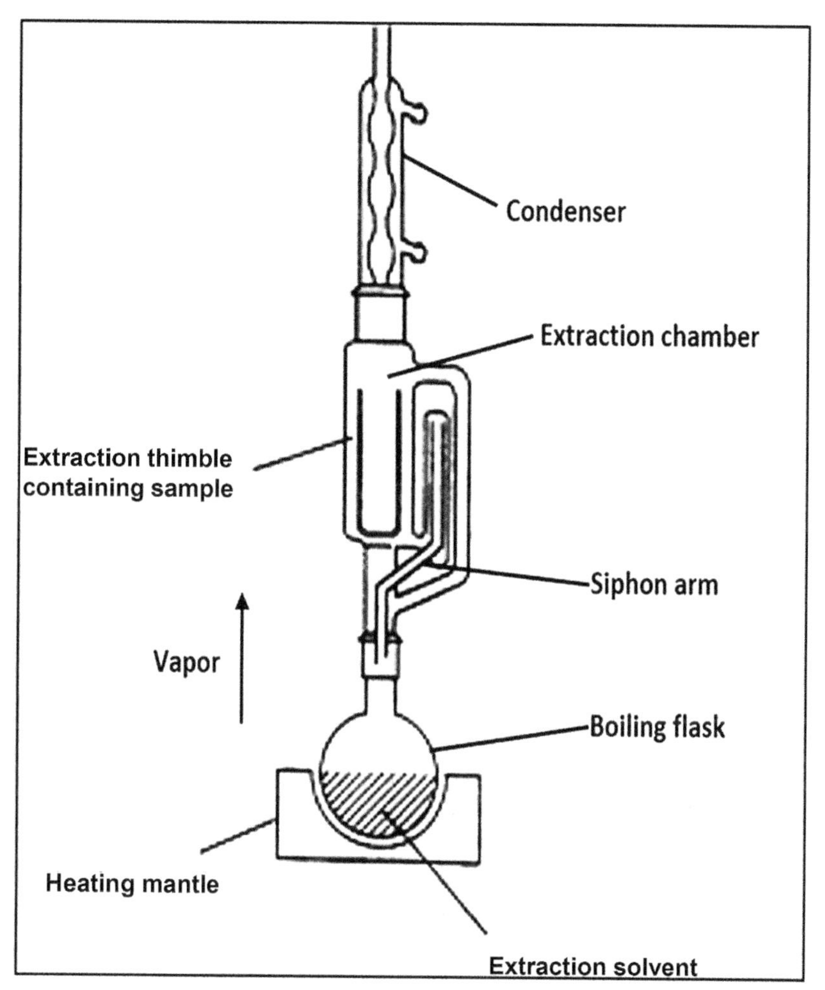

FIGURE 7.8 A typical soxhlet extractor.
Source: Modified from Ref. [28].

7.3.1.3 HYDRO DISTILLATION

Hydro-distillation does not make use of organic solvents and can be performed prior to dehydrating the plant materials. Hydro-distillation is of three types: water, water, and steam, and direct steam distillation. Plant materials are located in a static compartment. Steam and/or hot water extracts the bioactive compound out of the plant tissue [13]. Figure 7.10 represents the hydro-distillation apparatus.

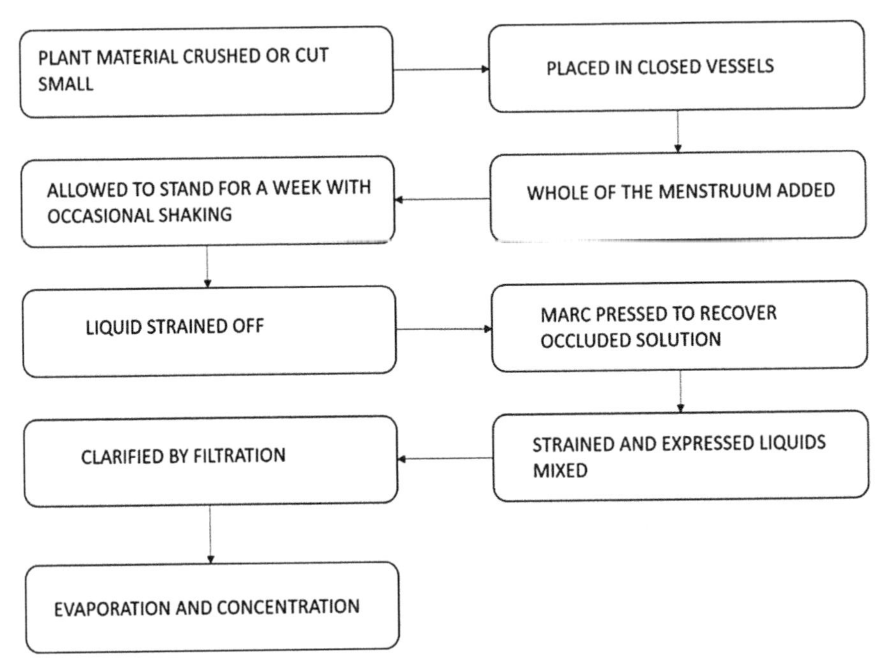

FIGURE 7.9 Sequential steps of maceration.
Source: Self-developed: Concepts were adapted from TNAU-agritech portal [46].

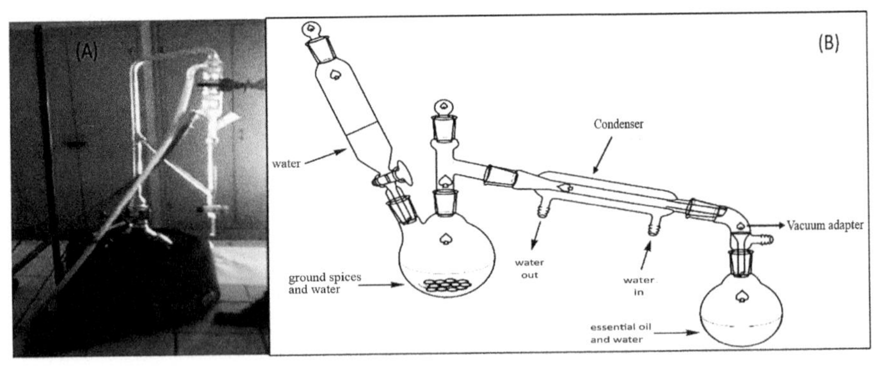

FIGURE 7.10 Small scale hydro-distillation apparatus (A) [modified from [41]; and a typical hydro-distillation apparatus (B) [modified from [44]].

7.3.2 *NON-CONVENTIONAL METHODS*

Huge extraction time, the requirement of expensive and high purity solvent, low extraction selectivity and thermal degradation of heat labile compounds restrict the usefulness of conventional extraction techniques [13]. To overcome these drawbacks, non-conventional methods with promising advantages are being preferred, and these are even eco-friendly.

7.3.2.1 ULTRASOUND ASSISTED EXTRACTION (UAE)

Ultrasound waves are sound waves with frequencies greater than 20 kHz [125], and pass through a medium producing compression and rarefaction to result in a phenomenon termed as cavitation. Cavitation includes generation, expansion, and crashing of bubbles. A huge quantity of energy is generated by channeling the kinetic energy of flux into heating of the contents of the bubble. UAE is based on this principle. Ultrasound energy aids the organic and inorganic compounds of a plant sample to leach out from the plant matrix. Degree of milling, size of the particle, moisture content of the sample, choice of the solvent, temperature, frequency, pressure, and duration of sonication are important factors governing the efficiency of UAE extraction [13].

7.3.2.2 PULSED ELECTRIC FIELD EXTRACTION (PEF)

PEF is based on the principle of destroying cell membrane structure to facilitate extraction. Electric potential passes through the membrane of living cells when they are suspended in an electric field. The electric potential partitions the charged membrane molecules based on their dipole nature. When the potential across the membrane surpasses a certain critical value, repulsion occurs between charged molecules. This repulsive force generates pores in the feeble areas of the membrane and drastically increases the permeability. Usually, exponentially decaying pulses are used in treating plant samples during PEF extraction. Efficacy of PEF treatment depends on the field strength, specific energy input, and number of pulses, etc., [13].

7.3.2.3 ENZYME ASSISTED EXTRACTION (EAE)

Many bioactives are present in the cell walls of plant cells (trapped in the polysaccharide network) and cannot be completely extracted using conventional extraction processes. Pre-treatment with specific enzymes (such as: cellulase, pectinase, and α-amylase, which break down cellulose, pectin, and starch, respectively) facilitate the release of bioactives by degrading the cell-wall components. EAE can be of two types: (1) Enzyme-assisted aqueous extraction (EAAE), used mainly for the extraction of oil from the seeds; and (2) Enzyme-assisted cold pressing (EACP), used to hydrolyze the seed cell wall [13].

7.3.2.4 MICROWAVE ASSISTED EXTRACTION (MAE)

Microwaves are electromagnetic waves with frequencies ranging from 300 MHz to 300 GHz and wavelengths ranging from 1 mm to 1 m [133]. Many plant materials can absorb part of the electromagnetic energy of microwaves and can convert it to heat. Like all electromagnetic waves, microwaves are also composed of mutually perpendicular electric and magnetic fields. The electric component changes 4.9×10^9 time $second^{-1}$. This produces a jumbled movement of polar molecules irradiated by microwaves, thus generating heat. MAE involves different steps: (a) separation of solutes from sample; (b) diffusion of solvent across sample matrix; and (c) discharge of solutes from sample matrix to solvents. MAE reduces the time and quantity of solvent required in the extraction process [13].

7.3.2.5 SUPERCRITICAL FLUID EXTRACTION (SFE)

SFE involves the separation of the extractant from the matrix using supercritical fluids as the solvent. A supercritical fluid is a substance at a temperature and pressure beyond its critical point, where distinctive liquid and gas phases do not exist. Even by modifying the temperature and pressure, a supercritical fluid cannot be liquefied [135]. They possess viscosity, surface tension, gas-like diffusion properties, liquid-like density, and solvation power. These properties can be altered by varying the temperature and pressure, thereby making supercritical fluids useful in the selective extraction. Carbon-dioxide, often mixed with methanol/ ethanol/dichloromethane, is the supercritical fluid (solvent) of choice for extraction. The critical temperature (T_c) of carbon-dioxide is 31°C with a critical pressure (P_c) of 74 bars. These factors allow carbon-dioxide to be used as a solvent under room temperature and moderate pressure. However, the low polarity of carbon-dioxide makes it suitable for the extraction of non-polar substances mainly. To increase the polarity of carbon-dioxide, chemical modifiers (such as: ethanol or methanol) are mixed. The efficiency of SFE is affected by particle size, temperature, pressure, moisture content of the sample, flow rate of carbon-dioxide, etc., [13].

7.3.2.6 PRESSURIZED LIQUID EXTRACTION (PLE)

PLE technique employs high pressure to keep the solvent in the liquid-form beyond their normal boiling point to facilitate the extraction of biomolecules from the crude extract. High temperature and pressure increase solubility of the analytes by decreasing surface tension and viscosity of the solvent. PLE is known by several names today, such as: High-pressure solvent extraction (HPSE), Accelerated fluid extraction (AFE), Enhanced solvent extraction (ESE) and Pressurized fluid extraction [13].

7.4 ROLE OF PHYTOBIOACTIVES

Phytobioactives (bioactive compounds) play an important role in keeping the human body healthy. They act as signaling components in various metabolic pathways and also epigenetically influence the transcription of many genes [51]. Phytobioactives mostly function by acting as active antioxidants [143].

7.4.1 MITIGATION OF OXIDATIVE STRESS

The term reactive oxygen species (ROS) describes a set of reactive molecules and free radicals, derived from molecular oxygen. These include superoxides, peroxides, alpha-oxygen, hydroxyl radical and singlet oxygen. The general chemical mechanism of ROS formation includes:

- Molecular oxygen (O_2) is reduced to superoxide anion (O_2^-).

$$O_2 + e^- \rightarrow O_2^-$$

- Superoxide anion acts as the precursor of other ROS species. Superoxide anion is disproportionated to produce hydrogen peroxide (H_2O_2).

$$2H^+ + 2\cdot O_2^- \rightarrow H_2O_2 + O_2$$

- Hydrogen peroxide is partially reduced to hydroxyl radical (OH).

$$H_2O_2 + e^- \rightarrow HO^- + .OH$$

- Singlet oxygen ($^1O_2$) is produced due to photosensitization.
- Alpha-oxygen is generated through an oxygen atom abstraction from nitrous oxide by alpha-Fe catalysts.

ROS can be generated both endogenously and exogenously. Endogenous ROS are produced mainly as by-products during oxidative phosphorylation.

Mitochondria, the powerhouses of the cell, generate ATP by the transport of protons through an electron transport chain across the inner mitochondrial membrane. The electrons are passed through series of proteins by redox reactions across the chain with oxygen as the terminal electron acceptor. Oxygen is usually reduced to water; however, sometimes a fraction of it is partially reduced to superoxide radical, which in turn continues to generate the other ROS [19].

Exogenous ROS are produced due to tobacco, smoke, ionizing radiations, xenobiotics, and other pollutants [102]. Ionizing radiation can react with water in a process termed as radiolysis to generate harmful intermediates [106]. The probability of radiolysis is greater under the presence of ionizing radiation because almost 70% of the human body is composed of water.

ROS hampers cognition, stimulates Deoxyribonucleotide acid (DNA) damage and aging, promotes Alzheimer's disease [72, 81, 142]. Since the generation of ROS is inevitable in aerobically respiring cells, therefore their detoxification is of prime importance for the survival of cells and the entire organism thereof. To counteract the harmful effects of ROS, the human body has developed an endogenous complex defense system of anti-oxidants. This defense mechanism comprises both enzymatic and non-enzymatic constituents, which work at several extents: arrest formation of ROS, erase them prior to causing damage, mend the damage done and eliminate the damaged molecules. If there is any disparity in the number of free radicals and antioxidants in the body, so that the number of ROS exceeds, then the phenomenon is known as oxidative stress.

Oxidative stress can affect either at the molecular level or systemic level. It breaks down the functionality of the body. Mutations in the genes, coding for components of this anti-oxidant defense system, lead to harmful consequences; sometimes even death. Also, the DNA sequence encoding genes vary with individuals. These small differences in sequences lead to the altered capability of fighting oxidative stress within different individuals. A knowledgeable diet can help to induce more antioxidant production. Apart from just combating ROS, the defense system also detoxifies xenobiotics to readily excretable forms. One of the major health benefits of phytobioactives is to combat oxidative stress. The major cytoprotective compounds of plants engaged in this mechanism include the glucosinolates and other sulfur-containing compounds. Glucosinolates induce the production of anti-oxidant enzymes, such as, Nicotinamide adenine dinucleotide phosphate (NADPH), quinone reductase (NQO1) and heme oxygenase 1 (HO1) [16].

Broccoli, onion, spinach, cabbage, and garlic are examples of plant sources rich in cytoprotective compounds. Glutathione secreted from the alveolar epithelial cells helps counteract hypochlorous acid (HOCl), which is released by neutrophils. It squeezes from the blood into the air spaces in persons, who practice smoking. Glutathione peroxidase catalyzes the reduction of hydrogen peroxide to water. Glutathione peroxidase and catalase prevent hydroxyl radical attack on DNA. Vitamin-E is a peroxyl radical scavenger to maintain the integrity of polyunsaturated fatty acids (PUFAs) in cell membranes by preventing their oxidation. A plant-based diet, with proper consumption of vegetables and fruits, helps to combat oxidative stress-induced diseases. It has been found that diets rich in antioxidants work better than antioxidant supplements [11].

Moreover, antioxidants cannot be replaced due to several reasons. For example, the effects due to the deficiency of Vitamin E cannot be mitigated by supplementation of Vitamin C in diet and vice-versa, although both are antioxidants with similar functions. This is because Vitamin C and Vitamin E have different subcellular locations. Moreover, their combination works as a better anti-oxidant than either of them alone [110].

Most of the bioactives obtained from plants are secondary metabolites. Phenolics are the most copious antioxidants in the diet. Phenolics can be subdivided into groups, such as, phenolic acids, stilbenes, flavonoids, and lignans. All of them have at least one phenol ring. Phenols are mainly present in beverages (tea, coffee) and fruits, less commonly in vegetables and cereals [53]. Table 7.1 provides the dietary sources of some phenolics.

TABLE 7.1 Dietary Sources of Some Common Phenolic Compounds and Their Sources

Phenolic Class	Subclass	Dietary Source
Flavonoids	Flavanols	Tea, onion, broccoli
	Flavones	Cereals, celery
	Flavanones	Citrus fruits
	Flavanols	Chocolate, red wine, apple
	Anthocyanidins	Berries, red wine
	Isoflavones	Soy products
Phenolic acids	Hydroxybenzoic acid	Onion, berries
	Hydroxycinnamic acid	Kiwi, plum
Lignans	—	Cereals, garlic
Stilbenes	—	Wine, blueberries

Source: Self-developed from information by Manach et al. [75].

Another important class of antioxidants includes carotenoids. Around 60 carotenoids are consumed by humans through a plant-based diet. Some important carotenoids and their dietary sources are listed in Table 7.2. Apart from these, other plant-derived antioxidants include allyl sulfides, curcumin, indoles, isothiocyanates, lignans, monoterpenes, phytic acid, saponins, etc.

Resveratrol stimulates the longevity assurance genes, the Sirtuins (*SirTs*) and extends life span using a protein called SIRT1, which is activated by NAD^+ provided by Pre-B cell colony-enhancing factor (PBEF). SIRT1 interacts with Forkhead box O-1 (FOXO1), which is an anti-aging transcription factor involved in various protective cellular activities, such as: DNA repair, the arrest of the cell cycle, induction of apoptosis and anti-oxidative stress [35].

TABLE 7.2 Examples of Carotenoids and Their Sources

Carotenoids	Dietary Sources	Carotenoids	Dietary Sources
Lutein	Spinach, broccoli	α-carotene	Carrot, pumpkin
Lycopene	Tomato, watermelon	β-carotene	Spinach, parsley
Zeaxanthin	Maize, spinach	B-cryptoxanthin	Avocado, papaya

Source: Self-developed from information by Higdon et al. [59].

7.4.2 HEALTH PROMOTING EFFECTS IN SOME DISEASES

7.4.2.1 CARDIOVASCULAR DISEASES

According to WHO [27], "*CVDs are disorders of the heart and blood vessels and include coronary heart disease, cerebrovascular disease, rheumatic heart disease, congenital heart disease, deep vein thrombosis, and pulmonary embolism and many other diseases.*" The consequences of coronary heart disease can range from angina to heart failure [92]. A stroke occurs when the supply of blood is cut off to a part of the brain. In a mini-stroke or transient ischemic attack, the blood flow is just transiently hindered to the brain. Stroke can cause damage to the brain and even death [27].

Generally, strokes, and heart attacks are severe incidents, which are caused due to some blockage of blood flow to the brain or heart (atherosclerosis) [12]. These blockages are caused by fatty deposits (mainly due to cholesterol) on the inner linings of blood vessels. The prime risk factors for CVDs are an unhealthy diet, lack of physical exercise, excessive use of tobacco and alcohol. However, 90% of CVDs are curable [131]. Carotid intima-media thickness (IMT) is closely associated with increased risks of atherosclerosis and coronary artery disease [118]. A daily intake of around 50 mL of pomegranate juice (containing high amounts of vitamin C, anthocyanins, and tannins) for 3 years has shown to reduce carotid IMT, blood pressure (BP) and low-density lipoprotein (LDL) oxidation in 19 patients with coronary artery disease [118].

Ischemic heart disease mortality, cardiovascular disease mortality and stroke mortality have been reported to be greatly reduced [68] due to intake of plant-based bioactives. Increased consumption of fruits and vegetables daily for a period of 3 months has resulted in reduced systolic and diastolic BP in hypertensive patients [117, 118].

Several polyphenols (such as resveratrol, phytoalexin, and curcumin) help to prevent CVDs. Resveratrol induces nitric oxide (NO) synthesis, which leads to ischemic reperfusion (restoring blood flow after a heart attack or stroke) of heart and brain. NO is synthesized by 3 nitric oxide synthase (NOS) isozymes, namely, endothelial NOS (eNOS), inducible NOS (iNOS), and neuronal NOS (nNOS) [134]. These isozymes convert L-arginine to L-citrulline, thereby generating NO free radical. Resveratrol also safeguards perfused hearts through increased iNOS generation and up regulation of the iNOS-vascular endothelial growth factor (VEGF)-KDR-eNOS signaling pathway [36].

Platelets are activated during thrombus formation. However, their aggregation can lead to vascular occlusion. Resveratrol prevents the aggregation of platelets, thereby preventing atherosclerosis. It also inhibits platelet activation by collagen, along with immobilization of Ca^{++} ions, the formation of thromboxane A2, breakdown of phosphoinositide and activation of Protein Kinase C (PKC) [35]. Resveratrol is present in red wines. Moderate drinkers suffer from less CVDs than teetotalers [70].

An anti-cancer drug named Adriamycin is potent to increase cardio-toxic side effects. Adriamycin has been demonstrated to cause an increase in heart rates, the elevation of the ST segment, reduction in activity and quantities of myocardial glutathione and glutathione peroxidase and increased levels of lipid peroxide in serum [126]. Curcumin prevents lipid peroxidation by free radical scavenging and also increases glutathione content. It also acts as a histone acetyltransferase (HATs) inhibitor and prevents cardiac injury by increasing levels of the protein, manganese superoxide dismutase [2]. Stabilization of the membranes of ischemic muscle cells and the lysosomal membranes prevents this enzymatic hydrolysis. Curcumin has been found to stabilize such membranes. It also promotes the release of endogenous corticoids, which stabilize lysosomal membranes [136].

Cardiac hypertrophy is the abnormal enlargement or thickening of the heart muscles, due to stress or myocardial infarction. Hypertrophy stimulates a number of signaling pathways that alter the gene expression pattern in cardiac muscle cells. Monocyte enhancing factor-2, AP-1, serum response factor, and GATA-4 are transcription factors mediating these changes. The p300, an adenovirus E1 A-associated protein, acts as a co-activator of these transcription factors. The p300 also acts as a HAT, thus promoting an active chromatin configuration. Over expression of p300 leads to cardiac hypertrophy. Curcumin has been reported to act as an inhibitor of p300 [86].

Curcumin also down regulates nuclear factor-κB (NF-κB), which in turn down regulates the expression of certain inflammatory factors, such as: tumor necrotic factor-α (TNF-α), interleukin-1 and interleukin-6 (IL-1, IL-6). It also prevents the mitogen-activated protein kinase (MAPK) pathway. These anti-inflammatory properties of curcumin play protective role against acute coronary heart syndrome, atherosclerosis, and atrial arrhythmia [29].

Heme oxygenase-1 (HO-1) is present in mammalian cells and breaks down heme to iron, carbon monoxide and biliverdin. It also inhibits growth of vascular smooth muscles and decreases arteriosclerosis. Curcumin increases HO-1 expression by activating Nrf2-dependent anti-oxidant response element in various cardiovascular cells, such as: vascular smooth muscle cells, vascular endothelial cells, and aortic smooth muscle cells. Curcumin, in a dose-dependent manner, also maintains calcium homeostasis in cardiac muscles by regulating Ca^{2+} ATPases of sarcoplasmic reticulum in heart cells [140].

7.4.2.2 CANCER

Cancer is caused due to multiple factors and accumulated damage to genes. Cancer is a result of mutations in genes that are crucial for regulating cell growth and differentiation. Thus, a cancer cell is an altered self-cell. There are two types of cancer genes present in humans: oncogenes and tumor suppressor genes. Oncogenes are generated from proto-oncogenes. Proto-oncogenes help cells grow normally; and have potential to cause cancer. The mutation of proto-oncogenes makes random cells divide, by bypassing the cell cycle checkpoints. Tumor suppressor genes suppress the formation of tumors under normal circumstances. Mutation (s) lead (s) to cessation of their activities thus increasing the risk of cancer [71].

Homeostasis is maintained by molecular signaling pathways, which in turn are influenced by exogenous factors and diet. The effect of bioactives on DNA expression, or epigenomics, regulates the spatial and temporal expression of specific sets of genes. Epigenetic modifications can be at the nucleotide level. The resulting changes induce modifications in chromatin structure or nucleotide sequence. Transcription of DNA at or around these modified areas is altered. These epigenetic modifications are also called 'marks' in the genome, which are sometimes passed down to the next generation [51].

The first type of mark, termed as histone modification, indirectly affects DNA transcription by changing the packaging or wrapping of DNA around the histone proteins. The second type of mark, characterized as DNA methylation, directly affects DNA transcription by modifying the DNA bases. There may be instances, where a phytobioactive can leave an epigenetic mark (say a DNA methylation event) on the genome of an individual. This mark may result in altered gene expression [51]. Some examples of such epigenetic modifications leading to altered gene expression are given in this section.

Folate, one of the vitamins B, is involved in the synthesis, repair, and methylation of DNA. In the human body, folate is converted to tetrahydrofolate, which remethylates homocysteine to methionine. This chemical reaction is important because methionine is the precursor of S-adenosyl-L-methionine (SAM), which acts as the primary methyl group donor in most methylation reactions. Folate deficiency leads to changes in DNA methylation patterns in proto-oncogenes, i.e., c-myc, c-Ha-ras, and c-fos, resulting in digestive neoplastic lesions or colon cancer [39].

The vitamin A, having been metabolized to its active form (retinoic acid), is translocated to the nucleus, where it binds to certain nuclear retinoic acid receptors (RARs). This complex then binds to corresponding response elements and alters the transcription of target genes. The downstream proteins regulate oncogenic transformation, DNA repair, DNA replication during S-phase and even changes in DNA methylation patterns that reduce certain types of cancer [18, 69, 101, 139].

Likewise, the active form of Vitamin D3 (calcitriol) exhibits a similar biological mechanism of action as vitamin A. The downstream target gene products catalyze a dephosphorylation event that is involved in inhibition of the AKT signaling pathway. It also exhibits a weak phosphatase activity involved in tumor suppression and changes in DNA methylation patterns preventing colon and prostate cancer [17, 110, 111].

Green tea polyphenols, curcumin, flavonoids, lupeol, and resveratrol have antineoplastic properties within their physiological limits [89].

Resveratrol induces apoptosis in cancer cells by stimulating loss of potential of mitochondrial membrane, thereby releasing cytochrome c, Smac/Diablo and caspases. It also activates pro-apoptotic Bax, p21 waf, and p53 while repressing anti-apoptotic Bcl-2, Bcl-xL, TNF receptor-associated factor, and cyclin D1. In case of breast cancer, resveratrol has been found to hinder anti-apoptotic phosphatidylinositol-3'-kinase (PI3K)/ Akt signaling pathway [35].

Multi-ethnic population-based cohort studies have shown that high intake of fruits and vegetables can significantly reduce risks of prostate cancer, breast cancer, rectal, and colon cancer [98].

One of the important signaling pathways leading to cancer is the Wnt/β-catenin pathway. Activation of this pathway leads to an increase in cytoplasmic β-catenin levels and regulates tumor progression [115]. Flavonoids inhibit this pathway by upregulating the expression of GSK-3β that phosphorylates β-catenin [7]. Curcumin hinders the transcription of Wnt target genes, such as: c-myc, c-jun, c-fos, and iNOS; thus prohibiting proliferation and inducing apoptosis in cancer cells [20].

7.4.2.3 DIABETES

When the blood glucose levels rise higher than normal, the phenomenon is termed as diabetes. Complex carbohydrates in food are digested into simple sugars, which are absorbed into the bloodstream. Insulin is a hormone, secreted by the endocrine part of the pancreas (i.e., Islets of Langerhans) and it helps in the uptake of blood glucose by the cells through many glucose transporters. When these endocrine cells do not produce enough insulin or the target cells become insulin irresponsive, the blood glucose levels tend to increase. Diabetes can affect anyone of any age group or gender.

Hyperglycemia causes epigenetic modifications of inflammatory genes. One of them is Nuclear Factor-κB (NF-κB), which interacts with HATs, thus leading to hyper-acetylation of target genes, such as: tumor necrosis factor (TNF) and cyclooxygenase-2 promoter. Also, prolonged hyperglycemia leaves epigenetic marks in pro-inflammatory promoters, thereby leading to long-term expression of these genes in spite of diabetic control [51].

Genistein, resveratrol, anthocyanins, and quercetin lead to glucose-stimulated insulin secretion in pancreatic β-cells. They also make the β-cells to proliferate, thereby increasing insulin production [98].

7.4.2.4 OBESITY

A person is called obese when his body mass index (BMI) exceeds 30 kgm^{-2}. It results when anabolism exceeds catabolism, i.e., when the intake of calories is greater than that burned by regular physical exercise [94]. Garnering excess body fat often leads to serious health complications, such as: CVDs, diabetes, sleep disorders, cancer, etc., [95]. Modern-day unhealthy diet and sedentary lifestyles lead to obesity [64, 97].

The etiology of obesity is complex [64], making it a serious health issue around the world. The adipocytes (chiefly), fibroblasts, and endothelial cells are constituents of adipose tissue that are linked by collagen fibers. Due to adipocyte hyperplasia (increase in count) or hypertrophy (increase in size), mainly in the preadipocytes, the adipose tissue expands resulting in obesity. White adipose tissue, apart from storing fat, also has endocrine functions [3].

It secretes hormone angiotensin II, which is involved in regulating BP and body fluid homeostasis. Besides, it secretes pro-inflammatory cytokines (termed as adipokines), interleukins (IL-1, IL-6, IL-8), tumor necrosis factor (TNF-α), monocyte chemo attractant protein (MCP-1), etc. In non-obese individuals, the adipocytes are of usual size, secrete anti-inflammatory cytokines and are sensitive to insulin. On the other hand, the adipocytes of obese individuals are larger than normal, with infiltration by many pro-inflammatory macrophages. They also secrete pro-inflammatory cytokines, which lead to insulin resistance [66, 74].

Various therapeutic drugs have been developed to prevent obesity. However, most of them have adverse side effects on the patients [67, 116]. Shift to a healthy diet and rigorous physical exercise are best methods to prevent obesity. Many of the phytobioactives help in mitigating obesity. Some of them are discussed in this section.

Turmerone and curcumin have anti-obesity properties. In adipocytes, curcumin enhances fatty acid oxidation and adenosine monophosphate-activated protein kinase (AMPK) activity [40]. It also decreases lipogenesis in the adipocytes by suppressing fatty acid synthase [144]. Curcumin has also been found to inhibit adipocyte differentiation by repressing MAP kinase phosphorylation. Even in the differentiated fat cells, it prevents further adipose tissue expansion via the Wnt/β-catenin pathway [4]. Adipose tissue inflammation, a hallmark of obesity, is reduced by curcumin by down regulating the secretion of pro-inflammatory cytokines [62].

The thiacremonone in garlic down-regulates the transcription factor peroxisome proliferator-activated receptor gamma (PPAR-γ), which prevents differentiation of adipocytes. Alliin in garlic ameliorates adipose tissue inflammation by suppressing ERK1/2 phosphorylation, which in turn decreases secretion of the pro-inflammatory cytokines [103]. Grape anthocyanins (like cyanidin-3-glucoside) repress the expression of inflammatory cytokines in adipocytes and also promote insulin sensitivity by induction of Glucose transporter 4 (GLUT-4) transporters [113].

Resveratrol decreases the expression of retinol binding protein 4 (RBP4) and resist in gene in fat cells, which reduce pro-inflammatory cytokine secretion, adipocyte differentiation and insulin resistance [83].

The α-linoleic acid (an ω-3 polyunsaturated fatty acid) reduces endoplasmic reticulum stress-associated genes (such as: X-box binding protein 1 or XBP1) expression in subcutaneous adipose tissue. The ω-3 fatty acids also enhance the biogenesis of mitochondria and β-oxidation of fatty acids, thereby reducing adiposity of the body [48, 63, 114]. They serve as ligands for certain members of the PPAR family of transcription factors (such as: PPAR-α and PPAR-γ), which regulate lipid metabolism [48, 49, 54]. They also suppress the secretion of inflammatory factors by inhibiting the activity of cyclooxygenase (COX) and down-regulate toll-like receptors [14]. Zingerone in ginger inhibits obesity-related adipose tissue inflammation [137].

7.4.2.5 HYPERPIGMENTATION

Skin pigmentation is caused due to a pigment called melanin. Although the term melanin is used loosely, yet it collectively refers to a group of pigments that are responsible for contributing skin color in humans, as also in other vertebrates. Melanin is synthesized by cells called melanocytes, by a multistep process (termed as melanogenesis). Melanocytes produce two types of melanin: eumelanin (black or brown) and pheomelanin (yellow or pink or red). The particular type of melanin expressed by the melanocytes determines the color of the skin. Tyrosinase-Related Protein-1 (TRP-1) and Tyrosinase-Related Protein-2 (TRP-2) regulate the quality of melanin. Although melanin serves to protect against the harmful ultra-violet B radiations, yet hyper-pigmentation is not desirable by most people especially in the Asian subcontinent. Hyper-pigmentation usually does not have many harmful effects on the body. However, the phenomenon is very unwanted. This has made the quest for efficient skin lightening products equally competitive and demanding [6]. In melanogenesis, the aromatic amino acid tyrosine is oxidized and then polymerized in organelles (called melanosomes). Pathway of melanin biosynthesis is shown in Figure 7.11.

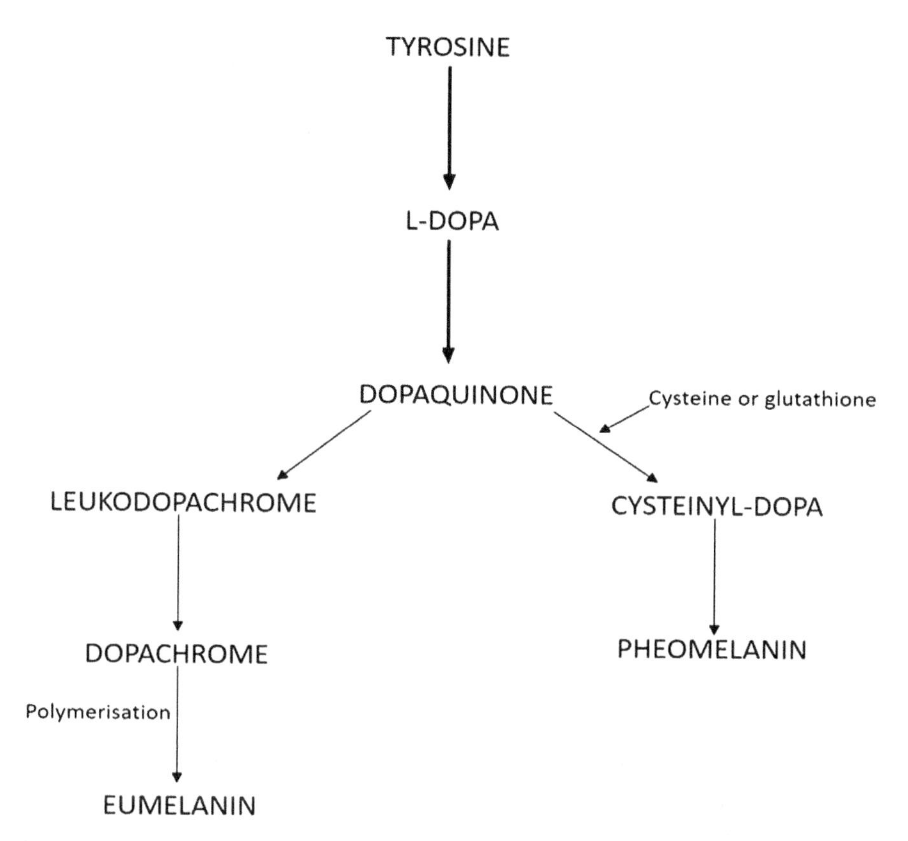

FIGURE 7.11 Melanin biosynthesis pathways.
Source: Modified from Ref. [120].

In Figure 7.11, the first two steps of melanin biosynthesis (i.e., tyrosine to L-3,4-dihydroxyphenylalanine (L-DOPA) and L-DOPA to Dopaquinone) are catalyzed by the enzyme tyrosinase (TYR). The inhibition of dermal pigmentation (or ceruloderma) involves hindering the melanin synthesis pathway or its products somehow. Many phytobioactives have been identified, which are involved in similar actions and hence act as significant skin lightening agents. These bioactives can be both be consumed or applied topically. Hyper-pigmentation can be obstructed at different levels:

- **By Inhibiting Expression of Tyrosinase:** The first mode of action to prevent excessive melanin synthesis is the modulation of transcription of gene-encoding tyrosinase enzyme. Many bioactives have been on the hunt to accomplish this. *Angelica daburica* has two active compounds, namely, imperatorin, and isoimperatorin. These can decrease the levels of messenger ribonucleic acids (m-RNAs) of the tyrosinase gene [30].

- **By Inhibiting the Activity of Tyrosinase:** Even if tyrosinase is produced, hampering its activity can still interfere with the melanin biosynthetic pathway. Green tea is a competitive inhibitor of the enzyme tyrosinase. The gallocatechin moiety of green tea flavonoids has been reported to be responsible for this action. Aloesin, a bioactive compound of *Aloe vera,* acts as a non-competitive inhibitor of enzyme tyrosinase and prevents DOPA from converting into melanin [99]. Glabridin from *Glycirrhiza glabra* [50, 60]and hesperidin from citrus fruits [77] also exhibit anti-tyrosinase activities in melanocytes.
- **By Inhibiting Dispersion or Translocation of Melanin:** After synthesis, melanin needs to be shifted from melanocytes to keratinocytes. Many studies have been conducted to identify the regulatory factors, involved in the movement of melanosomes during their transfer from melanocytes to keratinocytes. Niacin (or Vitamin B3) has a biologically active form, niacin amide, which is found in root vegetables. It inhibits translocation of melanosomes to keratinocytes from melanocytes [55, 84].
- **By Controlling the Degradation of Melanin and its Subsequent Removal:** If all of the above checkpoints fail to control pigmentation, then the excess melanin can be degraded and subsequently removed. Liquirtin from licorice can perform desquamation by shortening the cell cycle of melanized keratinocytes and subsequently removing them [141].

However, the beneficial effects of most phytobioactives are dosage-dependent, but the relationship is not linear. There is a certain range for phytobioactives, within which if taken, can prove to be effective. Beyond this range, they have negative impacts. The doses vary depending on several factors, such as: age, gender, pregnancy status, health, and pathological condition. For example, high consumption (ten times more than normal) of non-alcoholic beverages (such as: coffee) can increase methylxanthine concentrations in the body. This may cause genetic damage in sperm cells and also reduced sperm count in males [10]. Fetuses are also affected in uterus due to high methylxanthine consumption by the pregnant mother [9].

Of recent, over-dosage of bio-actives has been reported to epigenetically modify DNA, leading to alteration of gene expression and hence certain diseases. Obesity genes undergo reduced methylation, triggering obesity.

Promoters of genes are involved in energy balance. Overdosage of bioactives alters the balance between energy gained and spent by a body, shifting the flux towards obesity. Similar changes at the genetic and epigenetic level, due to biological transcendence of phytobioactives have also been found to be involved with CVDs, cancer, etc., [33, 34].

Overdoses of certain vitamins also impact negatively. Excessive intake of Vitamin C causes digestive distress [24]. Vitamin B6, if taken in large amounts (i.e., beyond the tolerable upper limit), can result in nerve damage [128]. A large intake of Vitamin D can lead to an increase in blood calcium levels [76] and kidney failure [129]. Vitamin A in excess reduces bone mineral density, which increases the risk of fractures [107]. Vitamin A toxicity can be acute or chronic depending on the duration of intake [56].

7.5 STRATEGIES FOR DELIVERING BIOACTIVE COMPOUNDS FOR HEALTH THERAPY

The best way to intake a variety of bio-actives is to eat sufficient quantities of nutritious food. Not all types of natural plant-based products are available in substantial amounts at all human habitations. Also, many people are sensitive to certain plant-based proteins [21]. Therefore, whole plant-based food products are not suitable for them. Hence, plant-based bio-actives need to be made available to them in purified form, free from plant protein contamination.

A potential bioactive compound, after its identification, must be evaluated first for its efficacy and safety of administration. Once established, the ease of its availability, synthesis, stability, and mode of transport require to be investigated. The cost of the ultimate commercial product is also a matter of concern. Therefore, novel procedures are always on the hunt to make the best out of what is available so that it is accessible to common people.

7.5.1 MICROBIAL DELIVERY

One such technique is delivery through microbes. We ingest a lot of bio-actives through food. Yet, the limitations of the human digestive system do not allow us to extract all components from the meal. Gut microbiota, especially the lactic acid bacteria (LABs), secrete proteolytic enzymes that completely digest food matrices. LABs are added to the processed food to ensure the extraction of encrypted bio-actives in food. This technology mimics the natural system of delivery of nutriments from food into the gut [52].

7.5.2 NANOTECH DELIVERY

Development in nano-technology has provided benefits of delivery of bioactives in nano-emulsions through nano-encapsulation. Quercetin, an anticancer agent, had its medical applications limited due to low aqueous solubility, rapid metabolism and clearance from the body (very little biological half-time), poor stability due to oral administration. These days, quercetin nano-formulations are being developed and explored to overcome most of these shortcomings [79].

7.6 NUTRIGENOMICS

Nutrigenomics refers to "*the science how bio-actives in foods and supplements alter the molecular expression and/or structure of an individual's genetic makeup*" [104]. Although the beneficial effects of food to the human body were known since early times, yet the scientific quest in finding out the actual interactions and relationship between active ingredients in food and the genome, started only with the discovery and isolation of DNA. This was followed by the cracking of the DNA structure and subsequent disclosure of the entire human genome. Since nutrigenomics deals with the diet-genome interaction, it brings together genomics, nutritional science, molecular medicine, computational biology, and bioinformatics at one platform to answer the complex etiology of chronic diseases and to cure them [104].

"*The new science of nutrigenomics teaches us what specific foods tell your genes. What you eat, directly determines the genetic messages your body receives. These messages, in turn, control all the molecules that constitute your metabolism: the molecules that tell your body to burn calories or store them. If you can learn the language of your genes and control the messages and instructions they give your body and your metabolism, you can radically alter how food interacts with your body, lose weight, and optimize your health,*" Mark Hyman [2006: 88].

Nutrigenomics involves interactions between active food ingredients and the genes that have been inherited. Certain diseases, which were previously broadly termed as "the inborn errors of metabolism" have traditionally always been treated by modifying or manipulating the patient's diet. Phenyl-ketonuria is an example of this. This disease is caused by mutation (a single nucleotide polymorphism) in the phenylalanine hydroxylase (PAH) gene that codes for phenylalanine hydroxylase. This enzyme catalyzes the conversion of phenylalanine to tyrosine. The mutation leads to build-up of phenylalanine in the body, leading to harmful effects on hyper-accumulation. A low phenylalanine diet is the only recommendation to keep good health [57].

Another example of nutrigenomics is provided by phenomenon of lactose intolerance. An enzyme called lactase breaks down lactose (the milk sugar: present in milk and milk products). Usually, the gene encoding for the enzyme lactase is turned off in humans shortly after the weaning period. Therefore, lactose intolerance is a normal thing to happen as one ages. However, a few thousand years ago, a single nucleotide polymorphism in the lactase encoding gene resulted in prolonged expression of lactase into adulthood of Europeans. This mutation was favored as the mutants can utilize milk and milk products (nutritionally rich products) in geographical areas, where the growing season of crops and fruits is short [88].

To date, nutrigenomics has mainly focused on single nucleotide polymorphisms (SNPs). SNPs account for more than nearly 90% of the DNA sequence variations (genetic variations) in humans. Certain SNPs transform the function of housekeeping genes and often lead to the predisposition of a particular disease. Bioactives in the diet have the potential to differentially modify the impact of SNPs, which decrease the risk of the disease. For example, the C677T polymorphism of methylenetetrahydrofolate reductase (MTHFR) gene involves a diet-SNP interaction. C677T variant slows down the MTHFR enzyme activity resulting in a decrease in the ability to utilize folic acid or folate to convert homocysteine to methionine and then to S-adenosyl methionine (SAM).

SAM is required for the preservation of cytosine methylation in DNA and the corresponding expression of genes. However, levels of another form of folate may be increased by the same enzyme variant that synthesizes the thymidine base. Thymidine prevents the mis-incorporation of uracil in DNA, which is mutagenic. Therefore, this shift in the pattern of DNA methylation explains why a low folate diet makes homozygous C677T individuals more decumbent to developmental defects. At the same time, it also provides protection against certain types of cancer [82].

Another case of gene polymorphism, predisposing their bearers to chronic pathological conditions, is provided by the *APOE* gene. This gene has three phenotypes, each with different probabilities of risk of CVD and different responses to environmental factors, lifestyle, and diet. Majority of the population in the United States bear the *APOE3* phenotype and responds well to low intake of fat in diet along with physical exercise. However, a substantial percentage of the United States' population carries a polymorphic *APOE-ε4* gene. This variant is associated with high cholesterol levels, increased risk of Alzheimer and diabetes [82].

From literature review in this section, it can be concluded that bioactives can be beneficial to our health right from the genetic level. Different bioactives affect the genomes of different individuals in uniquely in a dose-dependent manner. The beneficiary range of doses even varies from person to person. Different lifestyles and diet habits determine the type and quantity of bioactives necessary for an individual. Accordingly, these may play beneficiary or negative roles in their health. Moreover, subtle DNA changes exist in populations. The effect of bioactives on genomes thus also varies in individuals. However, nutrigenomics also explores the possibilities in the other way round, i.e., how deficiency or excess of nutriments can cause mutations in the genome, either at the level of DNA bases or chromosome.

The nutrigenomics views diet in terms of its genome protecting constituents and potential. Ben van Ommen hypothesizes that "*all diseases can be reduced to imbalances in four overarching processes: inflammatory, metabolic, oxidative, and psychological stress*" [82]. Kaput believes that "*nutrigenomics represents a major effort to improve our understanding of the role of nutrition and genomic interactions in at least the first three of these areas*" [82].

Over the course of human evolution, the metabolic abilities of humans have been severely changed due to dietary styles. This profound molding of diet over the years has sparked the emergence of modern-day chronic diseases. Diet is a limiting factor from the evolutionary point of view. Like other environmental clauses, it levies selective pressure on populations. Within a population, certain genotypes require high amount of nutrients. When the appropriate requirements are not fulfilled, selection occurs against those genotypes. If the requirements are met, then these genotypes persist in the population. Certain alleles of a gene provide some selective advantage over others. An optimal level of the required nutriment may increase the frequency of these alleles in a population [82]. Figure 7.12 represents how diet is intricately related to the genome and vice-versa.

7.7 FUTURE PERSPECTIVES

One of the major goals of nutrigenomics is to diagnose and prevent chronic diseases nutritionally based on genetic makeup. Lack of awareness about diet-SNP interaction is a considerable impediment in public health awareness. However, identifying these interactions is challenging. It is difficult to demonstrate a link between a bioactive ingredient and health/disease. Often, a "single active ingredient-single effect" relationship does not hold valid. Inside a human body, all pathways are connected in some manner. Therefore, it is difficult to demonstrate which one of the active ingredients is related with a particular health benefit. Many people are of the opinion that the activity of a bioactive ingredient is not due to the compound per se but due to the production of a metabolite from the compound by the host or the gut microbiota. For example, Vitamin D exhibits its antioxidant properties only after getting converted to its active form calcitriol [47].

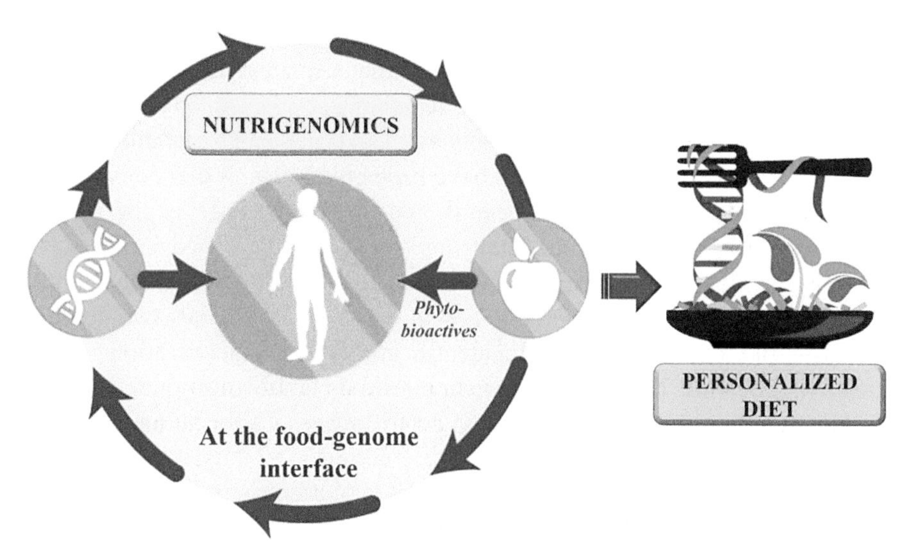

FIGURE 7.12 Interaction between diet and genome.
Source: Self-developed from concepts by Rugby Renegade [109]; and Nutritional sciences news and highlights [93].

It has also been established that human gut microbial diversity is greater in slim than obese objects. Weight-loss attempts are less effective in people with low gut microbial genetic diversity [37]. In order to accumulate evidence for the relation of bioactive components to health, biomarkers need to be developed. Bioactivity of food can be attributable to more than one component, which can make multiple tissues to respond at the same time. This makes it difficult to monitor the link between an active ingredient and health.

The study of nutrigenomics requires the use of high omics technologies, such as, genomics, proteomics, and metabolomics. Yet, the presently available technologies are not enough. Many attempts to identify the SNP-diet interactions have not been successful due to insufficient statistical data and other methodology problems. Kaput says that "*the diet-gene interactions are highly complex and hard to predict, thus demonstrating the need for completely controlled genotypes and environmental conditions that allow for identifying different regulatory patterns based on diet and genotype*" [82]. He adds that "*the challenges we now face may ultimately require a nutrigenomics project on a scale of the Human Genome Project to identify the genes that cause or promote chronic disease and the nutrients that regulate or influence the activity of these genes*" [82].

In the 21st century, chronic pathological conditions (such as: obesity, diabetes, CVDs, cancer, etc.), are widely prevalent. Their increased occurrence is also partly responsible because of the lazy lifestyle and diet of individuals. Thus, a modification in lifestyle habits and diet can be of substantial benefit in preventing or curing these maladies. Fruits and vegetables contain myriad bioactive components that have protective roles. A diet consisting of more fruits and vegetables have been shown to work better in keeping chronic diseases at bay than taking medical supplements as mentioned earlier.

Bio-fortification of staple food with these active ingredients is also required to make them available to common masses. In the last two decades, a substantial focus has been in the identification and characterization of these active ingredients. Their exact role in human body is still an enigma. This field holds exciting research prospects with promising results for human health.

7.8 SUMMARY

The healthy diet should include substantial quantities of fruits and vegetables as they consist of active ingredients, which have protective roles against chronic diseases. These phytobioactives have been found to interact with DNA and thus exert genetic effects in a dose-dependent manner. The field of nutrigenomics is concerned with the identification and characterization of these interactions. Currently, progress in this research is slow due to certain technological and statistical limitations. However, it is clear that phytobioactives are beneficial for health. We need to get a diet personalized, based on our genetic composition and abide by it. It is high time that we know our genome and start eating smart.

ACKNOWLEDGMENTS

The authors thank Rev. (Dr.) Dominic Savio, S. J., Principal, and Rector, St. Xavier's College (Autonomous), Kolkata, for constant support in this research.

KEYWORDS

- **alkaloids**
- **bioactives**
- **cardiovascular diseases**
- **epigenomics**
- **nutrigenomics**
- **phenolics**
- **terpenes**

REFERENCES

1. Achan, J., Talisuna, A. O., Erhart, A., Yeka, A., Tibenderana, J. K., Baliraini, F. N., Rosenthal, P. J., & D'Alessandro, U., (2011). Quinine-an old anti-malarial drug in a modern world: Role in the treatment of malaria. *Malaria Journal, 10*, 144.
2. Aggarwal, B. B., & Harikumar, K. B., (2009). Potential therapeutic effects of curcumin, the anti-inflammatory agent, against neurodegenerative, cardiovascular, pulmonary, metabolic, autoimmune, and neoplastic diseases. *International Journal of Biochemistry and Cell Biology, 41* (1), 40–59.
3. Ahima, R. S., & Flier, J. S., (2000). Adipose tissue as an endocrine organ. *Trends in Endocrinology and Metabolism, 11* (8), 327–332.
4. Ahn, J., Lee, H., Kim, S., & Ha, T., (2010). Curcumin-induced suppression of adipogenic differentiation is accompanied by activation of Wnt/β-catenin signaling. *American Journal of Physiology, 298* (6), 1510–1516.
5. Aldred, E. M., (2009). Terpenes. *Pharmacology—A Handbook for Complementary Healthcare Professionals* (pp. 167–174). Churchill Livingstone/Elsevier, Edinburgh, New York.
6. Ali, S. A., Choudhary, R. K., Naaz, I., & Ali, A. S., (2015). Understanding the challenges of melanogenesis: Key role of bioactive compounds in the treatment of hyper-pigmentary disorders. *Journal of Pigmentary Disorders, 2* (11), 1–9.
7. Amado, N. G., Predes, D., Moreno, M. M., Carvalho, I. O., Mendes, F. A., & Abreu, J. G., (2014). Flavonoids and Wnt/β-catenin signaling: Potential role in colorectal cancer therapies. *International Journal of Molecular Sciences, 15* (7), 12094–12106.
8. Ameh, S. J., Obodozie, O. O., Abubakar, M. S., & Garba, M., (2010). Current phytotherapy—An inter-regional perspective on policy, research, and development of herbal medicine. *Journal of Medicinal Plants Research, 4* (15), 1508–1516.
9. Andersson, H., Hallström, H., & Kihlman, B. A., (2005). *Intake of Caffeine and other Methylxanthines During Pregnancy and Risk for Adverse Effects in Pregnant Women and their Uses* (p. 387). Nordic Cooperation, Nord Pub, Copenhagen, Denmark: Nordic Council of Ministers.

10. Andersson, H. C., (2008). Bioactive plant compounds in social non-alcoholic drinks-a risk for human reproduction? In: *Bioactive Compounds in Plants-Benefits and Risks for Man and Animals* (pp. 144–158). Proceedings from a symposium held at The Norwegian Academy of Science and Letters, Oslo, Norway.
11. *Antioxidants in Depth.* NIH-National Center for Complementary and Integrative Health. https://nccih.nih.gov/health/antioxidants/introduction.htm (accessed on 5 August 2020).
12. *Arteriosclerosis/Atherosclerosis.* Mayo Clinic. https://www.mayoclinic.org/diseases-conditions/arteriosclerosis-atherosclerosis/symptoms-causes/syc-20350569 (accessed on 5 August 2020).
13. Azmir, J., Zaidul, I. S. M., Rahman, M. M., Sharif, K. M., Mohamed, A., Sahena, F., Jahurul, M. H. A., et al., (2013). Techniques for extraction of bioactive compounds from plant materials: A review. *Journal of Food Engineering, 117* (4), 426–436.
14. Bagga, D., Wang, L., Farias-Eisner, R., Glaspy, J. A., & Reddy, S. T., (2003). Differential effects of prostaglandin derived from ω-6 and ω-3 polyunsaturated fatty acids on COX-2 expression and IL-6 secretion. *Proceedings of the National Academy of Sciences of the United States of America, 100* (4), 1751–1756.
15. Balasundram, N., Sundram, K., & Samman, S., (2006). Phenolic compounds in plants and agri-industrial by-products: Antioxidant activity, occurrence, and potential uses. *Food Chemistry, Elsevier, 99* (1), 191–203.
16. Becker, T., & Juvik, J., (2016). The role of glucosinolate hydrolysis products from *Brassica* vegetable consumption in inducing antioxidant activity and reducing cancer incidence. *Diseases, 4* (4), 22, 23.
17. Ben-Eltriki, M., Deb, S., Adomat, H., & Tomlinson, G. E. S., (2016). Calcitriol and 20 (S)-protopanaxadiol synergistically inhibit growth and induce apoptosis in human prostate cancer cells. *Journal of Steroid Biochemistry and Molecular Biology, 158*, 207–219.
18. Benkoussa, M., Brand, C., Delmotte, M. H., Formstecher, P., & Lefebvre, P., (2002). Retinoic acid receptors inhibit AP1 activation by regulating extracellular signal-regulated kinase and CBP recruitment to an AP1-responsive promoter. *Molecular and Cellular Biology, 22* (13), 4522–4534.
19. Blomhoff, R., (2008). Role of dietary phytochemicals in oxidative stress. In: *Bioactive Compounds in Plants-Benefits and Risks for Man and animals* (pp. 52–70). Proceedings from a symposium held at The Norwegian Academy of Science and Letters, Oslo, Norway.
20. Bose, S., & Panda, A. K., (2015). Curcumin and tumor immune-editing: Resurrecting the immune system. *Cell Division, 10*, 6–9.
21. Breiteneder, H., & Radauer, C., (2004). A classification of plant food allergens. *Journal of Allergy and Clinical Immunology, 113* (5), 821–830.
22. Britannica. https://www.britannica.com/plant/Cinchona (accessed on 5 August 2020).
23. Britannica. https://www.britannica.com/science/alkaloid (accessed on 5 August 2020).
24. Bsoul, S. A., & Terezhalmy, G. T., (2004). Vitamin C in health and disease. *Journal of Contemporary Dental Practice, 5* (2), 1–13.
25. *Cancer*, WHO. https://www.who.int/news-room/fact-sheets/detail/cancer (accessed on 5 August 2020).
26. *Cardiovascular Diseases*, WHO. https://www.who.int/cardiovascular_diseases/en/ (accessed on 5 August 2020).
27. *Cardiovascular Diseases*, WHO. https://www.who.int/news-room/fact-sheets/detail/cardiovascular-diseases- (cvds)_(accessed on 5 August 2020).

28. Castro, M. D. L., & Ayuso, L. E. G., (2000). Environmental applications-soxhlet extraction. In: *Encyclopaedia of Separation Science* (pp. 2701–2709). New York: Academic Press.
29. Cho, J. W., Lee, K. S., & Kim, C. W., (2007). Curcumin attenuates the expression of IL-1beta, IL-6, and TNF-alpha as well as cyclin E in TNF-alpha-treated HaCaT cells; NF-kappaB and MAPKs as potential upstream targets. *International Journal of Molecular Medicine, 19* (3), 469–474.
30. Cho, Y. H., Kim, J. H., Park, S. M., Lee, B. C., Pyo, H. B., & Park, H. D., (2006). New cosmetic agents for skin whitening from *Angelica dahurica. Journal of Cosmet Science, 57*, 11–21.
31. *Coleus Forskohlii (Plectranthus barbatus) Herbal Monograph.* Brett Elliott Ultimate Herbal Health.https://www.brettelliott.com/detox-blog/coleus-forskohlii-plectranthus-barbatus-herbal-monograph_(accessed on 5 August 2020).
32. Cosme, F., Pinto, T., & Vilela, A., (2018). Phenolic compounds and antioxidant activity in grape juices: A chemical and sensory view. *Beverages, MDPI, 4*, 22–24.
33. Crescenti, A., Solá, R., Valls, R. M., Caimari, A., Del, B. J. M., Anguera, A., Anglés, N., & Arola, L., (2013). Cocoa consumption alters the global DNA methylation of peripheral leukocytes in humans with cardiovascular disease risk factors: A randomized controlled trial. *PLoS One, 8* (6), 224–226.
34. Crowe, K. M., & Allison, D., (2015). Evaluating bioactive food components in obesity and cancer prevention. *Critical Reviews in Food Science and Nutrition, 55* (5), 732–734.
35. Das, D. K., Mukherjee, S., & Ray, D., (2011). Erratum to: Resveratrol and red wine, healthy heart and longevity. *Heart Failure Reviews, 16* (4), 425–435.
36. Das, S., Alagappan, V. K., Bagchi, D., Sharma, H. S., Maulik, N., & Das, D. K., (2005). Coordinated induction of iNOS-VEGF-KDR-eNOS after resveratrol consumption: A potential mechanism for resveratrol preconditioning of the heart. *Vascular Pharmacology, 42* (5, 6), 281–289.
37. Davis, C. D., (2016). The gut microbiome and its role in obesity. *Nutrition Today, 51* (4), 167–174.
38. *Diabetes*, WHO. https://www.who.int/news-room/fact-sheets/detail/diabetes_(accessed on 5 August 2020).
39. Duthie, S. J., Narayanan, S., Sharp, L., & Little, J., (2004). Folate: DNA stability and colo-rectal neoplasia. *Proceedings of the Nutrition Society, 63* (4), 571–578.
40. Ejaz, A., Wu, D., Kwan, P., & Meydani, M., (2009). Curcumin inhibits adipogenesis in 3T3-L1 adipocytes and angiogenesis and obesity in C57/BL mice. *Journal of Nutrition, 139* (5), 919–925.
41. Elkacimi, I. M., Cherrah, Y., & Alaoui, K., (2018). The technological influence of the two different methods of extraction on the initial concentration of thymol in the essential oil of thyme (*Thymus satureioides*). *African Journal of Biotechnology, 17* (37), 1180–1187.
42. Elsamanoudy, A. Z., Neamat-Allah, M. A. M., Mohammad, F. A. H., Hassanien, M., & Nada, H. A., (2016). The role of nutrition related genes and nutrigenetics in understanding the pathogenesis of cancer. *Journal of Microscopy and Ultrastructure, 4* (3), 115–122.
43. (2012). *Epharmacognosy*. http://www.epharmacognosy.com/2012/07/phenylalanine-derived-alkaloids.html (accessed on 5 August 2020).
44. (2014). *Extracting Essential Oils from Spices Using Steam Distillation.* AZoM, https://www.azom.com/article.aspx?ArticleID=11355_(accessed on 5 August 2020).
45. *Extraction Techniques of Herbal Plants.* http://agritech.tnau.ac.in/horticulture/extraction_herbal_techniques%20_medicinal_plants.pdf (accessed on 5 August 2020).

46. *Extraction Techniques of Medicinal Plants.* TNAU Agritech Portal http://agritech.tnau.ac.in/horticulture/extraction_techniques%20_medicinal_plants.pdf_(accessed on 5 August 2020).
47. Feldman, D., Pike, J. W., & Adams, J., (2011). *Vitamin D* (3rd edn., p. 231). New York: Academic Press.
48. Flachs, P., Horakova, O., Brauner, P., Rossmeisl, M., Pecina, P., Franssen-van, H. N., Ruzickova, J., et al., (2005). Polyunsaturated fatty acids of marine origin up regulate mitochondrial biogenesis and induce β-oxidation in white fat. *Diabetologia, 48* (11), 2365–2375.
49. Flachs, P., Rossmeisl, M., Bryhn, M., & Kopecky, J., (2009). Cellular and molecular effects of n-3 polyunsaturated fatty acids on adipose tissue biology and metabolism. *Clinical Science, 116* (1), 1–16.
50. Fu, B., Li, H., Wang, X., Lee, F. S., & Cui, S., (2005). Isolation and identification of flavonoids in licorice and a study of their inhibitory effects on tyrosinase. *Journal of Agriculture and Food Chemistry, 53*, 7408–7414.
51. García, P. B., López, L. G., Esparza, B. G., & Rosas, C. D. C., (2017). Effect of bioactive nutriments in health and disease: The role of epigenetic modifications. In: *Functional Food-Improve Health through Adequate Food.* Intech Open; online article.
52. Ghosh, T., Beniwal, A., Semwal, A., & Navani, N. K., (2019). Mechanistic insights into probiotic properties of lactic acid bacteria associated with ethnic fermented dairy products. *Food Microbiology, Frontiers in Microbiology, 10* (502), 210–218.
53. Giada, M. L. R., (2013). Food phenolic compounds: Main classes, sources, and their antioxidant power. In: *Oxidative Stress and Chronic Degenerative Diseases—A Role for antioxidants*. IntechOpen; Online article.
54. González-Périz, A., Horrillo, R., Ferré, N., Gronert, K., Dong, B., Morán-Salvador, E., Titos, E., et al., (2009). Obesity-induced insulin resistance and hepatic steatosis are alleviated by ω-3 fatty acids: A role for resolvins and protectins. *The FASEB Journal, 23* (6), 1946–1957.
55. Hakozaki, T., Minwalla, L., Zhuang, J., Chhoa, M., Matsubara, A., Miyamoto, K., Greatens, A., et al., (2002). The effect of niacinamide on reducing cutaneous pigmentation and suppression of melanosome transfer. *British Journal of Dermatology, 147*, 20–31.
56. Hammoud, D., Haddad, B. E., & Abdallah, J., (2014). Hypercalcemia secondary to hypervitaminosis a in a patient with chronic renal failure. *West Indian Medical Journal, 63* (1), 105–108.
57. Healthline. https://www.healthline.com/health/phenylketonuria (accessed on 5 August 2020).
58. Herbs Natural-Medicinal Herbs. http://www.naturalmedicinalherbs.net/herbs/p/papaver-somniferum=opium-poppy.php (accessed on 5 August 2020).
59. Higdon, J., & Johnson, E. J., (2016). Carotenoids. *Phytochemicals, Dietary Factors.* Micronutrient Information Center, Linus Pauling Institute; https://lpi.oregonstate.edu/mic/dietary-factors/phytochemicals/carotenoids (accessed on 5 August 2020).
60. Holloway, V. L., (2003). Ethnic cosmetic products. *Dermatologic Clinics, 21*, 743–749.
61. Hong, L., Guo, Z., Huang, K., Wei, S., Liu, B., Meng, S., & Long, C., (2015). Ethnobotanical study on medicinal plants used by Maonan people in China. *Journal of Ethnobiology and Ethnomedicine, 11*, 32–35.
62. Jayarathne, S., Koboziev, I., Park, O. H., Oldewage-Theron, W., Shen, C. L., & Moustaid-Moussa, N., (2017). Anti-inflammatory and anti-obesity properties of food

bioactive components: Effects on adipose tissue. *Preventive Nutrition and Food Science, 22* (4), 251–262.

63. Kabir, M., Skurnik, G., Naour, N., Pechtner, V., Meugnier, E., Rome, S., Quignard-Boulangé, A., et al., (2007). Treatment for 2 Mo with n-3 polyunsaturated fatty acids reduces adiposity and some atherogenic factors but does not improve insulin sensitivity in women with type 2 diabetes: A randomized controlled study. *American Journal of Clinical Nutrition, 86* (6), 1670–1679.
64. Kalupahana, N. S., Claycombe, K., Newman, S. J., Stewart, T., Siriwardhana, N., Matthan, N., Lichtenstein, A. H., & Moustaid-Moussa, N., (2010). Eicosapentaenoic acid prevents and reverses insulin resistance in high-fat diet-induced obese mice via modulation of adipose tissue inflammation. *Journal of Nutrition, 140* (11), 1915–1922.
65. Kamohara, S., & Noparatanawong, S., (2013). A *Coleus forskohlii* extract improves body composition in healthy volunteers: An open-label trial. *Personalized Medicine Universe, 2*, 25–27.
66. Kanety, H., Feinstein, R., Papa, M. Z., Hemi, R., & Karasik, A., (1995). Tumor necrosis factor α-induced phosphorylation of insulin receptor substrate-1 (IRS-1):possible mechanism for suppression of insulin-stimulated tyrosine phosphorylation of IRS-1. *Journal of Biological Chemistry, 270* (40), 23780–23784.
67. Kang, J. G., & Park, C. Y., (2012). Anti-obesity drugs: A review about their effects and safety. *Diabetes and Metabolism Journal, 36* (1), 13–25.
68. Kaviratna, A. C., & Sharma, P., (1913). *The Charaka Samhita: 5 Volumes*. Sri Satguru Publications, New Delhi: India; http://www.shodhak.co.in/pdf/books/47-Part-B.pdf (accessed on 5 August 2020).
69. Kolb, E. A., & Meshinchi, S., (2015). Acute myeloid leukemia in children and adolescents: Identification of new molecular targets brings promise of new therapies. *Hematology, 2015* (1), 507–513.
70. Kris-Etherton, P. M., Hecker, K. D., Bonanome, A., Coval, S. M., Binkoski, A. E., Hilpert, K. F., Griel, A. E., et al., (2002). Bioactive compounds in foods: Their role in the prevention of cardiovascular disease and cancer. *American Journal of Medicine, 113* (9), 71–88.
71. Lodish, H., & Berk, A., (2013). Cancer. In: *Molecular Cell Biology* (7th edn., pp. 1113–1154). W.H. Freeman and Company, New York, USA.
72. Lovell, M. A., & Markesbery, W. R., (2007). Oxidative DNA damage in mild cognitive impairment and late-stage Alzheimer's disease. *Nucleic Acids Research, 35* (22), 7497–7504.
73. Mahomoodally, M. F., (2013). Traditional medicines in Africa: An appraisal of ten potent African medicinal plants. *Evidence-Based Complementary and Alternative Medicine (eCAM), 2013*, 14. Article ID 617459. http://dx.doi.org/10.1155/2013/617459 (accessed on 5 August 2020).
74. Makki, K., Froguel, P., & Wolowczuk, I., (2013). Adipose tissue in obesity-related inflammation and insulin resistance: Cells, cytokines, and chemokines. *ISRN Inflammation, 2013*, 12. Article ID 139239. http://dx.doi.org/10.1155/2013/139239 (accessed on 5 August 2020).
75. Manach, C., Scalbert, A., Morand, C., Rémésy, C., & Jiménez, L., (2004). Polyphenols: Food sources and bioavailability. *The American Journal of Clinical Nutrition, 79* (5), 727–747.
76. Mannheimer, B., Törring, O., & Nathanson, D., (2015). Vitamin D intoxication caused by drugs brought online. Sky high daily dosage for six months resulted in severe

hypercalcemia. *Läkartidningen, 112*, PMID: 26035535. https://europepmc.org/abstract/med/26035535 (accessed on 5 August 2020).
77. Mapunya, M. B., Hussein, A. A., & Rodriguez, B., (2011). Tyrosinase activity of *Greyia flanaganii* (Bolus) constituents. *Phytomedicine, 18*, 1006–1012.
78. Market Business News. https://marketbusinessnews.com/financial-glossary/food-biotechnology/ (accessed on 5 August 2020).
79. Martínez-Ballesta, M., Gil-Izquierdo, Á., García-Viguera, C., & Domínguez-Perles, R., (2018). Nanoparticles and controlled delivery for bioactive compounds: Outlining challenges for new "smart-foods" for health. *Foods, 7* (5), 72–75.
80. Maximum yield. https://www.maximumyield.com/definition/3034/bioactive-compounds (accessed on 5 August 2020).
81. Maynard, S., Fang, E. F., Scheibye-Knudsen, M., Croteau, D. L., & Bohr, V. A., (2015). DNA damage, DNA repair, aging, and neurodegeneration. *Cold Spring Harbor Perspectives in Medicine, 5* (10). Online article https://www.ncbi.nlm.nih.gov/pubmed/26385091 (accessed on 5 August 2020).
82. Mead, M. N., (2007). Nutrigenomics: The Genome-Food Interface. *Environmental Health Perspectives, 115* (12), 582–589.
83. Mercader, J., Palou, A., & Bonet, M. L., (2011). Resveratrol enhances fatty acid oxidation capacity and reduces resistin and Retinol-Binding Protein 4 expression in white adipocytes. *The Journal of Nutritional Biochemistry, 22* (9), 828–834.
84. Minwalla, L., Zhao, Y., Cornelius, J., Babcock, G. F., Wickett, R. R., Le Poole, I. C., & Boissy, R. E., (2001). Inhibition of melanosome transfer from melanocytes to keratinocytes by lectins and neoglycoproteins in an *in vitro* model system. *Pigment Cell Research, 14*, 185–194.
85. Mishra, A. P., Sharifi-Rad, M., Shariati, M. A., Mabkhot, Y. N., Al-Showiman, S. S., Rauf, A., Salehi, B., et al., (2018). Bioactive compounds and health benefits of edible Rumex species: A review. *Cellular and Molecular Biology, 64* (8), 27–34.
86. Morimoto, T., Sunagawa, Y., Kawamura, T., Takaya, T., Wada, H., Nagasawa, A., Komeda, M., et al., (2008). The dietary compound curcumin inhibits p300 histone acetyltransferase activity and prevents heart failure in rats. *Journal of Clinical Investigation, 118* (3), 868–878.
87. Munns, G. F., & Aldrich, C. A., (1927). Ephedrine in the treatment of bronchial asthma in children. *Jama Network, 88* (16), 1233–1236.
88. Neeha, V. S., & Kinth, P., (2013). Nutrigenomics research: A review. *Journal of Food Science and Technology, 50* (3), 415–428.
89. Niedzwiecki, A., Roomi, M. W., Kalinovsky, T., & Rath, M., (2016). Anticancer efficacy of polyphenols and their combinations. *Nutrients, 8* (9), 552–555.
90. NIH-National Cancer Institute. https://www.cancer.gov/research/progress/discovery/taxol (accessed on 5 August 2020).
91. NIH-National Center for Complementary and Integrative Health. https://nccih.nih.gov/health/ephedra (accessed on 5 August 2020).
92. Nordqvist, C., (2018). *Coronary Heart Disease: What you Need to Know*. Medical News Today. https://www.medicalnewstoday.com/articles/184130.php (accessed on 5 August 2020).
93. *Nutritional Sciences News and Highlights*. (2018). https://he.utexas.edu/ntr-news-list/entry/diet-and-genes (accessed on 5 August 2020).
94. *Obesity and Overweight*. WHO. https://www.who.int/news-room/fact-sheets/detail/obesity-and-overweight (accessed on 5 August 2020).

95. *Obesity*. Stanford Health Care. https://stanfordhealthcare.org/medical-conditions/healthy-living/obesity.html (accessed on 5 August 2020).
96. *Obesity*. WHO, Africa https://www.afro.who.int/health-topics/obesity (accessed on 5 August 2020).
97. *Obesity*. Wikipedia. https://en.wikipedia.org/wiki/Obesity (accessed on 5 August 2020).
98. Oh, Y. S., & Jun, H. S., (2014). Role of bioactive food components in diabetes prevention: Effects on beta-cell function and preservation. *Nutrition and Metabolic Insights, 7*, 51–59.
99. Picardo, M., & Carrera, M., (2007). New and experimental treatments of cloasma and other hypermelanoses. *Dermatology Clinics, 25* (3), 353–362.
100. Picotto, G., Liaudat, A. C., Bohl, L., & Talamoni, N. T. D., (2012). Molecular aspects of vitamin D anticancer activity. *Cancer Investigation, 30* (8), 604–614.
101. Pitha-Rowe, I., Petty, W. J., Kitareewan, S., & Dmitrovsky, E., (2003). Retinoid target genes in acute promyelocytic leukemia. *Leukemia, 17* (9), 1723–1730.
102. Pizzino, G., Irrera, N., Cucinotta, M., Pallio, G., Mannino, F., Arcoraci, V., Squadrito, F., et al., (2017). Oxidative stress: Harms and benefits for human health. *Oxidative Medicine and Cellular Longevity, 2017*. Epub ID: 8416763; doi: 10.1155/2017/8416763.
103. Quintero-Fabián, S., Ortuño-Sahagún, D., Vázquez-Carrera, M., & López-Roa, R. I., (2013). Alliin, a garlic (Allium sativum) compound, prevents LPS-induced inflammation in 3T3-L1 adipocytes. *Mediators of Inflammation, 2013*, 1–13.
104. Rana, S., Kumar, S., Rathore, N., Padwad, Y., & Bhushan, S., (2016). Nutrigenomics and its impact on life style associated metabolic diseases. *Current Genomics, 17* (3), 261–278.
105. Ray, P., Gupta, H. N., & Roy, M., (1980). *Suśruta Saṃhita-a Scientific Synopsis*. The Indian National Science Academy, Delhi, India; E-book available at: https://ia801900.us.archive.org/22/items/in.ernet.dli.2015.492978/2015.492978.Susruta-Samhita.pdf (accessed on 5 August 2020).
106. Reisz, J. A., Bansal, N., Qian, J., Zhao, W., & Furdui, C. M., (2014). Effects of ionizing radiation on biological molecules-mechanisms of damage and emerging methods of detection. *Antioxidants and Redox Signaling, 21* (2), 260–292.
107. Ribaya-Mercado, J. D., & Blumberg, J. B., (2007). Vitamin A: Is it a risk for osteoporosis and bone fracture? *Nutrition Reviews, 65* (10), 425–438.
108. (2015). ROSIN CERATE-Quirky quality science, http://www.rosincerate.com/2015/10/a-brief-history-of-opium-based.html (accessed on 5 August 2020).
109. *Rugby Renegade*. https://rugbyrenegade.com/rugby-renegade-podcast-episode-07/ (accessed on 5 August 2020).
110. Ryan, M. J., Dudash, H. J., Docherty, M., Geronilla, K. B., Baker, B. A., Haff, G. G., Cutlip, R. G., & Alway, S. E., (2010). Vitamin E and C supplementation reduces oxidative stress, improves antioxidant enzymes and positive muscle work in chronically loaded muscles of aged rats. *Experimental Gerontology, 45* (11), 882–895.
111. Saramäki, A., Diermeier, S., Kellner, R., Laitinen, H., Vaïsänen, S., & Carlberg, C., (2009). Cyclical chromatin looping and transcription factor association on the regulatory regions of the P21 (*CDKN1A*) gene in response to 1α,25-dihydroxyvitamin D_3. *Journal of Biological Chemistry, 284* (12), 8073–8082.
112. Sarmento, A., Barros, L., Fernandes, Â., Carvalho, A. M., & Ferreira, I. C., (2015). Valorization of traditional foods: Nutritional and bioactive properties of *Cicer arietinum L.* and *Lathyrus sativus L.* pulses. *Journal of the Science of Food and Agriculture, 95* (1), 179–185.

113. Sasaki, R., Nishimura, N., Hoshino, H., Isa, Y., Kadowaki, M., Ichi, T., Tanaka, A., et al., (2007). Cyanidin 3-glucoside ameliorates hyperglycemia and insulin sensitivity due to down regulation of retinol binding protein 4 expression in diabetic mice. *Biochemical Pharmacology, 74* (11), 1619–1627.
114. Sato, A., Kawano, H., Notsu, T., Ohta, M., Nakakuki, M., Mizuguchi, K., Itoh, M., et al., (2010). Anti-obesity effect of eicosapentaenoic acid in high-fat/high-sucrose diet-induced obesity. *Diabetes, 59* (10), 2495–2504.
115. Shang, S., Hua, F., & Hu, Z. W., (2017). The regulation of β-catenin activity and function in cancer: Therapeutic opportunities. *Oncotarget, 8* (20), 33972–33989.
116. Smith, S. R., Weissman, N. J., Anderson, C. M., Sanchez, M., Chuang, E., Stubbe, S., Bays, H., & Shanahan, W. R., (2010). Multicenter, placebo-controlled trial of Lorcaserin for weight management. *New England Journal of Medicine, 363* (3), 245–256.
117. Svendsen, M., Blomhoff, R., Holme, I., & Tonstad, S., (2007). The effect of an increased intake of vegetables and fruit on weight loss, blood pressure and antioxidant defense in subjects with sleep related breathing disorders. *European Journal of Clinical Nutrition, 61*, 1301–1311.
118. Svendsen, M., Ellingsen, I., & Tonstad, S., (2008). Compounds that may be responsible for reduced risk for cardiovascular diseases related to consumption of vegetables, fruits, and berries. In: *Bioactive Compounds in Plants-Benefits and Risks for Man and Animals* (pp. 71–77). Proceedings from a symposium held at The Norwegian Academy of Science and Letters, Oslo, Norway.
119. Taiz, L., & Zeiger, E., (2002). Secondary metabolites and plant defense. In: *Plant Physiology* (3rd edn., pp. 283–308). Sunderland: Sinauer Associates.
120. Taofiq, O., González-Paramás, A. M., Martins, A., Barreiro, M. F., & Ferreira, I. C. F. R., (2016). Mushrooms extracts and compounds in cosmetics, cosmeceuticals, and nutricosmetics: A review. *Industrial Crops and Products, Elsevier, 90*, 38–48.
121. Thakar, V. J., (2010). Historical development of basic concepts of Ayurveda from *Veda* up to *Samhita. Ayu, 31* (4), 400–402.
122. The Poison Garden Website. http://www.thepoisongarden.co.uk/atoz/taxus_baccata.htm (accessed on 5 August 2020).
123. TreeEbb. https://www.ebben.nl/en/treeebb/tabaccat-taxus-baccata/ (accessed on 5 August 2020).
124. Tsao, R., (2010). Chemistry and biochemistry of dietary polyphenols. *Nutrients, 2* (12), 1231–1246.
125. Ultrasonic technology. https://www.ndk.com/en/sensor/ultrasonic/basic01.html (accessed on 5 August 2020).
126. Venkatesan, N., (1998). Curcumin attenuation of acute adriamycin myocardial toxicity in rats. *British Journal of Pharmacology, 124* (3), 425–427.
127. Vimana Sthana (in Sanskrit); Chapter 28, pp. 225–226, verse 4–5.
128. Vitamin B6, NIH. https://ods.od.nih.gov/factsheets/VitaminB6-Consumer/ (accessed on 5 August 2020).
129. Wani, M., Wani, I., Banday, K., & Ashraf, M., (2016). The other side of vitamin D therapy: A case series of acute kidney injury due to malpractice-related vitamin D intoxication. *Clinical Nephrology, 11*, 236–241.
130. WebMD. https://www.webmd.com/vitamins-and-supplements/forskolin-uses-and-risks#1 (accessed on 5 August 2020).

131. Wikipedia. https://en.wikipedia.org/wiki/Cardiovascular_disease (accessed on 5 August 2020).
132. Wikipedia. https://en.wikipedia.org/wiki/Cinchona (accessed on 5 August 2020).
133. Wikipedia. https://en.wikipedia.org/wiki/Microwave (accessed on 5 August 2020).
134. Wikipedia. https://en.wikipedia.org/wiki/Nitric_oxide_synthase (accessed on 5 August 2020).
135. Wikipedia. https://en.wikipedia.org/wiki/Supercritical_fluid (accessed on 5 August 2020).
136. Wongcharoen, W., & Phrommintikul, A., (2009). The protective role of curcumin in cardiovascular diseases. *International Journal of Cardiology, 133* (2), 145–151.
137. Woo, H. M., Kang, J. H., Kawada, T., Yoo, H., Sung, M. K., & Yu, R., (2007). Active spice-derived components can inhibit inflammatory responses of adipose tissue in obesity by suppressing inflammatory actions of macrophages and release of monocyte chemoattractant protein-1 from adipocytes. *Life Sciences, 80* (10), 926–931.
138. Woodland Trust. https://www.woodlandtrust.org.uk/visiting-woods/trees-woods-and-wildlife/british-trees/a-z-of-uk-native-trees/yew/ (accessed on 5 August 2020).
139. Xia, Q., Zhao, Y., Wang, J., Qiao, W., Zhang, D., Yin, H., Xu, D., & Chen, F., (2017). Proteomic analysis of cell cycle arrest and differentiation induction caused by atpr, a derivative of all-trans retinoic acid, in human gastric cancer SGC-7901 cells. *Proteomics-Clinical Applications, 11* (7, 8). Epub: doi: 10.1002/prca.201600099.
140. Xiao, Y., Xia, J., Wu, S., Lv, Z., Huang, S., Huang, H., Su, X., et al., (2018). Curcumin inhibits acute vascular inflammation through the activation of heme oxygenase-1. *Oxidative Medicine and Cellular Longevity,* 12. Article ID 3295807.https://doi.org/10.1155/2018/3295807 (accessed on 5 August 2020).
141. Young, K. H., & Ortonne, J. P., (2009). Melasma update. *Actas Dermosifiliogr 100, 2,* 110–113.
142. Zawia, N. H., Lahiri, D. K., & Cardozo-Pelaez, F., (2009). Epigenetics, oxidative stress, and Alzheimer's disease. *Free Radical Biology and Medicine, 46* (9), 1241–1249.
143. Zhang, Y. J., Gan, R. Y., Li, S., Zhou, Y., Li, A. N., Xu, D. P., & Li, H. B., (2015). Antioxidant phytochemicals for the prevention and treatment of chronic diseases. *Molecules, MDPI, 20,* 21138–21156.
144. Zhao, J., Sun, X. B., Ye, F., & Tian, W. X., (2011). Suppression of fatty acid synthase, differentiation, and lipid accumulation in adipocytes by curcumin. *Molecular Cell Biochemistry, 351* (1), 19–28.

CHAPTER 8

ROLE OF DIETARY PHYTOCHEMICALS IN AMELIORATION OF ARSENIC-INDUCED CANCER: AN EMERGING ELIXIR

NILANJAN DAS, MANISHA GUHA, and XAVIER SAVARIMUTHU

ABSTRACT

Arsenic (As) is a documented carcinogen in persons who drink arsenic-contaminated groundwater. Ingestion of arsenic causes neoplasia in the lungs, urinary bladder, and skin. Phytotherapeutics from herbal source shows some efficacious and encouraging recovery from arsenic toxicity. Plant-based natural bio-compounds (such as: rutin, resveratrol, curcumin, polyphenols, etc.) have been found to reduce arsenic toxicity. These phytochemicals eliminate arsenic from the biological system and act as chemopreventive agents and help to reduce the dosage of drugs. This chapter is a comprehensive representation of the carcinogenic toxicity of arsenic and the application profile of phytochemicals that have the potential to detoxify and reduce its ruinous effects.

8.1 INTRODUCTION

Our deteriorating environment represents a prime contributor to the myriads of diseases and life-threatening syndromes [41]. An important theme of Environmental Sciences is to recognize and understand how the environment can affect and alter the basic biology of the human system and how the eco-toxicants kindle the progression of the disease [43]. Regular exposure to a huge number of eco-toxicants and pollutants can take a toll on our health and, in fact, is a prime challenge in the major public health. Certain grave diseases are initiated or exacerbated by exposure to environmental toxicants, including cancer, and neurodegenerative diseases [69].

With the advent of rapid industrialization and concomitant socio-economic changes, the patterns of resource utilization and management of disposed waste have become a serious problem for us. The scientific fraternity is making serious deliberations on developing strategies to conserve natural resources, prevent environmental pollution [54,] and look for natural ways to deal with diseases, caused due to environmental toxicity.

Human exposure to xenobiotic substances (chemicals that are not naturally metabolized in the body) is a matter of concern due to their lethal effects. Toxicity due to heavy metal exposure (such as: Cd, Cu, Co, Ni, and Pb) can prove fatal. Arsenic (As) is a metalloid that is mobilized from its natural sources by various actions, such as: smelting of metallic ores, indiscriminate development of irrigational facilities, application of As-containing weedicides or pesticides, the burning of fossil fuels and the e-wastes from electronics industry [50]. It is now established that As is a carcinogen agent [60] that exerts its genotoxicity by inducing single-strand breaks (SSBs) in deoxyribonucleic acid (DNA) and altering the methylation capacities [19].

Figure 8.1 outlines the contents of this chapter. Different anthropologic causes have led to heavy soil and water pollutions, and the As infestation. This chapter focuses on the use of herbal products against As toxicity to explore the potential of traditional therapeutic knowledge and infuse modern recent advancements in ecotoxicology.

8.2 ARSENIC MENACE

Arsenic ($^{74.92}As_{33}$) is a metalloid occurring in nature (i.e., an element whose properties are transitional between metals and non-metals), which is found ubiquitously in soil and ores of different metals. As, widely regarded in antiquity as the 'king of poison,' is a chemical toxicant that can have elevated health risks [48]. Most of the compounds of As are odorless, flavorless, and readily dissolve in aqueous media, which fosters an imposing health challenge to the people, drinking this As-contaminated water. Chronic subjection to inorganic compounds of As can lead to a plethora of health problems, including skin lesions (like hyperkeratosis, hyperpigmentation, and Black-foot disease), high blood pressure, disorders of peripheral blood vessels, diabetes, and increased risk of cancers of lungs, liver, urinary bladder and kidneys [94]. The United Nations' World Health Organization (WHO) brought down the permissible cut-off point of As titer in potable water from 50 $\mu g\ L^{-1}$ to 10 $\mu g\ L^{-1}$. Beyond this level, the water is regarded as unfit for human consumption [101].

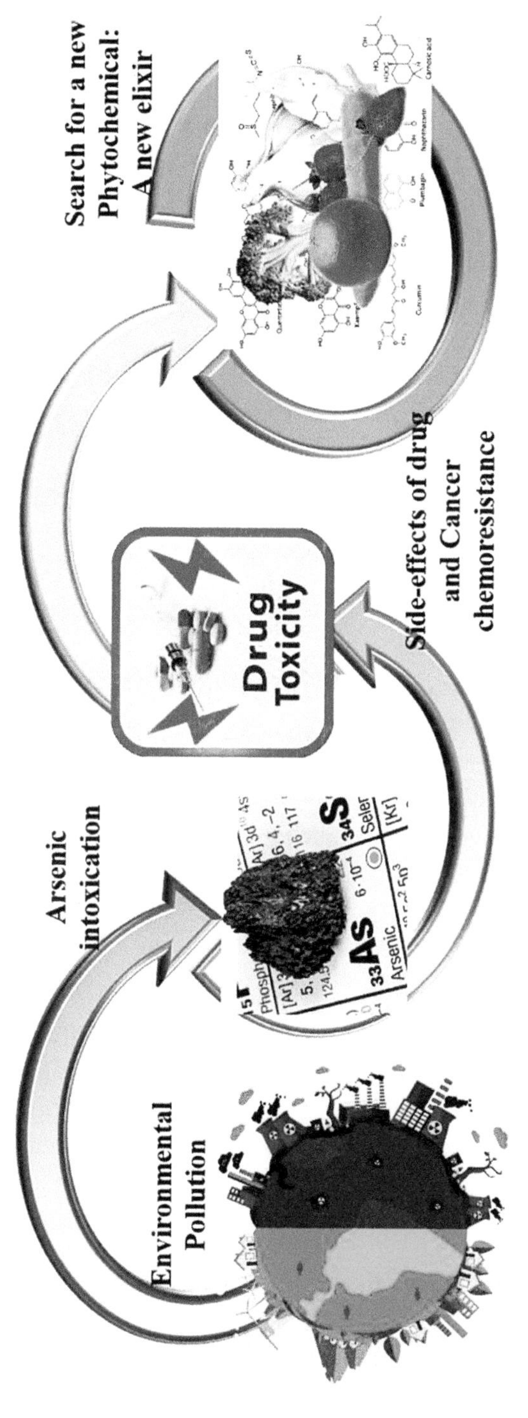

FIGURE 8.1 Challenge of arsenic toxicity and its mitigation by phytochemicals.
Source: Self-developed.

The elevated toxicity and extensive environmental contamination, different kinds of interdisciplinary research have been conducted throughout the globe, in connection with the site, attributes, environmental distribution, and remediating methods of As contamination. The geological and environmental data have depicted augmented As concentrations in water in many areas in the world (Figure 8.2). The fatal legacy is on the rise in growing in United States of America, China, Chile, Bangladesh, Republic of China (Taiwan), Mexico, Argentina, Japan, Ghana, and India [1].

FIGURE 8.2 Distribution of arsenic in countries in the world. Countries marked with blue color have arsenic-contaminated water.

Source: Self-developed by authors based on Refs. [1, 103].

The cause of the overtly augmented titer of As is a culmination of natural and anthropogenic sources of pollution.

8.2.1 ANTHROPOGENIC SOURCES

The sources of As pollution arising out of human activities are principally include irrigation, use of fossil fuels, mining activities, and the electrical burning of industrial and municipal wastes. The revolution in tube-well during the last three decades in the Gangetic plains of India and Bangladesh has unfolded this new reality of As contamination [92].

8.2.2 NATURAL SOURCES

In the biosphere, As finds its presence in sulfide ores in conjunction with certain other metals, such as: lead, silver, copper, and gold. The collected activities of mineral leaching and elevated evaporation rates in tropics release As compounds from these natural sources. This can also be caused by desorption of mineral oxides and reduction and dissolution of iron and manganese oxides [17].

Although As can exist as both organic and inorganic forms (Figure 8.3), yet the organic compounds of As (of a kind that are present in seafood) are lesser detrimental to human health than their inorganic counterparts (such as: As (III) and As (V)). Inorganic As is a documented carcinogen agent [103].

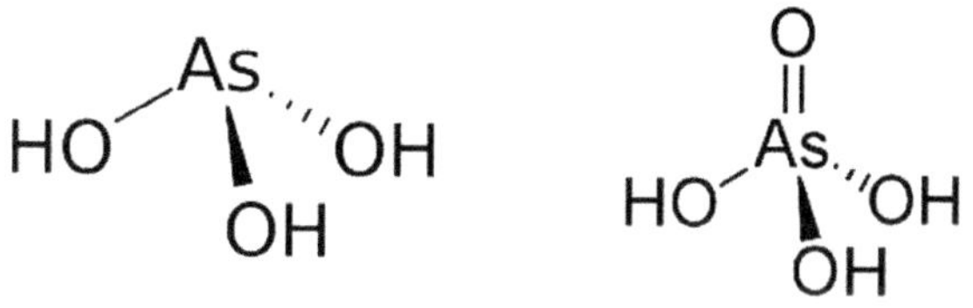

FIGURE 8.3 Structures of arsenic acid (left: H_3AsO_4) and arsenous acid (right: H3AsO3), which are typically encountered in arsenic-contaminated water.
Source: Self-developed.

8.3 ARSENICOSIS AND ARSENIC-INDUCED CANCER

The chronic exposure to As via potable water leads to arsenicosis (As poisoning) [102]. The health implications are often delayed and the best way out of this problem is drinking of potable water without or low concentration of As. As is not usually absorbed through the skin, and therefore exposure to As-infested water through hand-washing, laundry, etc., is not of great concern.

8.3.1 IMPACTS OF ARSENIC EXPOSURE ON HEALTH

8.3.1.1 ACUTE EFFECTS

Following acute As poisoning, instant clinical signs include abdominal pain, vomiting, and diarrhea. These are accompanied by tingling and numbness of hands and feet, muscle twitching and in some cases even death [46].

8.3.1.2 LONG-TERM EFFECTS

8.3.1.2.1 Keratosis and Skin Manifestations

Initial symptoms of chronic exposure to elevated concentrations of inorganic As (by drinking water and food route) are generally seen on the skin, including modifications in pigmentation, skin lesions, and hyperkeratosis (tough patches on the palms and feet sole). These skin disorders manifest at least five years after exposure and are regarded as preludes to skin cancer [13, 67].

8.3.1.2.2 Other Cancers

In addition to skin cancer, long-term As exposure can also trigger bladder and lung cancers. Different other cancers of the alimentary canal and urinary system (including kidney cancer) have also been reported. Thus, the As contaminated drinking water could be a potential source of carcinogenesis [13].

8.3.1.2.3 Systemic Disorders

Several other acrimonious health problems that could be associated with chronic intake of inorganic As, are: diabetes, developmental defects, breathing disorders (lung diseases) and cardiovascular disease. Often As-induced cardiovascular disease promotes myocardial infarctions. Heart attack due to As-induced toxicity is particularly a grave cause for death in the affected patients [13].

8.3.1.2.4 Blackfoot Disease

In the Republic of China (Taiwan), As poisoning has been related to "Blackfoot disease," a critical disorder of blood vasculature that gives rise to gangrene (a hypoxic condition in the tissues leading to the necrosis in the skin and soft tissues). The Blackfoot disease is not usually found in other As-affected areas of our planet, and it is assumed that such a phenotype may result from malnutrition coupled with As toxicity [13, 98].

8.3.1.2.5 Teratogenicity

As toxicity is also associated with developmental disorders and may result in negative results of pregnancy and effects on the health of children. In utero and early juvenescence exposure to As has been correlated with increased mortality among kids and adolescents. The affected individuals mostly succumb to multiple neoplasias, pulmonary syndromes, cardiovascular diseases, and kidney failure. Number of research studies have indicated that As exposure during growth can have adverse effects on fetal cognitive development, intelligence, and memory [13, 82].

The cutaneous manifestations become one of the most paramount elements for the diagnosis of Assis. They are regarded as one of the earliest As toxicity symptoms (Table 8.1). Figure 8.4 presents symptoms that are related to As toxicity in human life/health.

TABLE 8.1 Skin Manifestations as Diagnostic Symptoms of Arsenic toxicity

Condition	Characteristics
Melanosis	• It is the earliest and commonest. • Post-inflammatory hyper-pigmentation arises as brown macules or lesions, with a boundary that is scarcely defined, uneven in shape and generally with hardly any change in surface. • The melanosis may be dispersed or uneven, may take on a motif of rain-drop or leucomelanosis, and may occasionally cause mucosal pigmentation.
Keratosis	• It is most delicate indicator at an early stage to diagnose arsenicosis. • Depending on the density and size of keratotic lesions (papules, plaques, and nodules), keratosis may be ranked as benign, moderate, or serious.
Pre-malignant/ Malignantlesions	• On top of keratotic lesions, a dramatic rise in the size of the lesions, breaks, crevices, and inflammation suggest cancerous transformation. • Squamous cell carcinoma *in situ*: Bowen's multi-centric disease in zones subjected to non-sunlight. • Squamous cell carcinoma and Basal cell epithelial cancer.

Source: Self-developed with information from Refs. [13, 82].

8.3.2 *MECHANISM OF ARSENIC TOXICITY*

- As comes into contact with groups of thiol (–SH) and modulates the output of energy in cells. It interferes with ordinary metabolism by substituting metal co-factors like zinc and selenium in the holoenzymes from their binding locations [90].

- As poisoning elevates the level of reactive free radicals of oxygen and nitrogen (ROS and reactive nitrogen species (RNS)), *viz.*: nitric oxide, superoxide, and hydroxyl radicals that cause lesions in the DNA and proteins. It also impedes other cellular functions and alters cell architecture, membrane permeability, and cell survival and proliferation [90].
- As has been observed to switch on several downstream signaling pathways and further impedes on the normal cellular signaling, related to cell growth, cell division and cell death [90].
- As functions as a promoter of cancers and lead to tumorigenesis. As-induced tumorigenesis may result from the suppression of the p53 pathway, widely regarded as the guardian of the genome, which culminates into deregulated proliferation [90].
- Scientific trials found that there is a close relationship between exposure to As and enhanced activity of acetylcholine esterase [90].

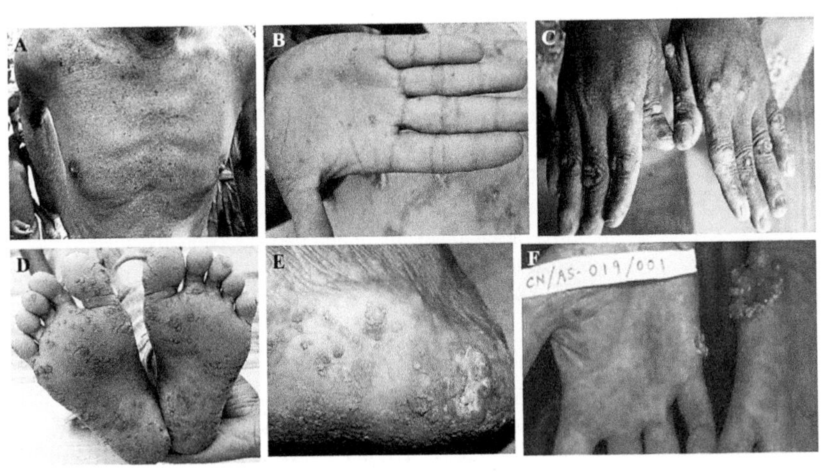

FIGURE 8.4 Cutaneous symptoms of arsenic toxicity.
Legend: Severe hyperpigmentation (A); Mild keratosis (B); Severe dorsal keratosis (C); Severe keratosis of the feet (D); Gangrene (cancer) in feet (E); and Gangrene (cancer) in hands (F).
Source: Self-developed based on Ref. [104].

8.3.3 CARCINOGENESIS FROM ARSENIC

Heiand his co-workers [39] reported that trivalent arsenite is a potent mutagen and clastogen (an agent that causes chromosomal aberration) in mammalian cell lines. They have designed an assay (A_L assay) to determine the clastogenicity [39]. As and its compounds are highly toxic to humans. According to the

International Agency for Research on Cancer (IARC) has explicitly declared that As-infested drinking water is highly carcinogenic to mankind [2]. As carcinogenesis involves genetic changes, inhibition of DNA repair, epigenome changes, and signal transduction. As's xenobiotic metabolism can play a part in this phase. The bio-transformation of inorganic As (V) into inorganic As (III) and its methylated conjugates play an important role in the genetic and epigenetic carcinogenicity of As [78, 81]. As genotoxicity can be generated either through direct contact with DNA or through indirect impacts resulting in DNA configuration mistakes (Figure 8.5). These indirect impacts may include: micronucleus formation, exchange of sister chromatids, or aberrations of chromosomes. Furthermore, enzymes are often inhibited in the DNA repair pathway, which may also lead to an indirect genotoxic mechanism [68].

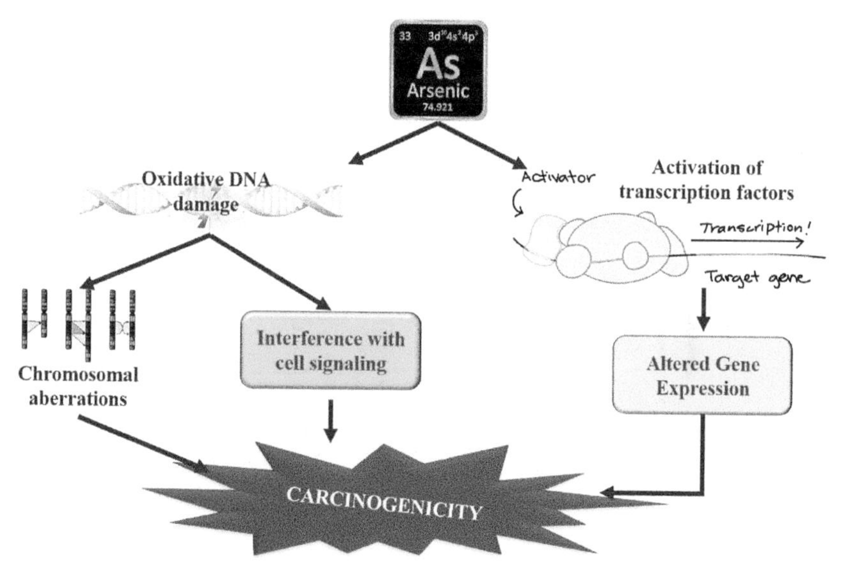

FIGURE 8.5 Mechanism of arsenic toxicity leading to carcinogenesis.
Source: Self-developed based on Ref. [91].

8.4 METHODS TO MITIGATE THE ARSENIC MENACE

As has been widely regarded as a '*silent killer.*' Mitigation of the As menace can warrant an integrated approach. A multipronged strategy may involve the supplying of As-free water to the masses and offering therapeutic means to treat patients with chronically exposed As via water and food. Figure 8.6 addresses various factors [58].

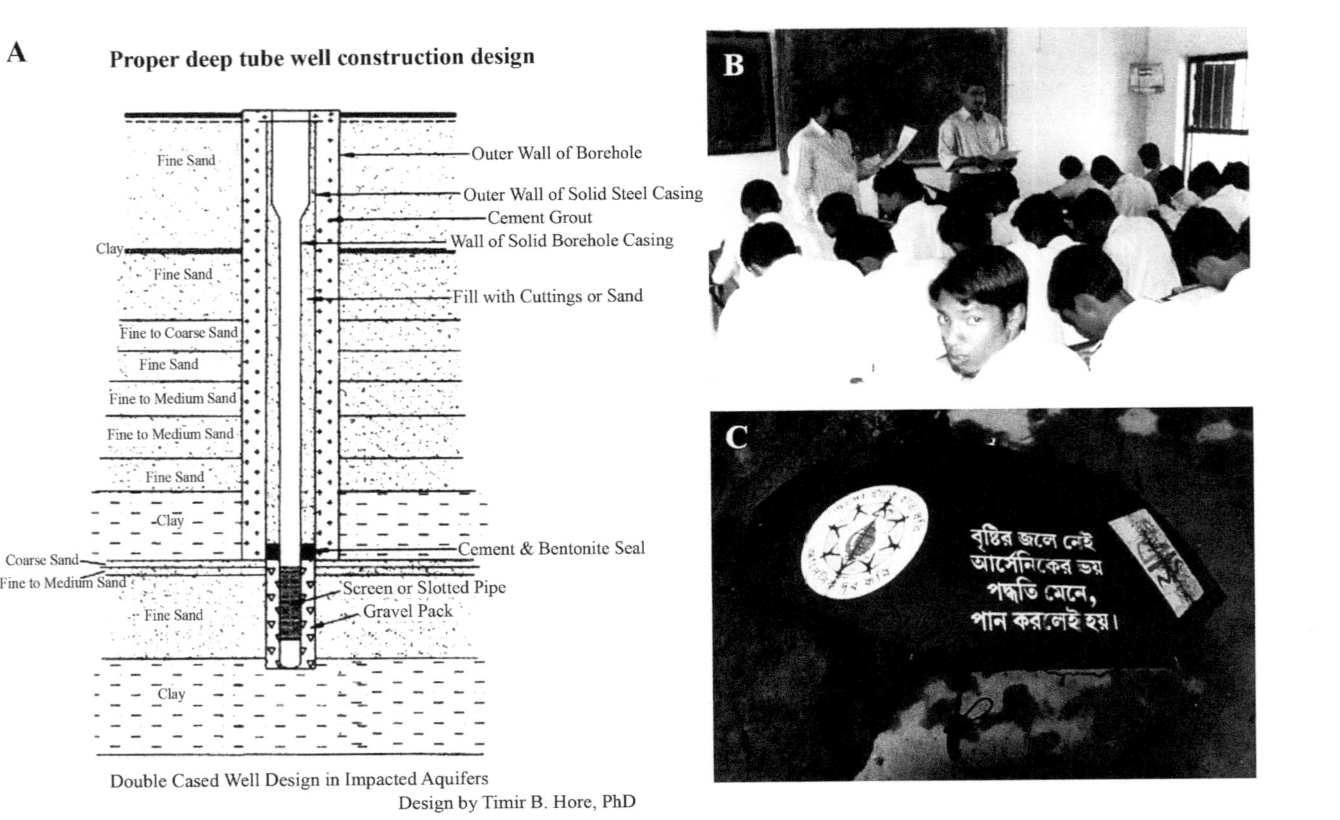

FIGURE 8.6 Ways of controlling arsenic menace: Designing of proper wells (A); conducting seminars to increase public awareness (B); and using posters and daily use items to increase public awareness (C).

Source: Self-developed based on Ref. [104].

- The first step to solve the As problem is to identify thc affected regions. Intensive monitoring of water quality from various surface and groundwater sources at different geographical locations is a must. The identification must be correlated with epidemiological studies to develop a bio-model of the problem.
- Following the identification and construction of the hit maps, emphasis must be placed on undertaking hydro-geological studies to evaluate the dynamic process of As seepage in groundwater and to identify new ways of taking groundwater through wells (Figure 8.6A).
- Delivery of safe drinking water in the affected areas of As-free or As. This could be accomplished by purifying surface water supply, setting up deep bore wells or tube wells to extract water from As-safe reservoirs, harvesting rainwater (domestic or community) and eliminating As from it.
- To generate consciousness and motivation among the impacted individuals, experts, and researchers must come forward. Training workshops need to be held periodically to make individuals conscious of the risk of As and the opportunities for As avoidance (Figures 8.6B and 8.6C).
- Biochemical transformation of inorganic As, using certain sulfate-reducing bacteria and fungal species, can convert them into a relatively non-toxic form. For instance, microbial methylation of As reduces it from As (V) to As (III), which in turn is methylated to dimethylarsine by coenzyme S-adenosylmethionine (SAM) [16].
- Last, research must focus on methods to treat the patients and to develop clinical modalities to cure As-induced cancers.

8.4.1 THERAPY FOR ARSENIC TOXICITY

Current therapeutic approach against As toxicity involves the use of a chelating agent. Chelation can be described as a metal ion being incorporated in a heterocyclic ring framework. This treatment contributes to a coordinating compound that connects more heavily toxic metal ions than cellular macromolecules and excretes the complex [24]. Some significant features should be the perfect chelating agent for clinical use in the treatment of metal ion poisoning: [26]

- It should be water-soluble;
- It must be resistant to metabolism;
- It must be accessible at the site of As accumulation;
- It must be able to displace As from biological carriers;
- It should form non-toxic complexes with metal;
- The As-chelating agent complex must be readily excretable.

Common chelating agents against As poisoning are: 2-3-dimercaprol (British Anti Lewisite, BAL), Meso 2,3-dimercaptosuccinic acid (DMSA): Monoisoamyl DMSA (MiADMSA), Sodium 2,3-dimercaptopropane-1-sulfonate (DMPS). All these are linked to toxicity and side effects, which are extremely warranted by safe treatment techniques.

8.5 PHYTOTHERAPY: PREVENTIVE AND CURATIVE HERBAL APPROACH

Plants are a rich source of natural organic compounds with various biological activities, which have offered medicinal benefits for centuries [44]. It is estimated that nearly a whopping 80% of the world's population throw their trust in traditional medicines as the first line of medication. Despite their cultural significance, only about 6% of the medicinal plants have been studied for their biological activities [21].

Phytotherapy is related to research efforts on the use of natural extracts as palliative and curative medicinal products and/or adaptogenic agents [99]. Phytotherapy represents a pragmatic approach of engaging evidence-based remedies and the emerging research-related practice model in biomedicine. Holistic phytotherapy seeks to exploit the healing potential of plant-derived compounds as a potent therapeutic modality.

Researchers are now focusing on natural antioxidant-containing plant products as a means of tackling As poisoning. Compared to synthetic drugs, plant-derived natural compounds offer some added benefits, such as: they are non-toxic, much more convincing to the public, commonly accessible, and relatively cheap. Owing to their antioxidant, anti-tumorigenic, healing, and chelating characteristics, various phytochemicals are now being suggested to have remedial potential against As toxicity. Cancer chemoprevention with herbal phytochemicals is an evolving approach for cancer prevention, delay or cure cancer [8].

8.5.1 CHEMOPREVENTION

Cancer chemoprevention is a mechanism of cancer control wherein one or more vitamins, dietary supplements, and medication prevent the onset of this disease. Prevention is the most desirable way to eradicate the effect of cancer on humans. Figure 8.7 shows the processes, where the chemopreventive agents act to suppress the development of cancer. To achieve this goal, the first set of strategies is to eliminate predisposing factors. The determinants of the majority of human cancers still are uncertain. When carcinogenesis is initiated, chemoprevention would involve the inhibition of the neoplastic microevolution process [75, 77]. Chemoprevention can be practiced through the following three methods:

- **Primary Chemoprevention:** Using a medication, vitamin, or supplement to prevent cancer in a healthy person;
- **Secondary Chemoprevention:** Using a medication, vitamin, or supplement to prevent a pre-cancerous area from becoming cancer;
- **Tertiary Chemoprevention:** Using a medication, vitamin, or supplement in a person who has already had cancer, to prevent them from developing another cancer [29].

The characteristics of an ideal chemoprevention agent are:

- The agent does not cause bothersome side effects that might affect the quality of life;
- The agent should be inexpensive and easily available;
- It should be non-toxic and safe to consume purely as a prophylactic measure;
- It should be potent to prevent cancer with a low dosage and its mechanism of action should be well-studied [29].

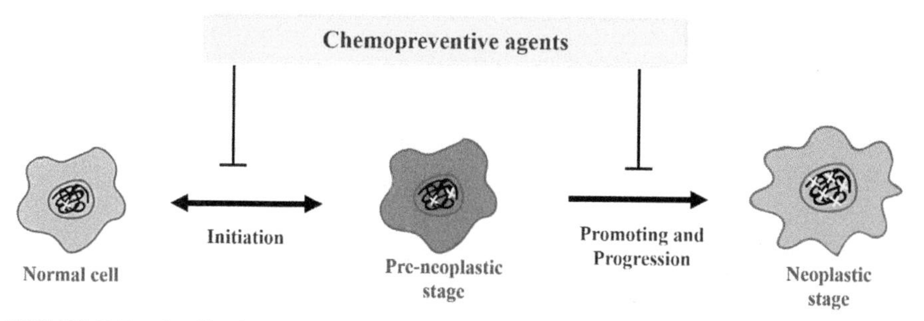

FIGURE 8.7 Application of chemoprevention in different stages of cancer.
Source: Self-developed based on Ref. [77].

8.5.2 TRADITIONAL MEDICINAL SYSTEMS

Traditional medicine refers to complementary or alternative medicine in many societies [47]. Herbal treatments (that is, rudimentary forms of phytotherapy) are the most popular form of traditional medicine.

8.5.2.1 THE INDIAN AYURVEDIC SYSTEM

The Ayurveda provides an orientation to the protection and management of various illnesses by medicinal products from whole plant or components of the plant. According to the Ayurveda principle, various plant products have the potential for health benefits. International organizations (such as: WHO, United Nations International Children's Emergency Fund (UNICEF), United States Agency for International Development (USAID)) have promoted the use of Ayurveda in medical research. Various such botanical products or phytomedicines have been used to maintain or augment health [21].

Rasayana (Sanskrit: rasa-essence; Ayana-going) involves the acquisition of nutrition, its circulation, absorption, and assimilation in the body tissues. The medicinal properties of these phytochemical drugs have been documented by the Ayurvedic practitioners and have been used for centuries to cure illness or as adaptogenic substances. Active principles that were extracted from the plants (such as: flavonoids, lignans, alkaloids, bisbenzyls, coumarins, and terpenes) have anti-oxidant properties. Ayurveda is not merely a collection of ancient alchemy but a comprehensive framework that vividly defines the symptoms of sickness and connotes them with a healing practice [32].

8.5.2.2 THE TRADITIONAL CHINESE SYSTEM

Traditional Chinese medical (TCM) practitioners resort to herbal drugs and different psychological and body practices (such as, acupuncture, and *tai chi*) to cure or prevent health issues. The ancient Chinese emperor, Shen Nung, is credited for the compilation of the first medicinal herbal literature, *Pentsao* in 2800 B.C. [40].

8.6 HERBS AND PHYTOCHEMICALS AGAINST ARSENIC-INDUCED CANCERS

Many of the available drugs for heavy metal poisoning are being accepted by patients due to their huge side effects. In contemporary times, plant-derived products have gained increasing interest to combat arsenic toxicity and As-induced cancer because of their safe, non-toxic, and efficacious anti-oxidative and chelating properties [5]. Natural resource antioxidant principles provide tremendous scope for correcting the body's pro-oxidant and antioxidant status imbalance.

There is an established positive correlation among dietary supplementation with various vegetables and plant components and the reduction of genotoxic and carcinogenic effects of various environmental contaminants, such as heavy metals and As. Natural antioxidants are found mostly in herbs and parts of the plant like wood, leaves, bark, roots, pods, etc. Some typical natural antioxidants are: vitamins, phenolic compounds, carotene, quercetin, cinnamic acid, flavonoids, and peptides [91]. In instances of As-induced cancer, research studies lead us to focus on the natural antioxidants to rectify vitiated homeostasis. The range of studies has yielded valuable outcomes causing a growing interest in the use of herbal extracts to tackle the problem of As toxicity [5].

8.6.1 BASIL (OCIMUM SANCTUM)

Ocimum sanctum (or *Ocimum tenuiflorum*: commonly called Tulsi in India) is a well-known aromatic medicinal plant [97]. Eugenol, present in the leaves, is a natural antioxidant as an efficient inhibitor of lipid peroxidation [28]. The essential oil of *O. sanctum* has antibacterial, anti-fungal, immunostimulatory, anti-carcinogenic, anti-diabetic properties and protects against radioactive damage. Tulsi also possesses anti-inflammatory and antioxidant properties. One of the mechanisms by which tulsi confers protection against cellular damage is via scavenging of free radicals [38].

Sharmila Banu et al. [89] reported protective effects of *O. sanctum* at a dose of 200 mg kg^{-1} per day in rats after chronic exposure to As (III). As intoxication inhibited the enzymatic action of δ-aminolevulinic acid dehydratase (ALAD), decreased the glutathione levels, and augmented the free oxygen radicals in the blood. These were reversed in case of Tulsi leaf extract. The drug showed a significant reversal of oxidative stress parameters and a marginally significant lowering of As titer in the blood and metabolic organs in the body. They recommended the use of Tulsi along with thiol chelator as adjuvant therapy.

8.6.2 ALOE VERA

Aloe vera (also known as the Aloe or Ghritakumari medicinal plant) is a succulent plant that is commonly grown in Asia, North Africa and other biogeographically tropical areas [73]. Laboratory studies have shown that *Aloe vera* possesses anti-ulcer, anti-diabetic, immunostimulant, anti-inflammatory, antiviral, anti-microbial, antibacterial, anti-tumor, anti-oxidant, and ROS scavenging activities [26].

Flora and his co-workers [26] investigated the preventive potential of *Aloe vera* against As-mediated lethality in rats and their outcomes suggest that the concomitant application of *Aloe vera* with As (III) avoids As-inflicted oxidative damage. However, it does not have any significant impact on blood and soft tissue As concentration. There was inhibition of the enzyme activity of ALAD, a slight decline in glutathione and an increase in blood zinc protoporphyrin. Leukocyte levels were lesser while most of the other blood parameters stayed unchanged after As poisoning.

Simultaneous administration of *Aloe vera* had great protective action on blood dehydratase activity inhibited by δ-aminolevulinic acid and helped restore the amount of blood glutathione. Although the signs of ROS scavenging were evident, yet there was not much effect on As titer in blood and liver. Protection was shown by most of the hepatic biochemical parameters that indicate ROS stress. There were also no significant changes in the renal biomarkers of As toxicity. The research, therefore, proposed that *Aloe vera* had very restricted preventive impacts on As toxicity [36]. Even though it showed some protection to As toxicity by virtue of its anti-oxidant properties, yet it does not support As clearance from the body. Thus, further investigation should be directed towards the use of *Aloe vera* as a candidate for adjuvant.

More recently, Zodape [108] attempted to understand the role of *Aloe vera* juice on As-induced toxicity on Rohu fish (*Labeo rohita*) and established effective hepato-protective and tissue-protective properties of *Aloe vera* against As toxicity. Fingerlings were subjected to sub-lethal concentration of a mixture of As and *Aloe vera* juice. The hepatic and muscular tissues showed a significant effect in decreasing the enzyme activities, like glutamate-oxaloacetate transaminase, glutamate-pyruvate transaminase, acid phosphatase, and alkaline phosphatase. Total protein, lipids, and glycogen levels were reduced and there was an increase in protease and free amino acid levels.

8.6.3 GARLIC (*ALLIUM SATIVUM*)

Allium sativum L. possesses ROS scavenging property [21, 49]. Recently, Flora et al. [23] reported As-induced hepatic manifestations mediated via generation of ROS, causing apoptosis through mitochondria-mediated pathways. These findings and the associated etiology also indicated that the administration of aqueous extract of *Allium sativum* prevented As poisoning by reducing As a burden, oxidative stress and hepatic apoptosis in mice.

8.6.4 ZINGER (*ZINGIBER OFFICINALE*) AND TURMERIC (*CURCUMA LONGA*)

Zingiber officinale is a herbaceous perennial plant [30] that originated from Island Southeast Asia. It was mainly grown in southern India and it was also carried by traders into the Middle East and the Mediterranean.

Curcuma longa is one of the oldest cultivated spice plants in the southeast Asian countries [52]. The plant is a perennial, rhizomatous, herbaceous plant that is rich in the phytochemical Curcumin [12].

Several herbs such as *Shunthi* (dried ginger root) and *Haridra* (turmeric) are used for As poisoning management in Ayurveda [6]. It has been described as a stimulant, analgesic, appetizer, anti-flatulent, antimicrobial, hypolipidemic, and anti-oxidant [45].

Biswas et al. [6] attempted to identify if the application of turmeric and ginger powder abet experimentally induced As toxicity in calves and found that the spice was significantly effective in eliminating As from the body. Ginger and turmeric also protect against potential harm from As exposure. For 90 days, the calves were given orally with sodium arsenite ($NaAsO_2$) at one mg kg body weight^{-1} per day. From 46th day onwards, 10 mg kg body weight^{-1} per day was performed orally with ginger and turmeric powder. The As content was estimated every 15 days in feces, hair, urine, and plasma, as well as different biochemical, hematological, and antioxidant parameters. Plasma and hair As concentrations were considerably decreased by ginger powder through enhanced excretion through feces, and urine. The amount of hemoglobin, erythrocyte, and leucocyte counts were also reduced, but these were considerably enhanced in communities treated with turmeric and ginger after 75th day. In experimental groups, blood urea nitrogen and plasma creatinine were also reduced considerably relative to the placebo group from 60 days onwards. After treating individuals with ginger and turmeric, the superoxide dismutase (SOD) and catalase activity were considerably restored. Thus, the phytotherapeutic option of chronic As amelioration could be ginger and turmeric.

8.6.5 TEA (*CAMELLIA SINENSIS*)

Camellia sinensis is used to make popular beverages, like teas and tea infusions. Tea has two principal varieties, namely, *C. sinensis* var. *assamica* and *C. sinensis* var. *sinensis*. While the Assamese variety of tea with large leaves is endemic to the Assamese belt, the sinensis variety of tea with short leaves is grown in the Sino-Tibetan region [80]. Tea has abundant bioactive compounds, such as: alkaloids, amino acids, and flavonols. A diverse group of compounds (such as: catechins, theaflavins, epicatechin, epigallocatechin, epicatechin gallate, epigallocatechin-3-gallate, etc.), are found in leaves of *C. sinensis* and these have antioxidant, anticancer, and anti-inflammatory potentials [71].

Sinha et al. [93] concluded that extracts of green and black tea varieties, as well as their principal polyphenols (e.g.: (–)-epigallocatechin gallate and theaflavin) effectively mitigated the cytotoxic impacts of As compounds *in vitro.* The study was relevant in not only proving the therapeutic potential value of tea phytochemicals but also stressing on the usefulness of tea intake to prevent adverse effects of exposure to As in developing countries, such as, Bangladesh.

8.6.6 MORINGA (MORINGA OLEIFERA)

Moringa oleifera is found in the Himalayas, African, and Arabian countries [56]. The leaves of *Moringa* are known to have elevated protein, mineral, and vitamin content, and sulfur-containing amino acids (such as: methionine and cysteine) [22]. High levels of essential amino acids, mineral ions and vitamins make *Moringa* an appropriate dietary additive. Taking its leaves (new or dried powder) regularly can help in avoiding anemia and malnutrition [26].

Moringa has been documented to possess antidiabetic, anti-gastric ulcers, antitumor, antimicrobial, antioxidant properties; and it offers protection from radiation and controls high blood pressure. Its roots are useful in hysteria, flatulence, and epilepsy. The leaves, flowers, roots, fruits, and pods are used for the therapy of ascites, rheumatism, venomous bites, inflammation, liver disease, cancer, and pulmonary, hematological, and renal dysfunction. The benefits of using this plant are: no side effects, wide availability, and cost-effectiveness [26].

The *M. oleifera* seed powder has coagulating property that has been used for multiple elements of water purification, including turbidity decrease, alkalinity, and hardness check, and estimation of total dissolved solids (TDS). Some investigations have also shown biosorption behavior for the removal of toxic metals from water bodies [3, 15].

Saijdu et al. [83] documented successful removal of heavy metal ions (such as: lead, As, iron, and cadmium) with the use of seed kernels of *M. oleifera.* Fourier transform infrared (FTIR) spectrometry was used to detect protein/amino acid-arsenic interactions that are involved in the biosorption power in the seeds of *M. oleifera.*

Gupta and his co-workers [37] reported that *M. oleifera* seed powder can reduce As burden in body and ameliorate arsenic-induced oxidative stress via enhancing antioxidant enzymes [37]. *M. oleifera* possesses numerous antioxidants (such as: ascorbate, β-carotene, vitamin E, vitamin A, β-sitosterol and phenolics (i.e., quercetin and kaempferol), flavonoids, and anthocyanins. The seed powder of *Moringa* has elevated amounts of protein-rich in cysteine and methionine, and antioxidative vitamins and phytochemicals. These chemicals have anti-arsenic toxicity effects.

Mishra et al. [64] delineated the protective role of *M. oleifera* and as an adjuvant with chelation therapy against As in mice. In As or most heavy metal chronic poisoning, metal distribution and redistribution constitute an important factor for therapeutic outcomes. The results have shown that the distinctive potential of *M. oleifera* in combination therapy with chelating agents is not only to recover from modified biochemical parameters linked to oxidative damage but also to mobilize As from soft tissue and blood, after exposure to As.

8.6.7 CENTELLA (*CENTELLA ASIATICA*)

Centella asiatica is grown in Sri Lanka, Northern Australia, Indonesia, Iran, Malaysia, Melanesia, New Guinea and other regions of Asia. It is also referred to as Brahmi Gotu Kola, Asiatic Pennywort, Indian Pennywort, Luei Gong Gen, Antanan, and Pegaga.

Centella asiatica revitalizes the brain and nervous system and helps to increase attention and concentration. It also has tranquilizing and anti-aging effects. It helps to overcome the negative effects of fatigue and stress. Furthermore, it also has a revitalizing effect on neurotransmitters (biogenic amines). It is a potent source of antioxidants and protects against membrane peroxidation [11, 70].

Saxena and Flora [86] discovered that *Centella asiatica* was efficient in providing more pronounced chelation therapy-related effects, particularly in recovering oxidative stress parameters. They also found that concomitant administration of *Centella asiatica* and MiADMSA was able to restore the altered neurotransmitters (norepinephrine, dopamine, and acetylcholine) levels to normal as compared with the effects of chelators alone. Although *Centella asiatica* is a significant medicinal plant with antioxidant activity, yet it has not been adequately studied for its potential in terms of toxicity regulation induced by As exposure.

Gupta and Flora [35] have shown that *Centella asiatica* aqueous extract at a concentration of 200 and 500 mg kg body weight^{-1} can significantly protect animals from arsenic-inflicted damage. They proposed that aqueous extract from *Centella asiatica* may be mediated to some extent against arsenic-induced toxicity, but its effects may not be as appreciable when used as a chelating agent.

8.6.8 GOOSEBERRY (PHYLLANTHUS EMBLICA)

Phyllanthus emblica is commonly known as Indian gooseberry or Amla [74]. Its fruits are rich source of ascorbic acid and are widely used in conventional Unani and Ayurvedic medicinal practices in India for the therapy of broad range of illnesses, including scurvy and ulceration [14]. *Phyllanthus emblica* is rich in natural antioxidants, like catalase, SOD, and glutathione peroxidase [26]. Research has shown that Amla is a potent immune-stimulator, as well as an adaptogen that can increase tolerance to stress. The fruits of this powerful *Rasayana* herb possess expectorant, cardiotonic, antioxidative, antiviral, and anti-emetic and antivenin activities. Dried fruit acts as a remedy for hemorrhage, diarrhea, and dysentery. It is used in conjunction with iron to treat iron deficiency, jaundice, and indigestion [26]. *Phyllanthus emblica* serves to protect against number of heavy metals. It has been shown to prevent DNA and cell damage induced by As poisoning to a significant extent [31].

Sharma et al. [88] recently reported strong protective role of the fruits of *Emblica officinalis* against As-induced hepatopathy in mice. Pre- and post-administration of *E. officinalis* extract at a dose of 500 mg kg body weight^{-1} per day for 30 days against As (4 mg kg body weight^{-1} per day) could significantly reverse the oxidative stress and liver function parameters. Biochemical results were supported by histopathological findings indicated the promising potential of *E. officinalis* in As-induced oxidative stress and its possible utilization for adjuvant therapy.

8.6.9 SEA-BUCKTHORN (HIPPOPHAE RHAMNOIDES)

Hippophae rhamnoides has immense therapeutic value in various oxidative stress disorders [55]. Potent medicinal essential oil from the fruit of *Hippophae rhamnoides* has been used in treating cardiovascular disorders, heal skin burns, eczema (atopic dermatitis) and radiation exposure. Furthermore, it can also be used for preventing gastrointestinal diseases [61]. The fruit is a good astringent and is used as a tonic. It contains essential fatty acids and antioxidants. Therefore, it is capable of reducing the incidence of cancer, flu, cardiovascular diseases, mucosal injuries, and skin disease. Furthermore, Sea Buckthorn also has protective functions on chromosomal structure, antibacterial, antiulcer, radioprotective, and immunomodulatory activities [26].

Gupta and Flora [34] recently reported that fruits of *Hippophae rhamnoides* are generally found near the shoreline, forming thickets on fixed dunes and sea rocks. *Hippophae rhamnoides* L. extract has an important shielding function against oxidative and tissue harm caused by As. Prognosis with different extracts of *H. rhamnoides* L. fruit offers significant protection against As inhibition of blood ALAD activity, which has shown decrease in blood concentrations of glutathione. Although the water extract of the Sea Buckthorn fruit does not chelate As and was unable to significantly decrease As concentration from the tissue, yet it has a promising role as a supplementing agent during chelation of As. Figure 8.8 indicates some common herbs with potential for reducing As-induced toxicity.

FIGURE 8.8 Promising plants that are rich sources of phytochemicals to ameliorate the arsenic-induced toxicity.

Legend: Wild like tulsi leaves (A); aloe vera (B); garlic (C); ginger (D); turmeric (E); tea (F);moringa (G);bramhi (H);amla (I); and sea-buckthorn (J).

Source: Self-developed.

8.6.10 *RHABDOSIA RUBESCENS*

Rabdosia rubescens has been used by Chinese physicians to treat inflammation of neck, insects, and snake bites, infection of the tonsils and cancer of the esophagus, stomach, liver, prostate, and breast [107]. Oridonin or rubesecens in A is an herbal diterpenoid compound isolated from *R. rubescens* that has been shown to protect against As toxicity and has been recognized as targeting a fresh pathway for its positive impact [72].

Similarly, many other plants from different biological distribution, like *Andrographis paniculata* (kalmegh), *Glycyrrhizae radix*, etc., have proven their effectiveness against As poisoning [21].

8.6.11 SOME ARGENTINIAN HERBS

Soria et al. [95] studied the effects of different solvent extracted biomolecules on As-inflicted oxidative stress in Vero-cells. Bioactive compounds were extracted from five Argentinian medicinal plants. Arsenite has been found to be pro-oxidative agent that operates in a time-dependent way. Extracts from *Eupatorium huniifolium* (dissolved in petroleum ether), *Lanlanagrise bacbii* (dissolved in petroleum ether and in water), *Mandevilla pentlandiana* (dissolved in petroleum ether and in water), and *Sebastiania commersoniana* (dissolved in dichloromethane, methanol, and in water) deterred the development of both aqueous and lipid hydroperoxides, but *Heterotbalamus alienus* avoided only lipid hydroperoxides. As a result, antioxidant extracts of these Argentinian herbs can possibly have anti-arsenite-induced renal damage protection. Amongst these, the *L. graisebacbi* decoction was suggested to be most suitable for human consumption, since it is traditionally added in tea as a herbal condiment.

8.7 PHYTOCHEMICALS AGAINST ARSENIC TOXICITY

The extracts or parts of medicinal herbs have been a part of many folk medicinal systems. However, the current research scenario relies on studying medicinal herbs as a source to isolate therapeutically promising molecules. Recent scientific reports suggest that phytotherapeutics of herbal origin show some efficacious and encouraging amelioration from As-mediated toxicity. Some natural compounds from plant sources, such as: rutin, resveratrol, curcumin, polyphenols, etc., have been identified. They are potential to reduce As-induced toxicity. Structures of some the phytochemicals involved in arsenic amelioration are represented in Figure 8.9.

8.7.1 RUTIN

Sárközi et al. [85] investigated the effects of flavonoid rutin in ameliorating As-induced toxicity. Wister rats were administrated As by gavage (@ 10 mg (kg body weight)$^{-1}$ of $NaAsO_2$ daily) on open field motility for 6 weeks.

In the experimental group, an aqueous solution of rutin (2 gL^{-1} daily) was co-administered to check whether the effects of evoked cortical and peripheral electrophysiological activity were able to increase body weight in rats. The functional neurotoxic impacts and overall toxicity of As exposure in rats could be significantly reduced by parallel oral administration of rutin. While the gain in body weight in the As-only group was considerably lowered the 4th week onwards, and the rutin in the experimental group mainly removed this impact. Rats treated with As alone showed reduction in open field motility; enhanced cortical potentials latency; and reduced peripheral velocity of the nervous impulse. Co-administration of rutin also counteracted these functional changes; and the antioxidant and chelating activities of rutin could have contributed to the ameliorative impact. These findings appear to promote natural agents' potential function in maintaining human health in a As-contaminated setting.

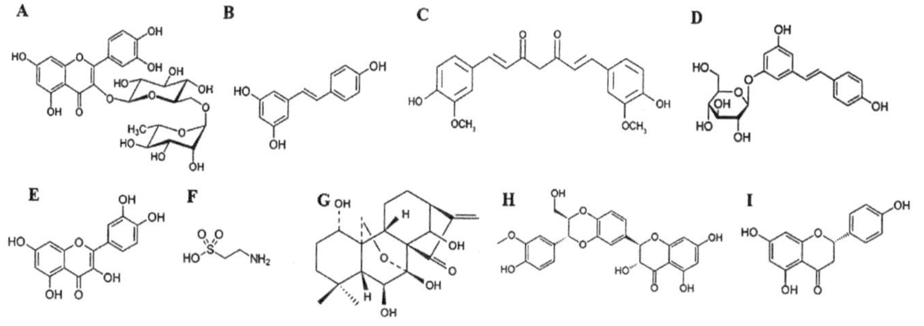

FIGURE 8.9 Structure of some the phytochemicals involved in arsenic amelioration: rutin (A), resveratrol (B), curcumin (C), polydatin (D), Quercetin (E), taurine (F), Oridonin (G), Silibinin (H) and Naringenin (I). (self-developed, figure compiled by authors).

8.7.2 RESVERATROL

Resveratrol is a phytoalexin having antioxidant and chemo preventive characteristics that are found in vegetables. Zhang et al. [106] accessed how resveratrol modulates As-induced cancer by using Arsenic trioxide (As_2O_3)-exposed male Wister rats as a model. In the placebo group, rats were treated with only As_2O_3 (@ 3 mg (kg boy weight)$^{-1}$ daily) and had significantly higher oxidative stress and As accumulation in the kidney. However, the experimental group (in which resveratrol (8 mg kg body weight^{-1} daily) was pre-administered one hour before As treatment) did not show these detrimental signs. In addition, the experimental group showed lower levels of blood urea nitrogen, creatinine, and insignificant renal tubular epithelial cell necrosis. The data clearly showed that resveratrol supplementation alleviated nephrotoxicity by enhancing antioxidant capacity and As efflux. These results indicate that use of resveratrol in communities exposed to As has the ability to safeguard against renal damage.

8.7.3 CURCUMIN

Curcumin from turmeric (*Curcuma longa*) issued to treat a broad variety of diseases. Khan et al. [53] investigated the effects of curcumin on apoptosis brought about by *in vitro* administration of $NaAsO_2$ in murine splenocytes. Cells were subjected to $NaAsO_2$ with and without curcumin. $NaAsO_2$ led to a decrease in cell viability and induction of apoptosis. These findings were consistent with increase in cell numbers with the production of free radicals (ROS), loss of mitochondrial transmembrane capacity, increased cell frequency with sub-G1 DNA content, and fragmentation of DNA. Conjunctive curcumin therapy (@ 5 or 10 $\mu g\ mL^{-1}$) with $NaAsO_2$ (@ 5 μM) resulted in cell viability recoveries and mitigation of molecular modifications induced by apoptosis. Substantial protection in murine splenocytes against apoptosis parameters proposed curcumin's protective efficacy.

8.7.4 POLYPHENOLS

Polyphenols (flavonoids, phenolic acids, lignans, and stilbenes) have antioxidant characteristics [100]. They have roles in preventing diseases like cardiovascular and neurodegenerative diseases and cancer [57]. They are also involved in the actions of enzymes and cell receptors [62].

Arslan-Acaroz et al. [4] investigated the impact of polydatin in As-exposed rats on free-radical overproduction. Thirty-five rats were randomly divided into five groups (7 rats per group). Whereas the control group was given just physiological saline, the second group was administered with only As (@ 100 mg/L) for 60 days through drinking water. The other groups were treated orally at various concentrations (@ 50, 100, and 200 mg (kg body weight)$^{-1}$ per day) of polydatin in addition to As. While As treatment enhanced malondialdehyde concentrations, glutathione levels in blood, liver, kidney, brain, lung, and heart of the rats were reduced. Moreover, As reduced the activity of SOD and catalase in erythrocytes, liver, kidney, brain, lung, and heart of rats. Besides, As therapy improved interleukin one beta (IL-1β), nuclear factor kappa-B (NFκB), p53, and tumor necrosis factor-α (TNF-α) expression in the lung, brain, kidney, and liver. However, application of polydatin alleviated As-exposed lipid peroxidation, antioxidant enzyme activity, DNA damage, gene expressions, and brain neuronal degeneration and focal gliosis.

Fatmi et al. [20] assessed the potential of green tea extracts, which are rich in polyphenolic content, as a natural means to counter the oxidative cellular damage inflicted by As. Authors found that green tea polyphenols can also prevent further generation of free radicals. In their studies, the treatment groups were given chronic exposure of $NaAsO_2$ (3 mg (kg body weight)$^{-1}$ daily) orally, accompanied by administration of *Camellia sinensis*@ 300 mg (kg body weight)$^{-1}$ daily for 6 weeks using gavage technique. Administration of fluid extricate of green tea had a noteworthy ameliorative impact on different biochemical parameters and lipid peroxidation levels. The defensive mechanism of these anti-oxidative agents was manifested in the form of progressive normalization of renal and hepatic functions and normal expression of enzymes and other markers.

8.7.5 QUERCETIN

Mishra and Flora [63] explored effectiveness of quercetin and monoisoamyl 2,3-dimercaptosuccinic acid (MiADMSA, thiol chelating agent), either separately or in conjunction, against oxidative stress caused by As and As clearance in the mouse. Animals were chronically subjected to 25 ppm of arsenite through their drinking water for 12 months, followed by MiADMSA therapy (@ 0.2 mmol (kg body weight)$^{-1}$ per day orally), quercetine (@ 0.2 mmol per day orally) either alone or in conjunction for 5 successive days. As exposure resulted in a substantial decrease in blood activity (ALAD), glutathione, white blood cells (WBC) and red blood cells (RBC). It also enhanced the number of platelets and caused substantial increase in erythrocyte ROS levels.

Activities of hepatic catalase and glutathione peroxidase showed depletion, whereas concentrations of thiobarbituric acid reactive substances (TBARS) increased exposure to As, thus suggesting oxidative stress in the blood and liver was caused by arsenite. Catalase activity in kidney showed deterioration, while As exposure increased TBARS concentrations. High concentration of As in blood, liver, and kidney accompanied these biochemical modifications. MiADMSA treatment was efficient in enhancing ALAD activity, while quercetin alone was ineffective. When co-administered with MiADMSA, Quercetin also provided no additional beneficial effect on the activity of blood ALAD but significantly altered the number of platelets.

Quercetin administration alone had profound beneficial impacts on the concentrations of hepatic oxidative stress and TBARS levels in the kidney. As and any of the treatments could not impact renal biochemical factors. Interestingly, mixed quercetin administration with MiADMSA had a notable impact on the depletion of complete blood and soft tissue As concentration.

8.7.6 SILYMARIN

Silymarin is a combination of certain flavonolignans that are extracted from milk thistle (*Silybum marianum* Gaertneri). Soria et al. [96] studied role of silymarin, and it was found that silymarin could not provide any beneficial effect. The study was conducted with mammary adenocarcinoma lines from human source, Michigan Cancer Foundation-7, (MCF-7) and ZR-75-1. Cells were *ex vivo* subjected to 200 μMNaAsO$_2$, 5 μM silymarin and/or 50 μM quercetin. Researchers observed some biomembrane parameters, such as sialic acid in gangliosides, γ-glutamyl transpeptidase (GGT) enzymatic activity, conjugated dienes, and free radical generation, to evaluate the arsenite-flavonoid interactions. Flavonoids in ZR-75-1 cells did not prevent time-dependent arsenite toxicity, while quercetin protected MCF-7 cells for 8 h.

Regarding the GGT activity, ZR-75-1 cells were shielded from stress only by quercetin. In MCF-7 cells, neither quercetin nor silymarin counteracted the arsenite-induced GGT activity. As and silymarin-treated As showed the decrease in the content of sialic acid in the MCF-7 membrane only. The membrane resistance in these cells to lipid oxidation was primarily due to GGT activity and sialylglycolipid content up-regulation.

8.7.7 TAURINE

Flora et al. [25] evaluated the efficacy of taurine in alleviating the harmful effects of As poisoning in rats, in conjunction with the chelator and mono isoamyl dimercaptosuccinic acid (MiADMSA). Male rats were chronically exposed to As (@ 25 ppm of NaAsO$_2$ in water for 24 weeks), following which they were treated with taurine (daily intraperitoneal injection of 100 mg (kg of body weight^{-1}) and MiADMSA (daily oral administration of 50 mg kg^{-1}), either individually or in combination for 5 days in succession. The evaluation of biochemical parameters indicated the presence of reactive oxygen species (ROS) along with elevated levels of As in the blood, liver, and kidney.

Taurine addition considerably decreased hepatic oxidative stress markers, but co-administering a greater dose of taurine and chelator medication offered improved outcomes in renal and hepatic antioxidant status. It also decreased body As stress, relative to individual MiADMSA or taurine therapy. The findings indicate that co-administration of taurine with MiADMSA may be preferred to obtain better impacts of chelation therapy.

8.7.8 ORIDONIN

Oridonin was isolated from *Rabdosia rubescens*, *Isodon japonicus* Hara, and *Isodon trichocarpus* [27], and it is a bitter tetracycline diterpenoid compound having certain anti-cancer characteristics. Du et al. [18] assessed the modulation by phytochemical oridonin of the cellular anti-oxidant regulator, nuclear-related factor 2 (Nrf2). The Nrf2 signaling pathway is cytoprotective and safeguards the cells from As-induced toxicity [51].

Pre-treatment of human urothelial cell line (UROtsa) with 1.4 μM oridonin caused the Nrf2 signaling pathway to be activated (at a low subtoxic dose of As, 30 uM As (III)). The Nrf2 protein was accumulated and stabilized due to the blocking of proteasomal degradation of Nrf2. The activation of the Nrf2-dependent pathway provided cytoprotective response. The cellular redox capacity was significantly enhanced, ROS levels were reduced, and the cells survival were alive upon As treatment.

8.7.9 SILIBININ

Silibinin is a principal flavonolignan compound that is found in milk thistle or silimarin (*Silybum marianum*). The spatial orientation of functional groups gives silibinin its potential for metal chelation and antioxidant potency [76].

Muthumani et al. [66] evaluated silibinin's ameliorative potential towards As poisoning affecting the liver in rats. As alone (@ 5 mg (kg body weight)$^{-1}$ per day) or silibinin (@ 75 mg (kg body weight)$^{-1}$ per day) were orally fed to rats for 4 weeks. The enhanced activity of serum liver-specific enzymes was used as an indicator for hepatotoxicity. Aspartate transaminase, alanine transaminase, alkaline phosphatase, γ-glutamyl transferase, lactate dehydrogenase enhanced concentration of lipid peroxidative marker.

The toxicity of As is also reflected in the reduced activities of membrane-bound ATPases, natural cellular anti-oxidants, like SOD, catalase, glutathione peroxidase, glutathione-S-transferase, glutathione reductase, and glucose-6-phosphate dehydrogenase along with non-enzymatic anti-oxidants, like glutathione, total sulfhydryl groups, ascorbate, and tocopherol. The dispensation of silibinin showed a substantial regression of As-induced toxicity in the liver tissue. Hepatic marker and bilirubin concentrations were reduced. Compared to As-treated rats, the concentrations of lipid peroxidative markers in rat liver were considerably lower. DNA lesions in hepatocytes were also decreased.

8.7.10 NARINGENIN

Naringenin is a flavanone, which is found in grapefruit, and other citrus fruits and plants (such as: tomato, oranges, lemons, and sweet lime).

Roy et al. [79] studied detoxifying potential of citrus-based naringenin against As-treated Swiss albino mice. In the placebo group, As trioxide (@ 3 mg kg^{-1}) was orally administered for continuous 15-days. In the experimental group, Naringenin (@ 5 mg kg^{-1} and 10 mg kg^{-1}) was orally given to mice at 30 min prior to the oral administration of As trioxide (@ 3 mg kg^{-1}). On the 15th day, number of physical (body weight and organ weights) and biochemical parameters (ALAD content, hemoglobin content, blood corpuscles counts, aspartate aminotransferase, alanine aminotransferase, alkaline phosphatase, serum bilirubin, total protein content, hepatic, and renal tissue anti-oxidative potential, glutathione levels and anti-oxidative enzyme activity) were observed. The oral feeding of nargininin As-exposed mice markedly brought down their body mass indices, organ weight, hematological, and blood serum profiles to normal levels and appreciably adjusted all of the hepatic and renal tissue biochemical parameters and extent of DNA fragmentation in As-exposed mice.

8.8 EMERGING ISSUES AND FUTURE PERSPECTIVES

For their anti-cancer characteristics, the plant-based bioactive compounds (or phytochemicals) are being widely investigated. Despite the excellent anticancer potential of various phytochemicals, low solubility in aqueous environment and insufficient bioavailability limit the efficacy of phytochemicals. This is among the major challenges in this area of food biotechnology.

According to U. S. Food and Drug Administration (FDA), bioavailability is "*the rate and extent to which the therapeutic moiety is absorbed and becomes available to the site of drug action*" [7]. This makes their action ineffective due to the lack of targeted action. Their absorption in the body system is often not satisfactory and large doses of such phytochemicals are excreted from the body [87].

Research reports suggest that combinatorial action of various phytochemicals can help in making them more bioavailable to the system. The emergence of nanotechnology is now trying to overcome this challenge.

There has been exhaustive research on nano-particle mediated drug delivery for cancer. In comparison with their free form, the phytochemicals are more soluble and bio-available when delivered by the nano-carriers and exhibit heightened action on tumor cells. Besides nanoparticles, the various carriers are used such as: micelles, dendrimers, and liposomes. Further, nanoparticle mediated delivery significantly reduces dosage of the phytochemical. Taken together, extensive scientific research on these phytochemicals delivered through nano-carriers foster the potential of "cancer nanotechnology," which will epitomizenano-carriers as more of a combatant in the improvisation of cancer chemotherapy by curtailing undesirable side-effects and boosting targeted drug delivery [33].

8.8.1 ENCAPSULATION AND IMMOBILIZATION OF PHYTOCHEMICALS

Three general drawbacks of the administration of phytochemicals are [65]:

- Phytochemicals do not have long-term stability and are light and heat-sensitive;
- Polyphenols are sparingly soluble in water and hence have a poor bioavailability; and
- Phytochemicals often have a very bitter taste, making it hard to administer orally.

Researchers have created number of delivery systems and stabilization techniques to avoid these detriments. Among the available methods of delivery systems, encapsulation is a promising and exciting option. Encapsulation involves coating the phytochemical of interest with natural, or artificial, biodegradable polymers, or lipids [9]. Numbers of encapsulation methods have been developed, such as:

- Physical methods, which include spray-drying, extrusion-spheronization, methods that use supercritical fluids, etc.

- Physicochemical methods, which include spray-cooling, hot melt coating, ionic gelation, solvent evaporation-extraction, etc.
- Chemical methods, which include interfacial polymerization, *in situ* polymerization, interfacial cross-linking, etc.

Recent advances include the design of Edible Nano-encapsulation Vehicles [105] for immobilization and delivery of phytochemicals. It greatly enhances the bioavailability of the phytochemicals, and also impacts their bio-efficacy by influencing their bio-distribution and renders them stability in the gut.

While a number of chemical substances have been used for encapsulation (such as: silica, magnetic compounds, etc.), number of natural proteins like soy protein have also been used [42, 84].

8.9 SUMMARY

Number of such herbal formulations are more prophylactic than therapeutic against As-mediated toxicity, indicating a role in chemoprevention or adjuvant therapy. Nevertheless, a long path remains to be followed in this direction of toxicological research. Further research needs to be done so that the mechanisms of action of these phytochemicals are clearly understood. Before these phytochemicals could be recommended as medicinal supplements, some aspects need to be reviewed. The biological interactions occur between the phytochemicals and other *in vivo* complexes, and the signals being generated need to be studied. Finally, the appropriate dosage of the phytochemicals that can be clinically recommended need to be established.

ACKNOWLEDGMENTS

Father Xavier Savarimuthu acknowledges the financial and infrastructural support by Fogarty International Center of NIEHS), the UC Berkeley International Training and Research Program in Environmental and Occupational Health (ITREOH). He also appreciates Dr. Ondine von Ehrenstein (Program Coordinator of Arsenic Research in India) for her help in the research work. The authors thank Rev. Dr. Dominic Savio, S. J., Rector, and Principal, St. Xavier's College (Autonomous), Kolkata, for support in all academic endeavors.

KEYWORDS

- **antioxidants**
- **arsenic toxicity**
- **chemoprevention**
- **ecotoxicants**
- **phytochemicals**
- **xenobiotics**

REFERENCES

1. Ahoulé, D. G., Lalanne, F., Mendret, J., Brosillon, S., & Maïga, A. H., (2015). Arsenic in African waters: A Review. *Water Air Soil Pollution, 226* (302), 1–13.
2. Ali, A., (2014). *Arsenic Toxicity*. Published in: Health & Medicine, https://www.slideshare.net/AyaAli3/as-toxicity-final (accessed on 5 August 2020).
3. Amagloh, F. K., & Benang, A., (2009). Effectiveness of *Moringa oleifera* seed as coagulant for water purification. *African Journal of Agricultural Research, 4* (2), 119–123.
4. Arslan-Acaroz, D., Zemheri, F., Demirel, H. H., Kucukkurt, I., Ince, S., & Eryavuz, A., (2018). *In vivo* assessment of polydatin, a natural polyphenol compound, on arsenic-induced free radical overproduction, gene expression, and genotoxicity. *Environmental Science and Pollution Research, 25* (3), 2614–2622.
5. Bhattacharya, S., (2017). Medicinal plants and natural products in amelioration of arsenic toxicity: A short review. *Pharmaceutical Biology, 55* (1), 349–354.
6. Biswas, S., Maji, C., Sarkar, P. K., Sarkar, S., Chattopadhyay, A., & Mandal, T. K., (2017). Ameliorative effect of two ayurvedic herbs on experimentally induced arsenic toxicity in calves. *Journal of Ethnopharmacology, 197*, 266–273.
7. Blume, H., Schug, B., Tautz, J., & Erb, K., (2005). Guidelines for the assessment of bioavailability and bioequivalence (in German). *Bundesgesundheitsblatt Gesundheitsforschung Gesundheitsschutz (Federal Health Gazette Health Research Protection), 48* (5), 548–555.
8. Bode, A. M., & Dong, Z., (2009). Cancer prevention research-then and now. *Nature Review Cancer, 9* (7), 508–516.
9. Bonifácio, B. V., Silva, P. B., Ramos, M. A., Negri, K. M., Bauab, T. M., & Chorilli, M., (2014). Nanotechnology-based drug delivery systems and herbal medicines: A review. *International Journal of Nanomedicine, 9*, 1–15.
10. (2017). *Centella asiatica (Asiatic pennywort)*. Invasive Species Compendium, CABI. https://www.cabi.org/isc/datasheet/12048 (accessed on 5 August 2020).
11. Chandraprabha, D., Annapurani, S., & Murthy, N., (1996). Inhibition of lipid peroxidation by selected medicinal plants. *The Indian Journal of Nutrition and Dietetics, 33*, 128–132.
12. (2018). *Curcuma longa L.* Plants of the World Online; Kew Science, Kew Gardens, Royal Botanic Gardens, Kew, England. http://www.plantsoftheworldonline.org/taxon/urn:lsid:ipni.org:names:796451-1 (accessed on 5 August 2020).

13. Das, N. K., & Sengupta, S. R., (2008). Arsenicosis: Diagnosis and treatment. *Indian Journal of Dermatology, Venereology, and Leprolopy*, *74*, 571–581.
14. Dasaroju, S., & Gottumukkala, K. M., (2014). Current trends in the research of *Emblica officinalis* (Amla): Pharmacological perspective. *International Journal of Pharmaceutical Sciences Review and Research*, *24* (2), 150–159.
15. Delelegn, A., Sahile, S., & Husen, A., (2018). Water purification and antibacterial efficacy of *Moringa oleifera* Lam. *Agriculture and Food Security*, *7*, 25–30.
16. Dey, T. K., Banerjee, P., Bakshi, M., Kar, A., & Ghosh, S., (2014). Groundwater arsenic contamination in West Bengal: Current scenario, effects, and probable ways of mitigation. *International Letters of Natural Sciences*, *13*, 45–58.
17. Díaz, J. A., Serrano, J., & Leiva, E., (2018). Review bioleaching of arsenic-bearing copper ores. *Minerals*, *8* (5), 215–220.
18. Du, Y., Villeneuve, N. F., Wang, X. J., Sun, Z., Chen, W., Li, J., Lu, H., et al., (2008). Oridonin confers protection against arsenic-induced toxicity through activation of the Nrf2-mediated defensive response. *Environmental Health Perspectives*, *116* (9), 1154–1161.
19. Faita, F., Cori, L., Bianchi, F., & Andreassi, M. G., (2013). Arsenic-induced genotoxicity and genetic susceptibility to arsenic-related pathologies. *International Journal of Environmental Research and Public Health*, *10* (4), 1527–1546.
20. Fatmi, N., Fatima, N., Shahzada, M. Z., Sharma, S., Kumar, R., Ali, M., & Kumar, A., (2018). Effect of aqueous extracts of green tea in arsenic induced toxicity in mice. *Open Journal of Plant Science*, *3* (1), 1–14.
21. Ferreira, P., Palmer, J., McKenna, E. B., & Gendron, G., (2015). Traditional elder's anti-aging cornucopia of North American plants Maria. Chapter 1; In: Watson, R., (ed.), *Foods*, and *Dietary Supplements in the Prevention and Treatment of Disease in Older Adults* (pp. 3–11). New York: Academic Press (Elsevier).
22. Ferreira, P. M. P., Farias, D. F., Oliveira, J. T. A., & Carvalho, A. F. U., (2008). *Moringa oleifera*: Bioactive compounds and nutritional potential. *Revista de Nutrição (Nutrition Magazine)*, *21* (4), 431–437.
23. Flora, S. J. S., Mehta, A., & Gupta, R., (2009). Prevention of arsenic-induced hepatic apoptosis by concomitant administration of garlic extract in mice. *Chemico-Biological Interactions*, *177*, 227–233.
24. Flora, S. J., & Pachauri, V., (2010). Chelation in metal intoxication. *International Journal of Environmental Research and Public Health*, *7* (7), 2745–2788.
25. Flora, S. J., Chouhan, S., Kannan, G. M., Mittal, M., & Swarnkar, H., (2008). Combined administration of taurine and monoisoamyl DMSA protects arsenic-induced oxidative injury in rats. *Oxidative Medicine and Cellular Longevity*, *1* (1), 39–45.
26. Flora, S. J. S., Mittal, M., Gupta, R., & Pant, S. C., (2010). Arsenic toxicity: Biochemical effects, mechanism of action and strategies for the prevention and treatment by chelating agents and herbal extracts. Chapter 25; In: Ray, A., & Gulati, K., (eds.), *Recent Advances in Herbal Drug Research and Therapy* (pp. 401–448). I.K. International Publishing House Pvt. Ltd., New Delhi, India.
27. Fujita, T., Takeda, Y., Sun, H. D., Minami, Y., Marunaka, T., Takeda, S., Yamada, Y., & Togo, T., (1988). Cytotoxic and antitumor activities of *Rabdosia* diterpenoids. *Planta Medica*, *54*, 414–417.
28. Geetha, K. R., & Vasudevan, D. M., (2004). Inhibition of lipid peroxidation by botanical extracts of *Ocimum sanctum*: *In vivo* and *in vitro* studies. *Life Science*, *76*, 21–28.

29. Ghosh, D., (2016). Seed to patient in clinically proven natural medicines, Chapter 64; In: Gupta, R. C., (ed.), *Nutraceuticals: Efficacy, Safety, and Toxicity* (pp. 925–931). New York: Academic Press (Elsevier).
30. (2016). *Ginger*. NCCIH Herbs at a Glance. US NCCIH. https://nccih.nih.gov/health/ginger (accessed on 5 August 2020).
31. Govindarajan, R., Vijayakumar, M., & Pushpangadan, P., (2005). Antioxidant approach to disease management and the role of *Rasayana* herbs of Ayurveda. *Journal of Ethanopharmacology, 2*, 165–178.
32. Greenlee, H., (2012). Natural products for cancer prevention. *Seminars in Oncology Nursing, 28* (1), 29–44.
33. Gunasekaran, T., Haile, T., Nigusse, T., & Dhanaraju, M. D., (2014). Nanotechnology: An effective tool for enhancing bioavailability and bioactivity of phytomedicine. *Asian Pacific Journal of Tropical Biomedicine, 4* (1), S1–S7.
34. Gupta, R., & Flora, S. J. S., (2006). Protection against arsenic-induced toxicity in Swiss albino mice by fruit extracts of *Hippophae rhamnoides*. *Human Experimental Toxicology, 25* (6), 285–295.
35. Gupta, R., & Flora, S. J. S., (2006). Effect of *Centella asiatica* on arsenic induced oxidative stress and metal distribution in rats. *Journal of Applied Toxicology, 26*, 213–222.
36. Gupta, R., & Flora, S. J. S., (2005). Protective value of *Aloe vera* against some toxic effects of arsenic in rats. *Phytotherapy Research, 19*, 23–28.
37. Gupta, R., Kannan, G. M., Sharma, M., & Flora, S. J. S., (2005). Therapeutic effects of *Moringa oleifera* on arsenic-induced toxicity in rats. *Environmental Toxicological Pharmacology, 20*, 456–464.
38. Halim, E. M., & Mukhopadhyay, A. K., (2006). Effect of *Ocimum sanctum* (tulsi) and vitamin E on biochemical parameters and retinopathy in streptozotocin induced diabetic rats. *Indian Journal of Clinical Biochemistry, 21* (2), 181–188.
39. Hei, T. K., Liu, S. X., & Waldren, C., (1998). Mutagenicity of arsenic in mammalian cells: Role of reactive oxygen species. *Proceedings of National Academy of Sciences, USA (Cell Biology), 95* (14), 8103–8107.
40. Hou, J. P., (1977). The development of Chinese herbal medicine and the Pen-ts'ao. *The Journal: Comparative Medicine East and West, 5* (2), 117–122.
41. http://www.theworldcounts.com/stories/environmental-degradation-facts (accessed on 5 August 2020).
42. https://quotefancy.com/quote/829233/Hippocrates-Nature-itself-is-the-best-physician (accessed on 5 August 2020).
43. https://study.com/academy/lesson/understanding-environmental-toxicology-epidemiology.html (accessed on 5 August 2020).
44. https://theministerofwellness.com/2018/04/09/let-food-be-thy-medicine-and-medicine-be-thy-food/ (accessed on 5 August 2020).
45. https://www.ayurvedacollege.com/articles/students/Ginger (accessed on 5 August 2020).
46. https://www.medicalnewstoday.com/articles/241860.php (accessed on 5 August 2020).
47. https://www.who.int/traditional-complementary-integrative-medicine/about/en/ (accessed on 5 August 2020).
48. Hughes, M. F., Beck, B. D., Chen, Y., Lewis, A. S., & Thomas, D. J., (2011). Arsenic exposure and toxicology: A historical perspective. *Toxicological Sciences: An Official Journal of the Society of Toxicology, 123* (2), 305–332.

49. Ilker, D., Bilal, A., Yusuf, A., Erdinc, D., Aslihan, A., Cetin, E., & Dervis, O., (2004). Effects of garlic extract consumption on plasma and erythrocyte antioxidant parameters in atherosclerotic patients. *Life Science, 75*, 1959–1966.
50. Jan, A. T., Azam, M., Siddiqui, K., Ali, A., Choi, I., & Haq, Q. M. R., (2015). Heavy metals and human health: Mechanistic insight into toxicity and counter defense system of antioxidants. *International Journal of Molecular Sciences, 16* (12), 29592–29630.
51. Jeong, W. S., Jun, M., & Kong, A. N., (2006). Nrf2: A potential molecular target for cancer chemoprevention by natural compounds. *Antioxidants and Redox Signaling, 8* (1, 2), 99–106.
52. Karlowicz-Bodalska, K., Han, S., Freier, J., Smolenski, M., & Bodalska, A., (2017). *Curcuma longa* as medicinal herb in the treatment of diabetic complications. *Acta Poloniae Pharmaceutica, 74* (2), 605–610.
53. Khan, S., Vala, J. A., Nabi, S. U., Gupta, G., Kumar, D., Telang, A. G., & Malik, J. K., (2012). Protective effect of curcumin against arsenic-induced apoptosis in murine splenocytes *in vitro*. *Journal of Immunotoxicology, 9* (2), 148–159.
54. Lambertini, M., (2018). *Technology Can Help us Save the Planet*. But more than anything, we must learn to value nature. https://www.weforum.org/agenda/2018/08/here-s-how-technology-can-help-us-save-the-planet/ (accessed on 5 August 2020).
55. Li, T. S. C., & Schroeder, W. R., (1996). Sea buckthorn (*Hippophae rhamnoides* L.): A multipurpose plant. *Horticulture Technology, 6* (4), 370–380.
56. Mahmood, K. T., Tahira-Mugal, T., & Haq, I., (2010). *Moringa oleifera*: A natural gift-A review. *Journal of Pharmaceutical Sciences and Research, 2* (11), 775–781.
57. Manach, C., Scalbert, A., Morand, C., Rémésy, C., & Jiménez, L., (2004). Polyphenols: Food sources and bioavailability. *The American Journal of Clinical Nutrition, 79* (5), 727–747.
58. Mani, R., (2008). *Mitigation of Arsenic Contamination in Groundwater in West Bengal: Need Experiences and Case Studies in Mitigation*. India Water Portal. https://www.indiawaterportal.org/questions/mitigation-arsenic-contamination-groundwater-west-bengal-need-experiences-and-case-studies (accessed on 5 August 2020).
59. Margaret Mead Quotes, (2019). BrainyQuote.com, Brainy Media Inc. https://www.brainyquote.com/quotes/margaret_mead_157496 (accessed on 5 August 2020).
60. Martinez, V. D., Vucic, E. A., Becker-Santos, D. D., Gil, L., & Lam, W. L., (2011). Arsenic exposure and the induction of human cancers. *Journal of Toxicology, 2011*, 431287.
61. Matthews, V., (1994). The new plants man. *Royal Horticultural Society, 1*, 1352–4186.
62. Middleton, E., Kandaswami, C., & Theoharides, T. C., (2000). The effects of plant flavonoids on mammalian cells: Implications for inflammation, heart disease, and cancer. *Pharmacological Reviews, 52*, 673–751.
63. Mishra, D., & Flora, S. J., (2008). Quercetin administration during chelation therapy protects arsenic-induced oxidative stress in mice. *Biological Trace Element Research, 122* (2), 137–147.
64. Mishra, D., Gupta, R., Pant, S. C., Kushwah, P., Satish, H. T., & Flora, S. J. S., (2009). Coadministration of mono isoamyl dimercaptosuccinic acid and *Moringa oleifera* seed powder protects arsenic-induced oxidative stress and metal distribution in mice. *Toxicology Mechnisms and Methods, 19*, 169–182.
65. Munin, A., & Edwards-Lévy, F., (2011). Encapsulation of natural polyphenolic compounds; a review. *Pharmaceutics, 3* (4), 793–829.

66. Muthumani, M., & Prabu, S. M., (2012). Silibinin potentially protects arsenic-induced oxidative hepatic dysfunction in rats. *Toxicology Mechanisms and Methods, 22* (4), 277–288.
67. National Research Council Committee on Medical and Biological Effects of Environmental Pollutants US. (1977). Arsenic: Medical and biologic effects of environmental pollutants. In: *Biological Effects of Arsenic on Man* (p. 6). Washington (DC): National Academies Press (US). Online; Available from: https://www.ncbi.nlm.nih.gov/books/NBK231021/ (accessed on 5 August 2020).
68. National Research Council Subcommittee on Arsenic in Drinking Water (US). (1999). Arsenic in drinking water. In: *Mechanisms of Toxicity* (p.7) Washington (DC): National Academies Press (US). Available from: https://www.ncbi.nlm.nih.gov/books/NBK230889/; (accessed on 5 August 2020).
69. Ostrosky-Wegman, P., & Gonsebatt, M. E., (1996). Environmental toxicants in developing countries. *Environmental Health Perspective, 104* (3), 599–602.
70. Padma, P. R., & Bhuvaneswari, V. S., (1998). The activities of enzymatic antioxidants in selected green leaves. *The Indian Journal of Nutrition and Dietetics, 35*, 1–3.
71. Passos, V. F., Melo, M. A. S., Marçal, F. F., Costa, C. A. G. A., Rodrigues, L. K. A., & Santiago, S. L., (2018). Active compounds and derivatives of camellia sinensis responding to erosive attacks on dentin. *Brazilian Oral Research, 32*, E40–E50.
72. Peng, M., Liu, B., & Mao, M., (2018). Study on the Application of Chinese Patent Drug and Chinese Formula of *Rabdosia rubescens. IOP Conference Series: Materials Science and Engineering, 301* (1), 012060–012070.
73. Perkins, C. (2020). *Is Aloe a Tropical Plant?* https://homeguides.sfgate.com/aloe-tropical-plant-67510.html (accessed on 5 August 2020).
74. *Phyllanthus Emblica*. From Biodiversity of India; https://www.biodiversityofindia.org/index.php?title=Phyllanthus_emblica (accessed on 5 August 2020).
75. https://europepmc.org/article/med/10519377, Prevention of cancer in the next millennium. *Cancer Res., 59* (19), 4743–4758.
76. Pubchem. Milk thisle extract. https://pubchem.ncbi.nlm.nih.gov/compound/Milk-thistle-extract (accessed on 5 August 2020).
77. Rather, R. A., & Bhagat, M., (2018). Cancer chemoprevention and piperine: Molecular mechanisms and therapeutic opportunities. *Frontiers in Cell and Developmental Biology, 6* (10), E-collection.
78. Ren, X., McHale, C. M., Skibola, C. F., Smith, A. H., Smith, M. T., & Zhang, L., (2011). An emerging role for epigenetic dysregulation in arsenic toxicity and carcinogenesis. *Environmental Health Perspectives, 119* (1), 11–19.
79. Roy, A., Das, A., Das, R., Haldar, S., Bhattacharya, S., & Haldar, P. K., (2014). Naringenin, a citrus flavonoid, ameliorates arsenic-induced toxicity in Swiss albino mice. *Journal of Environmental Pathology, Toxicology, and Oncology, 33* (3), 195–204.
80. Roy, S., Roy, L., & Das, N., (2019). Peeping into the kettle: Review on the microbiology of made tea. *International Journal of Pharmacy and Biological Sciences, 9* (1), 842–849.
81. Sage, A. P., Minatel, B. C., Ng, K. W., Stewart, G. L., Dummer, T., Lam, W. L., & Martinez, V. D., (2017). Oncogenomic disruptions in arsenic-induced carcinogenesis. *Oncotarget, 8* (15), 25736–25755.
82. Saha, K. C., (2003). Diagnosis of arsenicosis. *Journal of Environmental Science and Health. Part A, Toxic/Hazardous Substances and Environmental Engineering, 38* (1), 255–272.

83. Sajidu, S. M., Henry, E. M. T., Kwamdera, G., & Mataka, L., (2005). Removal of lead, iron, and cadmium ions by means of polyelectrolytes of the *Moringa oleifera* whole seed kernel. In: Cunha, D. C., Brebbia, C. A., (eds.), *Water Resources Management III* (pp. 251–258). Wit Press, London: U.K.
84. Šaponjac, V. T., Canadanovi´c-Brunet, J., Cetkovi´c, G., Jakišic, M., Djilas, S., Vuli´c, J., & Stajcic, S., (2016). Encapsulation of beetroot pomace extract: RSM optimization, storage, and gastrointestinal stability. *Molecules*, *21*, 584–590.
85. Sárközi, K., Papp, A., Máté, Z., Horváth, E., Paulik, E., & Szabó, A., (2015). Rutin, a flavonoid phytochemical, ameliorates certain behavioral and electrophysiological alterations and general toxicity of oral arsenic in rats. *Acta Biologica Hungarica*, *66* (1), 14–26.
86. Saxena, G., & Flora, S. J. S., (2006). Changes in brain biogenic amines and hemebiosynthesis and their response to combined administration of succimers and *Centella asiatica* in lead poisoned rats. *Journal of Pharmacy and Pharmacology*, *58*, 547–559.
87. Selby-Pham, S., Miller, R. B., Howell, K., Dunshea, F., & Bennett, L. E., (2017). Physicochemical properties of dietary phytochemicals can predict their passive absorption in the human small intestine. *Scientific reports*, *7* (1), 1931–1937.
88. Sharma, A., Sharma, M. K., & Kumar, M., (2009). Modulatory role of *Embilica officinalis* fruit extract against arsenic induced oxidative stress in Swiss albino mice. *Chemio-Biological Interactions*, *180* (1), 20–30.
89. Sharmila, B. G., Kumar, G., & Murugesan, A. G., (2009). Effects of leaves extract of *Ocimum sanctum* L. on arsenic-induced toxicity in Wister albino rats. *Food and Chemical Toxicology*, *47* (2), 490–495.
90. Singh, A. P., Goel, R. K., & Kaur, T., (2011). Mechanisms pertaining to arsenic toxicity. *Toxicology International*, *18* (2), 87–93.
91. Singh, R., Gautam, N., Mishra, A., & Gupta, R., (2011). Heavy metals and living systems: An overview. *Indian Journal of Pharmacology*, *43* (3), 246–253.
92. Singh, V., Brar, M. S., Sharma, P., & Malhi, S. S., (2010). Arsenic in water, soil, and rice plants in the Indo-Gangetic plains of Northwestern India. *Communications in Soil Science and Plant Analysis*, *41* (11), 1350–1360.
93. Sinha, D., Roy, M., Dey, S., Siddiqi, M., & Battacharya, R. K., (2003). Modulation of arsenic induced cytotoxicity by tea. *Asian Pacific Journal of Cancer Prevention*, *4*, 233–237.
94. Smith, A., Hopenhayn-Rich, C., Bates, M., Goeden, H., Hertz-Picciotto, I., Duggan, H., Wood, R., et al., (1992). Cancer risks from arsenic in drinking water. *Environmental Health Perspective*, *97*, 259–267.
95. Soria, E. A., Goleniowski, M. E., Cantero, J. J., & Bongiovanni, G. A., (2008). Antioxidant activity of different extracts of Argentinean medicinal plants against arsenic-induced toxicity in renal cells. *Human and Experimental Toxicology*, *27*, 341–346.
96. Soria, J. C., Tan, D. S. W., Chiari, R., Wu, Y. L., Paz-Ares, L., Wolf, J., Geater, S. L., et al., (2017). First-line ceritinib versus platinum-based chemotherapy in advanced ALK-rearranged non-small-cell lung cancer (ASCEND-4): A randomized, open-label, phase-3 study. *Lancet*, *389* (10072), 917–929.
97. Staples, G., & Kristiansen, M. S., (1999). *Ethnic Culinary Herbs* (p. 73). University of Hawaii Press, Hawaii, USA.

98. Tseng, C. H., (2005). Blackfoot disease and arsenic: A never-ending story. *Journal of Environmental Science and Health, C: Environmental Carcinogenesis and Ecotoxicology Reviews*, *23* (1), 55–74.
99. Wang, H., Khor, T. O., Shu, L., Su, Z., Fuentes, F., Lee, J. H., & Kong, A. N. T., (2012). Plants against cancer: Review on natural phytochemicals in preventing and treating cancers and their drug ability. *Anticancer Agents Med. Chem., 12* (10), 1281–1305.
100. Megan W. (2017). *Why Are Polyphenols Good For You?* https://www.medicalnewstoday.com/articles/319728.php (accessed on 5 August 2020).
101. World Health Organization (WHO), (1993). Geneva. *Guidelines for Drinking-Water Quality* (2nd edn., Vol. 1) Recommendations; https://www.who.int/water_sanitation_health/dwq/2edvol1i.pdf?ua=1 (accessed on 5 August 2020).
102. World Health Organization (WHO), (2003). Regional Office for South-East Asia, New Delhi. *Arsenicosis: Case-Detection, Management, and Surveillance* (p. 26). Report of a Regional Consultation.
103. World Health Organization (WHO). *Fact Sheet on Arsenic*. https://www.who.int/news-room/fact-sheets/detail/arsenic (accessed on 5 August 2020).
104. Xavier, S., (2005). *Demographic Status and Epidemiological Pattern: A Study on Environmental Health Impacts of Arsenic Affected Population in West Bengal*. PhD Thesis; Kalyani University, Kalyani, India; pages 20–55.
105. Xiao, J., Cao, Y., & Huang, Q., (2017). Edible nanoencapsulation vehicles for oral delivery of phytochemicals. *Journal of Agriculture, Food, and Chemistry, 65* (32), 6727–6735.
106. Zhang, W., Liu, Y., Ge, M., Jing, J., Chen, Y., Jiang, H., Yu, H., et al., (2014). Protective effect of resveratrol on arsenic trioxide-induced nephrotoxicity in rats. *Nutrition Research and Practice*, *8* (2), 220–226.
107. Zhou, G. B., Chen, S. J., Wang, Z. Y., & Chen, Z., (2007). Back to the future of oridonin: Again, compound from medicinal herb shows potent anti-leukemia efficacies *in vitro* and *in vivo*. *Cell Research*, 17, 274–276.
108. Zodape, G. V., (2010). Effect of *Aloe vera* juice on toxicity induced by arsenic in *Labeo rohita* (Hamilton). *Journal of Applied and Natural Science*, *2* (2), 300–304.

PART IV

Plant-Based Phytochemicals: Extraction, Isolation, and Healthcare

CHAPTER 9

PLANT SECONDARY METABOLITES: COMMERCIAL EXTRACTION, PURIFICATION, AND HEALTH BENEFITS

TSEGA Y. MELESSE, TESFAYE F. BEDANE, and FRANCESCO DONSÌ

ABSTRACT

Plant secondary metabolites include bioactive compounds, such as terpenes, alkaloids, polyphenols, glucosinolates, and carotenoids. Apart from conventional extraction methods (e.g., soxhlet and maceration), several novel technologies with basic extraction processes have been developed to extract secondary metabolites from various plant parts and sources. Such extraction methods are high pressure-assisted (HPA), negative pressure cavitation-assisted (NPCA), high-pressure homogenization-assisted (HPHA), high voltage electrical discharge-assisted (HVEDA), microwave-assisted (MA), ultrasound-assisted (UA), moderate electric field-assisted extraction (MEFA) and pulsed electric field-assisted (PEFA), enzyme-assisted (EA), and chemical pretreatment methods. Therefore, applications of plant secondary metabolites in pharmaceuticals, foods, and cosmetics are noteworthy. This chapter focuses on the characteristics of selected plant secondary metabolites, technologies for their industrial production, and applications.

9.1 INTRODUCTION

Historically, plants or herbs are used as a functional food, source of pharmaceuticals or medicine, and ingredients of cosmetics and fragrance [60, 101]. Primary metabolites of plants include low molecular weight sugar, polysaccharides, amino acids, proteins, intermediates in the Krebs cycle, and nucleic acids produced by glycolysis, Krebs cycle, photosynthesis, and associated pathways.

Examples of plant secondary metabolites are terpenes, alkaloids, polyphenols, carotenoids, glucosinolates, etc., [77]. These are stored at intra- and inter-cellular spaces in leaf, bark, root, stem, seed, etc. Isolation and extraction of these secondary metabolites has received significant attention due to their several health benefits [5]. Figure 9.1 indicates various metabolic pathways in plants.

However, plant secondary metabolites are important but are not essential for the growth of a plant. They are useful as nutrients for their growth under adverse situations, protect from abiotic stresses, often used for defense purpose and impart characteristics to plant, such as, the development of fragrance, color of flower and fruit, etc. Furthermore, secondary plant metabolites, such as, hormones, are important to regulate and signal the metabolic pathways and take a significant role in the overall plant development. However, there is no fixed rule to classify these plant metabolites.

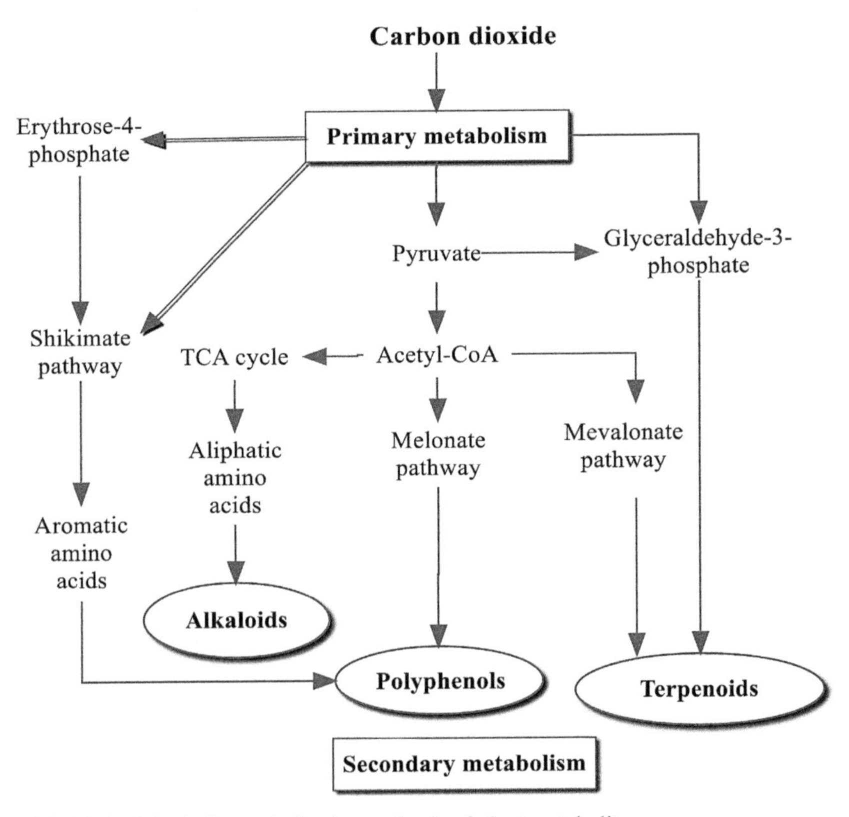

FIGURE 9.1 Metabolic cycle for the synthesis of plant metabolites.
Source: Self-developed with concepts from Ref. [77].

Based on the origin of biosynthesis, plant secondary metabolites are grouped into four major classes, such as: (a) terpene, (b) alkaloid, (c) polyphenol (d) carotenoid and (d) sulfur-containing compound, i.e., glucosinolates. Either plant secondary metabolites in plant extracts or as a pure compound are able to cure diseases and reduce risk of several health hazards [114]. Several unique biological activities of plant secondary metabolites are: antioxidant, anti-inflammatory, antibacterial, antiviral, antiallergic, antithrombotic, antiaging, and carcinogenicity, etc., [23].

This chapter focuses on selected plant secondary metabolites (such as: terpenes, alkaloids, polyphenols, carotenoids, and glucosinolates); their industrial production, purification, and applications in food, biopharmaceuticals, and cosmetics.

9.2 CLASSES OF PLANT SECONDARY METABOLITES

9.2.1 TERPENE

The term 'terpene' is originated from the word 'turpentine,' which also signifies' resin acids.' Chemically all terpenes are derived from 5-carbon isoprene units and are assembled in different ways. Terpenes are classified based on the number of isoprene units in a molecule, such as: (a) hemiterpenes, (b) monoterpenes, (c) sesquiterpenes, (d) diterpenes, (e) sesterterpenes, (f) triterpenes, (g) sesquiterpenes, (h) tetraterpenes, (i) polyterpenes, and (j) norisoprenoids.

Among all plant secondary metabolites, terpenes are the most diverse [19]. All living organisms including plants produce terpenes as a product of certain essential physiological function. It has been proven that selected terpenes have antitumor activity with slight or no side effects [84]. Furthermore, terpenoids can be utilized as pesticides and fungicides in agriculture and horticulture due to their strong antimicrobial and insecticidal properties [71].

9.2.2 ALKALOIDS

The word 'alkaloids' is derived from the Latin word 'alkali.' Alkaloids are natural organic compounds that contain mostly basic nitrogen atoms. In addition, they may also contain carbon, hydrogen, sulfur, oxygen, and rarely other elements (such as: bromine, chlorine, and phosphorus). Alkaloids are classified as: (a) true alkaloids, originated from amino acids and consisted nitrogen in the heterocycle, (b) protoalkaloids, originated from amino acids and contained nitrogen, but not the nitrogen heterocycle, (c) polyamine alkaloids, derivatives of putrescine, spermidine, and spermine, (d) peptide and cyclopeptide alkaloids, and (e) pseudo alkaloids (not originated from amino acids but alkaloid-like compounds). Alkaloids are useful for the development of pharmaceuticals, related to the nervous system and as a painkiller [6, 8, 9]. In addition, some alkaloids have been used as insecticides in agriculture.

9.2.3 POLYPHENOLS

The word 'polyphenol' is derived from the Latin word 'polus,' which also signifies 'many' or 'much.' Polyphenols are most abundant antioxidants that can be classified based on the hydroxyl groups in benzene rings and the number of phenol rings. In polyphenols, more than one hydroxyl group is attached to benzene rings [86, 103]. The main classes of polyphenols are phenolic acids and flavonoids. Furthermore, phenolic acids are classified as hydroxybenzoic and hydroxycinnamic acids. Flavonoids include anthocyanidins, aurones, catechins (including proanthocyanidins), dihydrochalcones, isoflavones, flavanones, flavans, flavonols, and leukoanthocyanidins [81]. These are produced as a by-product along with metabolic abnormality in the plant. According to research studies, these molecules have antioxidant activity and destructive behavior of reactive oxygen and nitrogen species [56, 118]. Epidemiologic reports indicate a significant role of polyphenols to protect against the development of diseases including cardiovascular, cancer, both Type I- and Type II-diabetes, inflammation, aging, asthma, etc., [56, 120].

9.2.4 CAROTENOIDS

Carotenoids are pigments soluble in fat having antioxidant properties [58] and additional physiological role in immunostimulantion [72]. Different classes of carotenoids are: (a) carotenes (hydrocarbons without oxygen, such as β-carotene, α-carotene and lycopene) and (b) xanthophylls (oxygen-containing compounds, such as lutein and zeaxanthin). In the plant, carotenoids take part in photosynthesis, absorb heat and protect the plant. Four different types of carotenoids (including α-carotene, β-carotene and γ-carotene, and β-cryptoxanthin) are considered as precursors of vitamin A in our body. Reaction of carotenoids with radical species in a biological system reduces the stress [58]. Research has shown the possibility of lowering chronic illness by consuming diets, rich or fortified with carotenoids [29].

9.2.5 GLUCOSINOLATES

Glucosinolates are plant secondary metabolites, contain sulfur and nitrogen, found in several vegetables (such as: cabbage, broccoli, mustard, capers, watercress, horseradish, and radishes). These molecules are derived from amino acid and glucose [95] and are well known for their insecticide, antimicrobial, antioxidant, anti-inflammatory, and cytoprotective activities [51].

In some cases, based on the chemical composition, plant secondary metabolites are classified as (a) alkaloids, (b) alkamides, (c) amines (d) amino acids, (e) cyanogenic glycosides, (f) glucosinolates, (g) lectins, peptides, and polypeptides, (h) steroids and saponins, (i) terpenes, (j) flavonoids and tannins, (k) phenylpropanoids, lignins, coumarins, and lignans, (l) polyacetylenes, waxes, and fatty acids, (m) polyketides and (n) carbohydrates and organic acids. The sources and biological activities of different plant secondary metabolites are shown in Table 9.1.

TABLE 9.1 Sources and Beneficial Activities of Plant Secondary Metabolites

Plant Secondary Metabolites	Sources	Beneficial Activities	References
Carotenoids	Plastids of roots, stem, flowers, fruits, and seeds	Antioxidant and prooxidant effects, provitamin activities, therapeutic effects	[75]
Glucosinolates	Cruciferous families Brassicaceae, Capparaceae, and Caricaceae	Antimicrobial, antioxidant, anti-inflammatory, and cytoprotective attributes	[51]
Polyphenols	Fruits, vegetables, cereals, and beverages; Tea, red wine, coffee, vegetables, leguminous plants, and cereals.	Antioxidants, protection against the development of chronic diseases. Antioxidant activity, development of functional foods	[32, 56]
Terpenes	Lamiaceae, coniferae, compositae, leguminoseae, taxaceae	Antitumor, anti-inflammatory, immunomodulatory, cardiac, and antithrombotic, endocrine	[3]

9.3 ISOLATION AND RECOVERY OF PLANT SECONDARY METABOLITES

Among all types of unit operations, extraction is widely used as a unit operation in food and biotechnology industries for the production and purification of several medicinal or functional compounds from the crude matrix. The target plant secondary metabolites are generally found inter- or intra-cellular positions with a complex microstructure. Thus, increase in the permeability of cell-wall and membrane is required to improve the extraction rate and yield of secondary metabolites from the plant tissue. Extraction has been identified as an essential step to recover the desired chemical components from the plant materials for further separation [6, 8, 9, 104]. In general, the extraction technologies of secondary metabolites can be divided into two subclasses: (a) conventional technology and (b) emerging or novel technology. These technologies (both conventional and novel) for isolation and purification of secondary metabolites from the plant matrix are shown in Table 9.2.

TABLE 9.2 General Processes and Operations for the Recovery of Plant Secondary Metabolites

Steps	Processes	Operations
First	Extraction	Conventional organic solvent-based extraction: Soxhlet, Maceration, EA, PEFA, HVEDA, HPA, NPCA, HPHA, MA, UA, MEFA
Second	Separation/Fractionation	Membrane-based separation: nanofiltration, ultra-filtration, microfiltration.
		Chromatographic process: ion-exchange chromatography, membrane chromatography, silica-gel column chromatography
Third	Concentration	Distillation, evaporation, freeze-drying, membrane processes with reverse osmosis, forward osmosis.

Source: Self-developed with information from Refs. [6, 8, 9, 104].

9.3.1 CONVENTIONAL TECHNOLOGIES

9.3.1.1 SOLVENT-BASED EXTRACTION

Conventional extraction of plant secondary metabolites consists of a solid-liquid and liquid-liquid extraction process. Soxhlet and maceration are commonly used technologies on a small laboratory-scale extraction process. Soxhlet extraction is performed with organic solvents. For any target plant secondary metabolite, the extraction yield significantly depends on the chemical nature of the solvent and characteristics of the targeted compound [10]. Hence, the selection of solvent and extraction technique are chosen according to the specific nature of plant secondary metabolite [62].

Criteria for the selection of a suitable solvent for a particular purpose include: less toxicity, volatility, and high distribution coefficient. Accordingly, hexane, methanol, chloroform, ethanol, water, ether, and acetone are commonly used solvents during the extraction of plant secondary metabolites. Properties of the solvent during extraction processes can affect further downstream processing. Although conventional solvent-based extraction technology is effective for the extraction of plant secondary metabolites, yet in some cases, it may cause the degradation of heat-sensitive compounds and offer a chance to contaminate with a toxic solvent. The method requires longer processing times, higher power, and the amount of sample and solvent [99]. Some examples of solvent-based extraction of plant secondary metabolites are shown in Table 9.3.

TABLE 9.3 Extraction of Plant Secondary Metabolites by Using Different Solvents

Solvent	Plant Secondary Metabolite
Acetone	Flavonoid
Chloroform	Flavonoid, terpenoid
Dichloromethanol	Flavonoid
Ethanol	Tannin, alkaloid, polyphenol, flavanol, terpenoid
Ether	Alkaloid, terpenoids
Methanol	Anthocyanin, flavones, terpenoid, saponins, tannins, polyphenols
Water	Anthocyanin, terpenoids, tannins, saponins

Source: Self-developed using information from Refs. [74, 104].

9.3.2 NOVEL TECHNOLOGIES

In recent years, the use of emerging and innovative technologies for the extraction of plant secondary metabolites have shown significant advancements [36]. Several pretreatment methods, such as, physical, biological, and chemical treatments are capable to improve cell structure and to increase the permeability of membrane during the extraction of plant secondary metabolites.

Extraction with different physical methods, e.g., HPA [33, 53], NPCA [65, 98, 111], high pressure homogenization-assisted (HPHA) [7, 34, 46], microwave-assisted (MA) process [27, 41, 47, 121, 122, 124], UA [1, 85, 101, 116, 119], MEFA [40, 63, 88, 102], pulsed electric field-assisted (PEFA) [30, 66, 94, 99, 113] and HVEDA [66, 99] have received significant attention. Beside these physical methods, extraction process with different enzymes are used to break for the cell structure and to release the secondary metabolites from cell-matrix [24, 39, 59, 64, 92]. In a similar manner, extraction with chemical pretreatment has been used to increase the inner or outer membrane permeabilization and to decrease the overall internal resistance for releasing plant secondary metabolites from matrix [57, 91].

Emerging technologies are considered as a hallmark in the platform of 'process intensification.' Novel extraction techniques are applicable for the extraction of different classes of plant secondary metabolites, with specific requirements in terms of low-temperature processing [36]and lower use of organic solvents [59]. These techniques have been used to enhance the mass transfer and increase of extraction yield and reduce the energy consumption during the extraction of plant secondary metabolites [36]. Modifications of the extraction methods are still under development according to the choice of researcher for a specific purpose [9].

9.3.2.1 EXTRACTION WITH ENZYMATIC TREATMENT

In enzymatic extraction, hydrolytic enzymes with highly specified catalytic activity are employed in the disruption of plant matrix before the extraction stage. It provides better quantity of yield compare to primitive extraction process [24, 39, 59, 64, 92]. In this context, commonly used hydrolytic enzymes are proteases, cellulases, lipases, and amylases. In Figure 9.2, the enzymatic extraction process for the extraction of plant-based secondary metabolites is represented.

The study has reported that significant improvement in the release of phenolic compounds from the hydro-distilled residual leaves after enzymatic pretreatment [18]. The enzymatic extraction has also been effective to increase the extraction efficiency of polyphenols from citrus peel [64], blackcurrant [61], ginger [76] and carotenoids from marigold flower [14].

9.3.2.2 EXTRACTION ASSISTED WITH PULSED ELECTRIC FIELD (PEFA)

PEFA extraction has significant potential to improve cell permeability of membrane through a phenomenon of irreversible electroporation. In this method, the application of moderate electric field strength with 0.1 to 40 kV/cm for microseconds increases the pore size in cell membranes and disrupts the cellular membrane, which increase the permeabilization of plant metabolites from matrix [37, 68]. Figure 9.3 shows PEFA extraction process for the extraction of plant secondary metabolites.

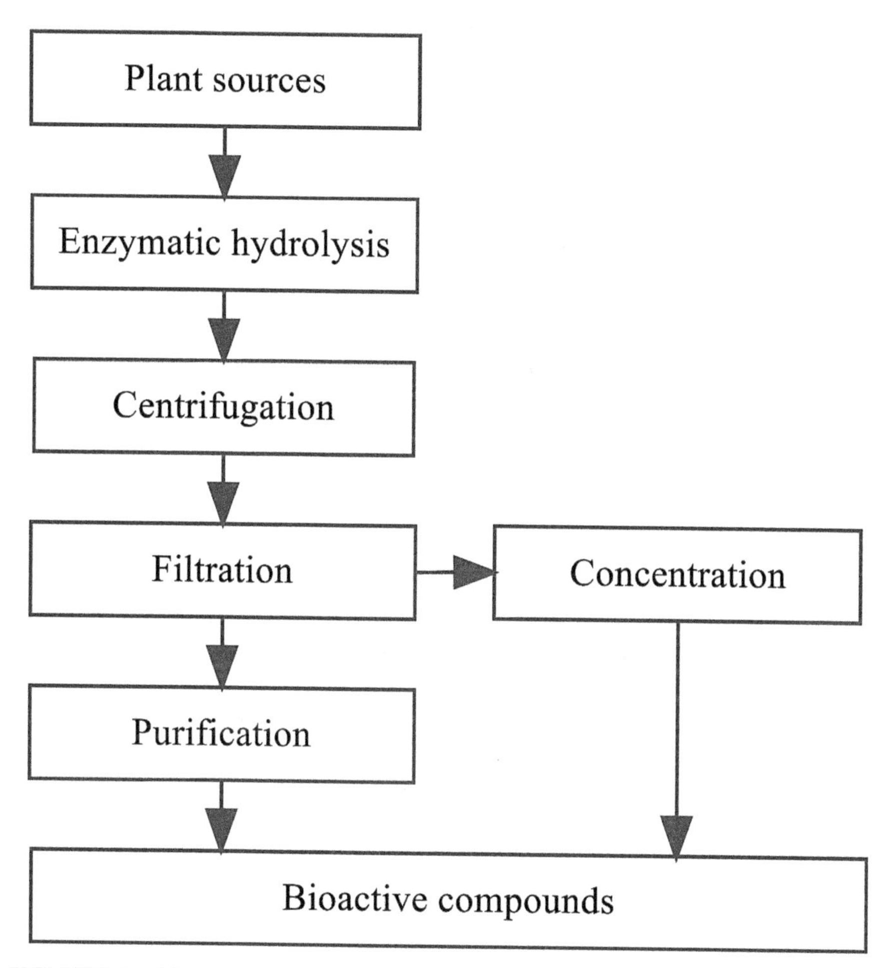

FIGURE 9.2 EA extraction process for the recovery of secondary metabolites from plants.
Source: Self-developed with concepts from Refs. [24, 92].

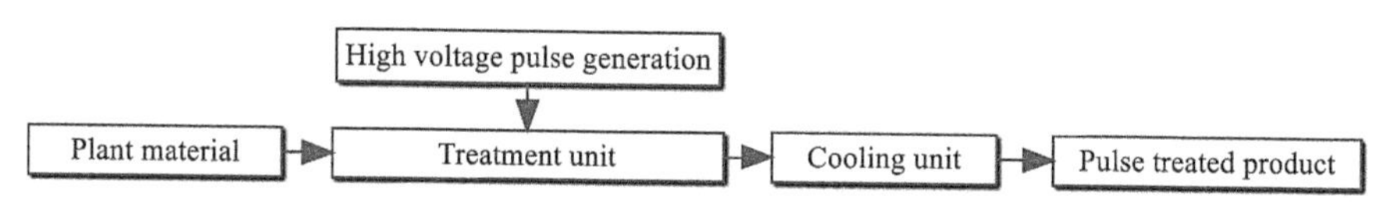

FIGURE 9.3 PEFA extraction process for the recovery of secondary metabolites from plants.
Source: Self-developed with concepts from Refs. [37, 68].

According to a recent study, the PEFA extraction process has been effectively utilized in winemaking process to increase the polyphenol content, antioxidant activity and color intensity in red wine [30]. Likewise, the PEFA extraction process was satisfactorily applied in brown rice to extract antioxidant compounds, such as, phenolic acids, polyphenols, γ-oryzanol, and saturated and unsaturated fatty acids [94].

9.3.2.3 EXTRACTION USING HIGH VOLTAGE ELECTRICAL DISCHARGE (HVEDA)

HVEDA extraction is a non-thermal treatment to disrupt cells in liquid samples and it subsequently increases the recovery of valuable components from the plant matrix. During this process, energy is released directly into the medium through the submerged electrodes. It is an extremely destructive treatment, where electric shock converts to mechanical energy and disintegrates cell-walls and cell-membranes [30, 66, 94, 99, 113]. Figure 9.4 shows the HVEDA extraction process for the extraction of secondary metabolites from plants.

This method has been satisfactorily employed to boost the recovery of polyphenol compounds from white grape pomace [66]. Moreover, it has been used to enhance the extraction efficiency of polyphenols and proteins with more effective energy inputs than the ultrasound and PEFA methods [99].

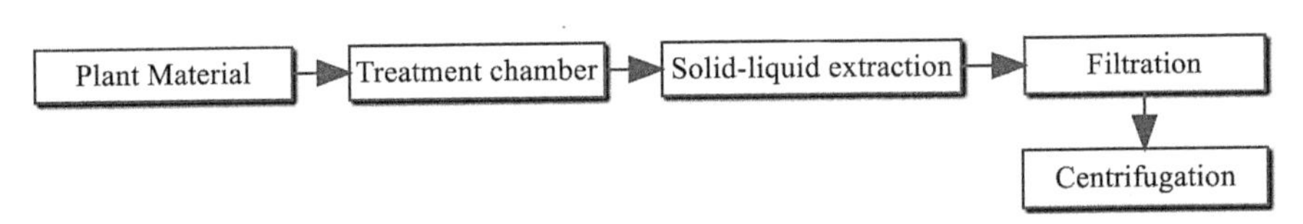

FIGURE 9.4 HVEDA extraction process for the recovery of secondary metabolites from plant matrix.
Source: Self-developed with concepts from Refs. [99].

9.3.2.4 HIGH-PRESSURE-ASSISTED EXTRACTION (HPA)

Although HPA is considered for the inactivation of pathogens in the food industry, yet this process has received attention for rupturing plant cell-walls for efficient extraction of plant secondary metabolites. The HPA extraction process involves compression, holding of the sample at the specific pressure and release of pressure. During the application of HPA, the sample should be dried prior to size reduction, because the presence of moisture may reduce the extraction efficiency. In this process, generally, pressure between 100–600 MPa is applied [53]. Figure 9.5 presents HPA extraction process for the extraction of secondary metabolites from plant matrix.

It has been reported that high hydrostatic pressure has a positive effect on the recovery of polyphenols and anthocyanins from various red fruit-based products. Furthermore, HPA treatment at moderate temperature increases the extraction efficiency of colored pigments and polyphenol content from fruits [33].

9.3.2.5 NEGATIVE PRESSURE CAVITATION-ASSISTED EXTRACTION (NPCA)

The NPCA extraction process is eco-friendly and efficient technology for the recovery of plant metabolites, where cavitation is formed by negative pressure with continuous introduction into the liquid-solid system to increase the turbulence, mass transfer and collision between the solvent and plant tissue. During this process, millions of tiny vapor bubbles are formed and collapsed.

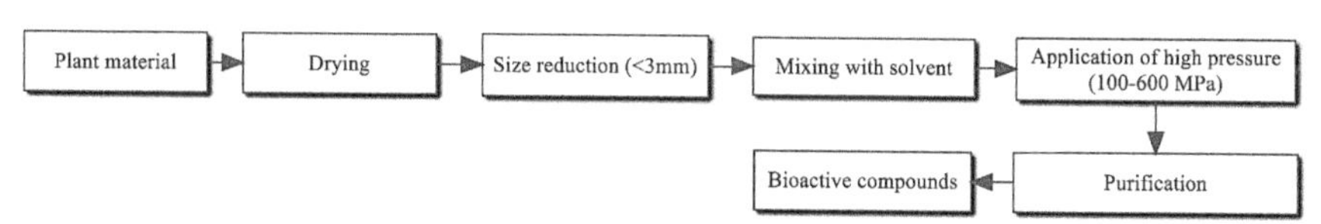

FIGURE 9.5 HPA extraction process for the recovery of plant secondary metabolites.
Source: Self-developed with concepts from Refs. [33, 53].

This collapse causes the release of high energy, which raises the local temperature and pressure at reaction sites. It increases the reaction rate in the system. Through continuous addition of water to the system, turbulence increases the mass transfer and collision between the matrix and solvent. This facilitates the permeabilization of target secondary metabolites from the plant matrix [50, 65, 111]. This method is effective in improving the mixing of substrate and enzyme. The NPCA extraction method has also been used to extract alkaloids, polyphenols, polysaccharides, and flavonoids from different plant parts [67]. Figure 9.6 shows the NPCA extraction process to extract secondary metabolites from the plant tissue.

The research study has demonstrated the effectiveness of NPCA extraction in combination with other methods, such as: microwave, homogenization, enzyme, ionic liquid solvents and deep eutectic solvents [98].

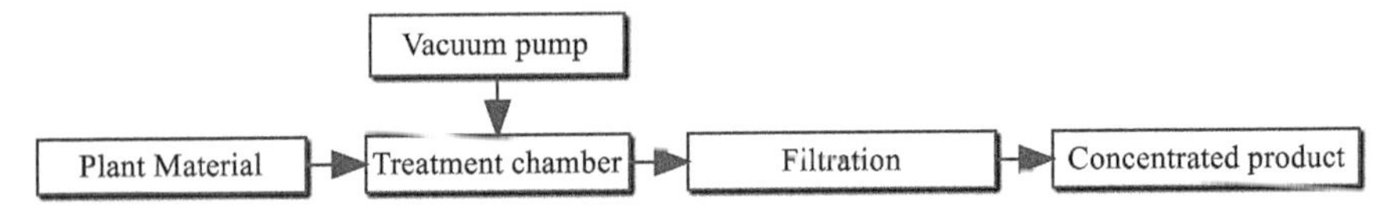

FIGURE 9.6 NPCA extraction process for the recovery of secondary metabolites from plants.
Source: Self-developed with concepts from Ref. [50].

9.3.2.6 EXTRACTION USING HIGH-PRESSURE HOMOGENIZATION (HPH)

High-pressure homogenization is a non-thermal process, where the cell membrane is disrupted using high-intensity fluid-mechanical stresses. In this process, the flow of the process fluid with high pressures (between 50–400 MPa) through homogenization is used. Samples are forced to pass through a narrow gap in the homogenizing valve, and high shear and turbulence create acceleration, compression, and pressure drop. Moreover, it results in the deformation and subsequent disruption of cells and macromolecules in the fluid [7, 34, 50]. Figure 9.7 shows the HPHA extraction for recovery of plant secondary metabolites is represented.

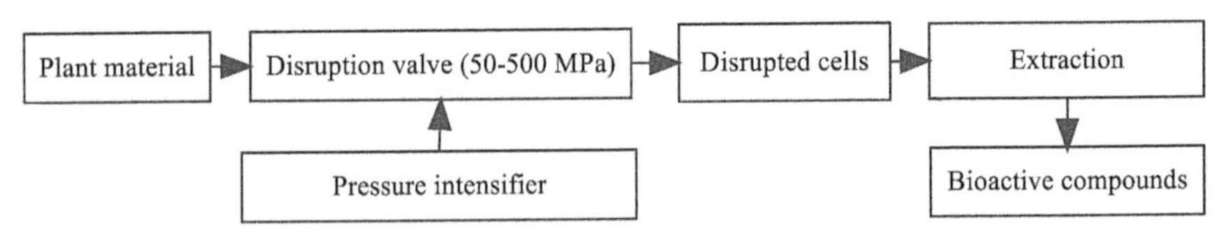

FIGURE 9.7 HPHA extraction process for the recovery of plant secondary metabolites.
Source: Self-developed with concepts from Ref. [89].

Recently, researchers have found that HPHA extraction has potential for the recovery of secondary metabolites from the plant. High-pressure homogenizer or ball mill was the most efficient for the recovery of carotenoids from microalgae (*Haemotococcus pluvialis, Chromochloris zofingiensis,* and *Chlorella sorokiniana*) [46, 110]. It has been found that HPHA extraction has an effect on changing the microstructure and bioaccessibility of carotenoids from tomato pulp. *In the same way,* HPHA extraction decreased the particle size of tomato because of matrix disruption. It also increased the consistency of the product [48].

9.3.2.7 MICROWAVE-ASSISTED EXTRACTION (MVA)

Microwaves are electromagnetic waves, which are used at 2.45 GHz and interact with polar molecules in the sample to generate heat [122]. During MA extraction, evaporation of the moisture from the cells enhances the porosity of the cell-matrix. It successively improves the penetration of a solvent into the matrix [47]. The elevated temperature can also improve solubility and yield of extraction. Compared to conventional extraction methods, MVA extraction has advantages, such as, higher recovery yield, less solvent consumption and shorter extraction time [122].

The change can be observed with light microscopy and scanning electron microscopy. The application of microwave during extraction provides a greater extent of cell disruption of the plant materials, which increase the rate of extraction [38]. However, microwave irradiation can change some target components due to chemical reactions [41, 121, 124]. Thus, MA extraction may improve extraction yield by modifying the structures of the target compounds [102]. In Figure 9.8, the MA extraction process for the extraction of plant secondary metabolites is represented.

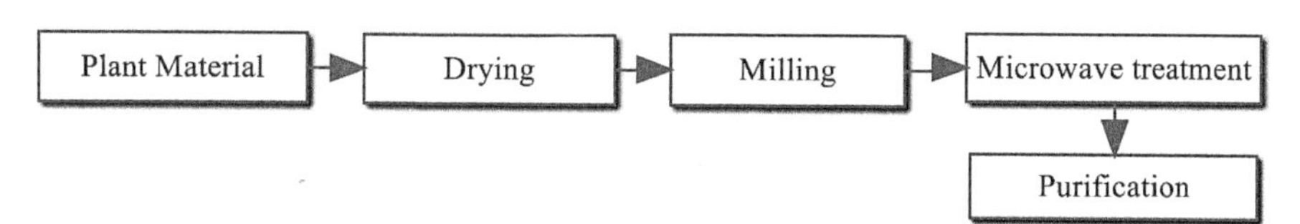

FIGURE 9.8 MA extraction process for the recovery of plant secondary metabolites.
Source: Self-developed with concepts from Refs. [122, 124].

9.3.2.8 ULTRASOUND-ASSISTED EXTRACTION (UA)

Ultrasound can improve the recovery process of plant metabolites through the interaction of acoustic waves with solvent to plant tissue. Due to the application of waves, dissolved gas creates bubbles to generate heat [8, 119]. This process induces significant cavitation phenomena, which is responsible for cell-wall disruption [42, 85]. UA extraction can enhance the extraction yield and aqueous extraction processes without using solvents, therefore, it provides the opportunity to use green solvents. Ultrasound treatment improves the extraction performance and extraction of heat-sensitive plant secondary metabolites [119]. UA extraction is an efficient technique to reduce the loss of plant secondary metabolites during extraction by decreasing extraction time [101, 116]. In Figure 9.9, the UA extraction process for the extraction of plant secondary metabolites is presented.

UA extraction of polyphenols from the bark [42], amino acids from grape [21] and isoflavones from soybean [85] have been reported.

9.3.2.9 MODERATE ELECTRIC FIELD-ASSISTED EXTRACTION (MEFA)

MEFA extraction uses an alternating current electric field usually from 1–1000 V cm^{-1} through two highly conductive electrodes to the biological material placed between them. It is a green method. The term moderate electric field can be defined as a process where electric field strength includes thermal treatment (ohmic heating) or excludes the heating process (electro-permeabilization) [105]. Unlike HVEDA, MEFA extraction involves direct use of the electric current in the form of alternating current at considerably lower strength than the pulsed electric field [40]. The use of an alternating current reduces the effect of electrolysis and the formation of undesired compounds around the electrodes. It can be operated at high temperature due to ohmic heating or at low temperature to minimize thermal effects. The method can reduce extraction time and energy with high quality and yield. The heating rate highly depends on the nature of biological materials [102].

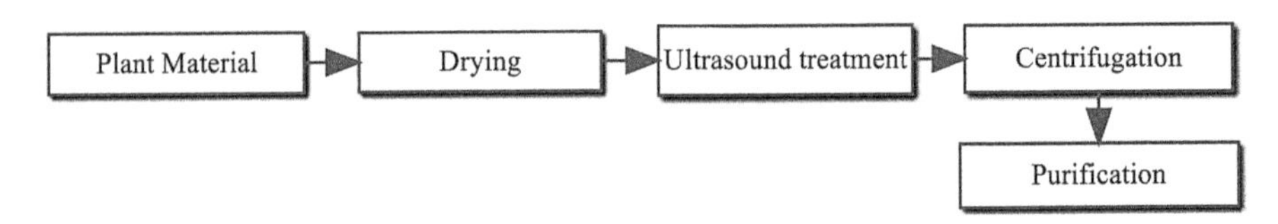

FIGURE 9.9 UA extraction process for the recovery of plant secondary metabolites.
Source: Self-developed with concepts from Ref. [85].

In the MEFA extraction process, quality, and yield of the products is affected by electric field strength, process time, frequency, and conductivity of the extraction medium and biological material. The types of electrode materials in the process also play a significant role in the performance of

MEFA extraction. The most commonly used electrode materials are stainless steel, titanium, platinum, and platinized titanium [40].

The MEFA extraction enhances permeabilization because of triggering of trans-membrane potential difference across the cell membrane (non-thermal effect) and high temperature [63]. It can be applied without solvent (by direct pressing of electrodes) or by using a conductive extraction medium. Salt solutions have also been used as extraction medium (solvent) during the MEFA extraction process. For example, the use of potassium chloride [88] or sodium chloride solution (0.05%) [63] has been reported in literature. However, in MEFA extraction process, energy consumption is reduced and the extraction yield is enhanced by enhancing the mass transfer. Process optimization is important to enhance the quality and quantity of the specifically extracted biomolecule and to minimize the electrochemical reactions, which may occur during the extraction process [40, 102]. In Figure 9.10, MEFA extraction process for recovery of plant secondary metabolites is presented.

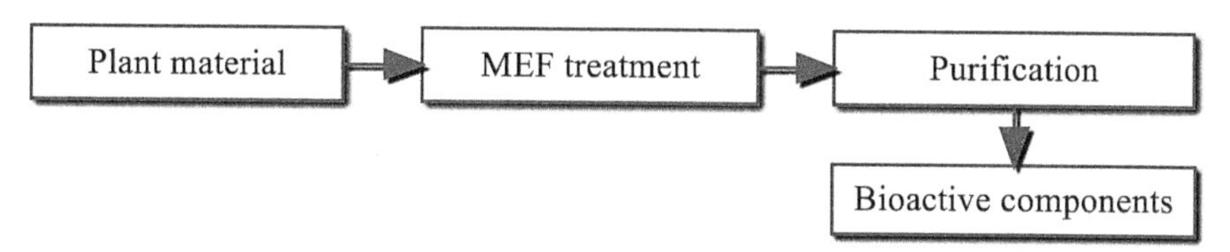

FIGURE 9.10 MEFA extraction process for the recovery of plant secondary metabolites.
Source: Self-developed with concepts from Refs. [102, 105].

The extraction of oils using distilled water as a medium has been used during the extraction of phenolic compounds [63]. MEFA extraction has also been successfully used to extract plant metabolites from potato tissue using different temperatures and treatment times. The total energy consumption in the conventional extraction process was about 300 kJ kg^{-1}, which was significantly reduced to 28 kJ kg^{-1} during MEFA extraction of phenolic compounds from colored potato [88].

In Table 9.4, some examples of the extraction of plant secondary metabolites from different biological materials by MEFA process are presented.

Research studies were conducted using emerging technologies to increase the recovery efficiency of secondary metabolites from the plant matrix. In Table 9.5, some examples for the extraction of plant secondary metabolites with emerging technology are represented.

TABLE 9.4 Examples of Extraction of Plant Secondary Metabolites from Different Biological Materials by the MEFA Process

Sample Matrix	Extracted Compounds	References
Beetroot	Betanin	[63]
Colored potato	Phenolic compounds	[88]
Gac (*Momordica cochinchinensis*)	β-carotene and lycopene	[1]
Green microalgae (*Chlorella vulgaris*)	Carotenoids and chlorophyll	[35]
Microalga (*Heterochlorella luteoviridis*)	Carotenoids	[49]
Tomato	Phenolic compounds	[43]
White Bran	Phenolic compounds	[4]

TABLE 9.5 Recovery of Plant Secondary Metabolites from Plant Matrix with Emerging Technologies

Plant Material	Extraction Method	Bioactive Components	References
Red grape skins Green tea leaves	HPA	Anthocyanin Caffeine	[28] [54]
Marigold flowers Bay leaves	EA	Carotenoids Essential oil	[14] [18]
Brown rice	PEFA	γ-oryzanol, polyphenols, and phenolic acids, saturated, and unsaturated fatty acids	[4]
Pigeon pea leaves	NPCA	Flavonoids	[67]
Microalgae Tomatoes	HPHA	Carotenoid Carotenoid	[110] [87]
Tea Tobacco leaves	MA	Phenols Solanesol	[108] [124]
Grapes Spruce wood bark	UA	Amino acids Polyphenols	[21] [42]
Pigeon pea leaves Radix Astragali (*Astragalus*)	NPCA	Flavonoids Polysaccharides	[66] [50]
Olive kernel	HVEDA	Polyphenols and proteins	[99]

9.3.2.10 SEPARATION AND PURIFICATION OF SECONDARY METABOLITES

Efficient separation and purification is another important issue for the recovery of plant metabolites. Plant secondary metabolites are often mixed with a wide range of compounds with similar polarities and structures. Due to this fact, the separation of plant secondary metabolites is considered a great challenge. A number of different separation and purification methods, such as membrane-based separation process [13] and chromatographic process [12, 93] are commonly used to obtain pure plant secondary metabolites [12, 13]. Apart from that, non-chromatographic techniques (such as distillation, evaporation, freeze-drying, adsorption, etc.), can be used for separation and purification of secondary metabolites from plant matrix.

9.3.2.10.1 Membrane Separation

Membranes are selective barriers, which are used for the separation of desired molecules. The membrane separation process is a pressure-driven process, where particles with larger sizes than the membrane pores are retained and smaller molecules move through the membrane pore. Based on the pore sizes, membranes can be ordered as microfiltration (MF), ultrafiltration (UF), nanofiltration (NF) and reverse osmosis (RO) [13]. In the industrial process, both cross-flow and dead-end membrane modules are used for purification and fractionation of plant secondary metabolites. In Figure 9.11, different types of membranes and their principle of separation process are presented.

The UF membrane has been used to concentrate polyphenols from grape seeds [66]; and to recover phenolic compounds from extracts of black tea [112]. Furthermore, the membrane separation process was successfully employed to concentrate bioactive components from kola-nut extract [80]. Although the membrane separation process is important to concentrate and recover plant metabolites, yet its performance is sometimes poor due to concentration polarization on the membrane surface and fouling of membrane [16].

9.3.2.10.2 Chromatographic Purification Techniques

Chromatography-related separation is used for the separation of a biomolecule from a mixture of different chemical compounds with unique polarity and molecular weight [6]. Different types of chromatographic processes (such as: adsorption chromatography, partition chromatography, ion exchange-chromatography, size-exclusion chromatography, thin-layer chromatography, paper chromatography, gas chromatography and high-performance liquid chromatography) are used to isolate secondary metabolites from plant tissues [12, 17]. In Figure 9.12, the classifications of chromatographic techniques are represented.

The separation efficiency of target metabolites is highly dependent on their adsorption affinity to the stationary phase. Subsequently, different spectroscopic techniques (such as: infrared nuclear magnetic resonance spectroscopy, ultraviolet (UV)-visible spectroscopy and tandem mass spectroscopy) are used to identify the separated and purified biological compounds [90].

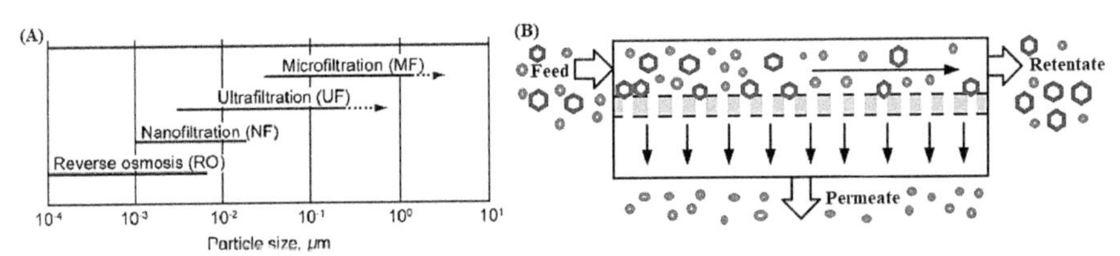

FIGURE 9.11 (A) Different types of membranes depending on their molecular weight cut-off (pore size), (B) Principle of membrane separation process.
Source: Self-developed with concepts from Ref. [13].

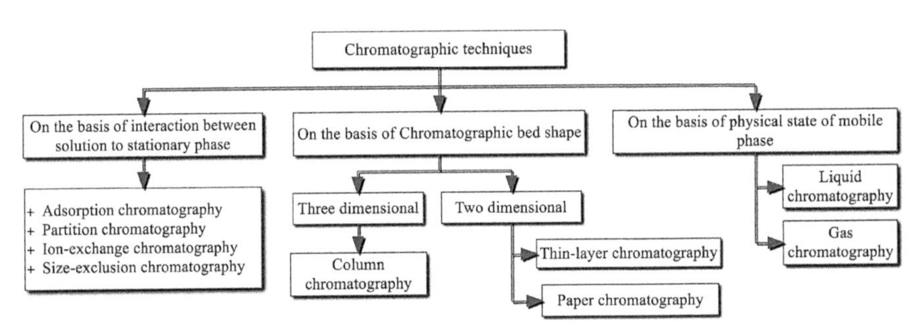

FIGURE 9.12 Classifications of chromatographic techniques.
Source: Self-developed with concepts from Refs. [12, 17].

The separation of secondary metabolites from the herbal matrix is not an easy task due to the presence of a wide variety of phytochemicals. This problem can be overcome by changing polarity of mobile phase. Therefore, in chromatographic process, gradient elution mode is frequently used instead of isocratic elution mode for separation of bioactive compounds. As plant metabolites have highly variable properties (polarity, chemical structure, glycosidic linkages and spectral characteristics), no single technique is appropriate for universal application of separation of plant secondary metabolites [73].

9.4 SECONDARY METABOLITES FROM HERBS: HEALTH BENEFITS

Plant secondary metabolites are known for their several beneficial effects on human health, such as: digestive stimulant, antimicrobial, anti-inflammatory, anti-carcinogenic, anti-diabetic, and anti-oxidant activities, etc. Therefore, the demand for natural bioactive components in food and consumatic industries have increased dramatically. These can improve the quality of processed foods and drugs by increasing the shelf-life, preserving organoleptic decoration and maintaining nutritional properties. They possess potential for reducing risks of several health hazards [115]. Therefore, plant secondary metabolites have received today more attention in food, pharmaceutical, and cosmetic industries [31, 52, 97, 109].

9.4.1 PHARMACEUTICS

In recent years, the majority of people shifted their attention towards herbal drugs and the indigenous system of medicine. Similar to synthetic drugs, several herb drugs as antitussives (codeine), analgesics (morphine), antihypertensives (reserpine), antineoplastics (vinblastine and taxol), cardiotonic (digoxin), anti-malarial (quinine and artemisinin), anticancer, anti-neuro disorders have become popular among the common population [70, 96]. Herbal extracts, in particular, essential oils contain many plant metabolites, such as, hydrocarbons (terpenes and sesquiterpenes) and oxygenated compounds (alcohols, ketones, aldehydes, acids, oxides, phenols, ethers, lactones, and esters). Due to the presence of these bioactive compounds, they are found to have inhibitory activities against several Gram-negative and Gram-positive pathogenic bacteria. Studies have shown the positive impact of flavonoids, especially their glycosides on human health. Furthermore, antioxidant, anti-inflammatory, anticancer, antiviral, and anti-diabetes activities of phenolic compounds are well known [118, 120]. It has been reported that extracts from different parts of Shatavari plant have a significant role in the treatment of female reproductive problems and diseases, and nervous disorders, dyspepsia, cough bronchitis, tuberculosis, and throat infections, etc., [107].

9.4.2 FOODS AND FOOD PRODUCTS

Plant secondary metabolites are important for food preservation technologies [44, 69, 79, 82]. Apart from preservative roles, plant secondary metabolites can be used as functional foods because they offer physiological benefits with a reduction of chronic diseases along with basic nutritional functions. Many plant metabolites (such as: polyphenols and flavonoids) have important health benefits against chronic health problems, such as, inflammation, cancer, and atherosclerosis [25]. The research study by Manach et al. represents the effectiveness of a plant-based diet for the reduction of risk of chronic diseases [70].

The potential use of plant secondary metabolites in fish farming has been also reported. Currently, due to stricter environmental legislation, we must reduce the use of synthetic antibiotics in aquaculture due to risk of chemical residues in food [22]. According to Badri et al. [11], plant metabolites can regulate microbial composition of the soil. Thus, it affects root exudates and plant productivity [11].

9.4.3 COSMETICS

Before the use of synthetic cosmetics, plant secondary metabolites with similar properties were the main source of all cosmetics in daily life. The use of secondary metabolites offers benefits as skincare products [31, 52, 109]. Plant secondary metabolites offer several advantages, such as: inhibition of tyrosinase, increase antioxidant capacity and antimicrobial activity. Those biological activities can be important for the prevention of several skin conditions [45, 97].

9.4.4 OTHERS

Apart from their role in food, cosmetics, and pharmaceutical items, plant secondary metabolites can be also used in dye production [25] and as active compounds (anti-oxidant and anti-microbial) in packaging systems [115]. Applications of selected plant secondary metabolites in pharmaceutics, food, cosmetic, and packaging system are summarized in Table 9.6.

9.5 SUMMARY

Plants or herbs produce different secondary metabolites, such as: terpenes, alkaloids, carotenoids, polyphenols, and glucosinolates. Besides conventional extraction, several novel technologies with the basic extraction

TABLE 9.6 Applications of Plant Secondary Metabolites in Pharmaceutics, Food, and Cosmetic Products

Application	Herbal Bioactive	Impacts/Benefits	References
Pharmaceutics	Polyphenols	Anti-inflammatory effect	[106]
	Flavonoids	Antioxidative effects, impede genetic mutation and carcinogenesis	[78]
	Glucosinolates		[20]
		Anti-aging effect	[55]
		Antioxidative effect	[51]
		Anti-Inflammatory, Immunomodulatory Agents,	
		Anti-cancer and Cardiovascular Disease	
	Terpenes	Anti-inflammatory	[83]
	Carotenoids	Antioxidant	[75]
Food	Carvone, limonene, and (E)-anethole	Effective against pests	[69]
	Terpenes or sterols	Antimicrobial activity	[2, 82]
	Phenolic compounds,	Antimicrobial activity	[79]
	Cinnamaldehyde,	Antifungal activity	[44]
	Cichoric acid, polysaccharides, Alkamides, and glycoproteins		
Cosmetics	Amino acids	Skincare/cleansing action	[15]
	Gallic acid, mannitol, glucose, fats, resin, traces of an alkaloid and mucilage;	Dandruff treatment	[45]
	Saponin	Collagen synthesis and anti-aging	[117]
Others	Polyphenolic extract;	Dyeing silk fabrics	[26]
	Polyphenolics	Active packaging	[115]

process have been developed. These technologies enhance the permeabilization of the plant cell-wall and facilitate the recovery of target components from inter or intra-cellular spaces. Plant secondary metabolites offer several health benefits. They have importance in pharmaceutical, food, and cosmetic industries.

KEYWORDS

- **herbal extracts**
- **high pressure-assisted**
- **microfiltration**
- **moderate electric field-assisted**
- **pharmaceuticals**
- **secondary metabolites**

REFERENCES

1. Aamir, M., & Jittanit, W., (2017). Ohmic heating treatment for gacaril oil extraction: Effects on extraction efficiency, physical properties and some bioactive compounds. *Innovative Food Science and Emerging Technologies, 41*, 224–234.
2. Abdalla, A. E., Darwish, S. M., & Ayad, E. H., (2007). Egyptian mango by-product: Antioxidant and antimicrobial activities of extract and oil from mango seed kernel. *Food Chemistry, 103*, 1141–1152.
3. Aldred, E. M., (2008). *Pharmacology: A Handbook for Complementary Healthcare Professionals* (pp. 167–175). Amsterdam: Elsevier Health Sciences.
4. Al-Hilphy, A. R. S., (2014). Practical study for new design of essential oils: Extraction apparatus using ohmic heating. *International Journal of Agricultural Sciences, 4*, 351–366.
5. Ali, P., Chen, Y. F., & Sargsyan, E., (2014). Bioactive molecules of herbal extracts with anti-infective and wound healing properties. In: Kon, K., & Rai, M., (eds.) *Microbiology for Surgical Infections Diagnosis, Prognosis, and Treatment* (pp. 205–220). Amsterdam, Elsevier.
6. Altemimi, A., Lakhssassi, N., & Baharlouei, A., (2017). Phytochemicals: Extraction, isolation, and identification of bioactive compounds from plant extracts. *Plants, 6*, 42–46.
7. Augusto, P. E. D., Ibarz, A., & Cristianini, M., (2012). Effect of high-pressure homogenization (HPH) on the rheological properties of a fruit juice serum model. *Journal of Food Engineering, 111*, 474–477.
8. Azmir, J., Zaidul, I. S. M., Rahman, M. M., & Sharif, K. M., (2013). Techniques for extraction of bioactive compounds from plant materials: A review. *Journal of Food Engineering, 117*, 426–436.

9. Azwanida, N. N., (2015). Review on the extraction methods use in medicinal plants, principle, strength, and limitation. *Medicinal and Aromatic Plants*, *4* (196), 3–8.
10. Baby, K. C., T. V., (2013). Ranganathan. Enzyme-assisted extraction of bioingredients. *Chem Week*, *59*, 213–224.
11. Badri, D. V., Chaparro, J. M., Zhang, R., Shen, Q., & Vivanco, J. M., (2013). Application of natural blends of phytochemicals derived from the root exudates of arabidopsis to the soil reveal that phenolic-related compounds predominantly modulate the soil microbiome. *Journal of Biological Chemistry*, *288*, 4502–4512.
12. Bailey, J. E., & Ollis, D. F., (1986). *Biochemical Engineering Fundamental* (2nd edn., pp. 753–764). Singapore: McGraw-Hill International Edition.
13. Bailey, J. E., & Ollis, D. F., (1986). *Biochemical Engineering Fundamentals* (2nd edn., pages 764–767). Singapore: McGraw-Hill International Edition.
14. Barzana, E., Rubio, D., & Santamaria, R. I., (2002). Enzyme-mediated solvent extraction of carotenoids from marigold flower (*Tagetes erecta*). *Journal of Agricultural and Food Chemistry*, *50* (16), 4181–4186.
15. Basmatker, G., Jais, N., & Daud, F., (2011). *Aloe vera*: A valuable multifunctional cosmetic ingredient. *International Journal of Medicinal and Aromatic Plants*, *1*, 338–341.
16. Belleville, M. P., & Vaillant, F., (2015). *Membrane Technology for Production of Nutraceuticals* (pp. 217–234). Chapter 8. online; https://www.researchgate.net/profile/Fabrice_Vaillant/publication/283851802_Membrane_Technology_for_Production_of_Nutraceuticals/links/565f0d1308aefe619b27dc97/Membrane-Technology-for-Production-of-Nutraceuticals.pdf (accessed on 6 August 2020).
17. Blanch, H. W., & Clark, D. S., (1996). *Biochemical Engineering* (1st edn., pp. 502–511). Boca Raton, FL: CRC Press.
18. Boulila, A., Hassen, I., Haouari, L., & Mejri, F., (2015). Enzyme-Assisted extraction of bioactive compounds from bay leaves (*Laurus nobilis L.*). *Industrial Crops and Products*, *74*, 485–183.
19. Breitmaier, E., (2006). *Terpenes: Flavors, Fragrances, Pharmaca, Pheromone* (pp. 1–9). Weinheim, Germany; Wiley-VCH Verlag GmbH & Co. KGaA
20. Cao, G., Booth, S. L., Sadowski, J. A., & Prior, R. L., (1998). Increases in human plasma antioxidant capacity after consumption of controlled diets high in fruit and vegetables. *The American Journal of Clinical Nutrition*, *68*, 1081–1087.
21. Carrera, C., Ruiz-Rodríguez, A., Palma, M., & Barroso, C. G., (2015). Ultrasound-assisted extraction of amino acids from grapes. *Ultrasonics Sonochemistry*, *22*, 189–505.
22. Chakraborty, S. B., & Hancz, C., (2011). Application of phytochemicals as immunostimulant, antipathogenic and antistress agents in finfish culture. *Reviews in Aquaculture*, *3*, 103–119.
23. Chandrasekara, A., & Shahidi, F., (2018). Herbal beverages: Bioactive compounds and their role in disease risk reduction: A review. *Journal of Traditional and Complementary Medicine*, *8*, 451–458.
24. Cheng, X., Bi, L., Zhao, Z., & Chen, Y., (2015). Advances in Enzyme-assisted extraction of natural products. *3rd International Conference on Material, Mechanical, and Manufacturing Engineering (IC3ME 2015)* (pp. 371–375). Guangzhou, China: Atlantis Press;
25. Chhikara, N., Devi, H. R., Jaglan, S., Sharma, P., & Gupta, P., (2018). Bioactive compounds, food applications and health benefits of *Parkia speciosa* (Stinky Beans): A review. *Agriculture and Food Security*, *7* (46), 1–9.

26. Chhikara, N., Kushwaha, K., Sharma, P., Gat, Y., & Panghal, A., (2019). Bioactive compounds of beetroot and utilization in food processing industry: A critical review. *Food Chemistry, 272,* 192–200.
27. Christen, P., & Kaufmann, B., (2002). Recent extraction techniques for natural products: Microwave-assisted extraction and pressurized solvent extraction. *Phytochemical Analysis: An International Journal of Plant Chemical and Biochemical Techniques, 13,* 105–113.
28. Corrales, M., García, A. F., Butz, P., & Tauscher, B., (2009). Extraction of anthocyanins from grape skins assisted by high hydrostatic pressure. *Journal of Food Engineering, 90,* 415–421.
29. Diplock, A. T., Charuleux, J. L., & Crozier-Willi, G., (1998). Functional food science and defense against reactive oxidative species. *British Journal of Nutrition, 1998,* 77–112.
30. Donsì, F., Ferrari, G., Fruilo, M., & Pataro, G., (2011). Pulsed electric fields—assisted vinification. *Procedia Food Science, 1,* 780–785.
31. Dureja, H., Kaushik, D., Gupta, M., Kumar, V., & Lather, V., (2005). Cosmeceuticals: An emerging concept. *Indian Journal of Pharmacology, 37,* 155.
32. El-Gharras, H., (2009). Polyphenols: Food sources, properties, and applications: A review. *International Journal of Food Science and Technology, 44,* 2512–2518.
33. Ferrari, G., Maresca, P., & Ciccarone, R., (2010). The application of high hydrostatic pressure for the stabilization of functional foods: Pomegranate juice. *Journal of Food Engineering, 100,* 245–253.
34. Floury, J., Bellettre, J., Legrand, J., & Desrumaux, A., (2004). Analysis of a new type of high-pressure homogenizer: A study of the flow pattern. *Chemical Engineering Science, 59,* 843–853.
35. Fraccola, G., Fernandes, B. D. O., & Geada, P., (2016). *Enhancing Extraction of Food Grade Pigments from the Microalgae Chlorella Vulgaris through Application of Ohmic Heating* (p. 2). Poster sess. *Present.* Meeting, Chicago; Institute of Food Technology.
36. Galanakis, C. M., (2013). Emerging Technologies for the production of nutraceuticals from agricultural by-products: A viewpoint of opportunities and challenges. *Food and Bioproducts Processing, 91,* 575–579.
37. Gamli, F., (2014). Review of application of pulsed electric field in the production of liquid/ semi-liquid food materials. *Advance Research in Agriculture and Veterinary Science, 1* (2), 54–61.
38. Gao, M., Huang, W., Roy Chowdhury, M., & Liu, C., (2007). Microwave-assisted extraction of scutellarin from erigeron breviscapus hand-mazz and its determination by high-performance liquid chromatography. *Analytica Chimica Acta, 591,* 161–166.
39. Gaur, R., Sharma, A., Khare, S. K., & Gupta, M. N., (2007). Novel Process for extraction of edible oils. Enzyme assisted three phase partitioning (EATPP). *Bioresource Technology, 98,* 696–699.
40. Gavahian, M., Chu, Y. H., & Sastry, S., (2018). Extraction from food and natural products by moderate electric field: Mechanisms, benefits, and potential industrial applications. *Comprehensive Reviews in Food Science and Food Safety, 17,* 1040–1052.
41. Ghani, S. B. A., Weaver, L., & Zidan, Z. H., (2008). Microwave-assisted synthesis and antimicrobial activities of flavonoid derivatives. *Bioorganic and Medicinal Chemistry Letters, 18,* 518–522.
42. Ghitescu, R. E., Volf, I., Carausu, C., & Bühlmann, A. M., (2015). Optimization of ultrasound-assisted extraction of polyphenols from spruce wood bark. *Ultrasonics Sonochemistry, 22,* 535–541.

43. Gil, F., Custódia, F., Emir, A., Annette, R., & Manuel, A., (2017). Polyoxometalates as inhibitors of P-type ATPases and the role of polyphenols. *AISANH (Archives of the International Society of Antioxidants in Nutrition and Health), 5*, 17–20.
44. Giuseppe, C. R., Coda, R., De, M. A., & Di, R. C., (2009). Long-term fungal inhibitory activity of water-soluble extract from *Amaranthus* Spp. Seeds during storage of gluten-free and wheat flour breads. *International Journal of Food Microbiology, 131*, 189–196.
45. Grimes, P., Nordlund, J. J., & Pandya, A. G., (2006). Increasing our understanding of pigmentary disorders. *Journal of the American Academy of Dermatology, 54* (5), 255–261.
46. Gupta, A., Naraniwal, M., & Kothari, V., (2012). Modern extraction methods for preparation of bioactive plant extracts. *International Journal of Applied and Natural Sciences, 1*, 8–16.
47. Ho, K. K. H. Y., Ferruzzi, M. G., & Liceaga, A. M., (2015). Microwave-assisted extraction of lycopene in tomato peels: Effect of extraction conditions on all-trans and Cis-isomer yields. *LWT-Food Science and Technology, 62*, 160–168.
48. Ho, K. H. V. H., De Boer, B. C., & Tijburg, L. B., (2000). Carotenoid bioavailability in humans from tomatoes processed in different ways determined from the carotenoid response in the triglyceride-rich lipoprotein fraction of plasma after a single consumption and in plasma after four days of consumption. *The Journal of Nutrition, 130*, 1189–1196.
49. Jaeschke, D. P., Merlo, E. A., & Rech, R., (2017). Moderate electric field influence on carotenoids extraction time from hetero chlorella *Luteo viridis*. *World Academy of Science, Engineering, and Technology International Journal of Nutrition and Food Engineering, 11*, 1–9.
50. Jiao, J., Wei, F. Y., Gai, Q. Y., Wang, W., Luo, M., Fu, Y. J., & Ma, W., (2014). Pilot-scale homogenization-assisted negative pressure cavitation extraction of *Astragalus* polysaccharides. *International Journal of Biological Macromolecules, 67*, 189–194.
51. Johnson, T. L., & Dinkova-Kostova, A. T., (2015). Glucosinolates from the *Brassica* vegetables and their health effects. In: *Encyclopedia of Food and Health* (pp. 248–255) New York: Elsevier Inc.
52. Joshi, L. S., & Pawar, H. A., (2015). Herbal cosmetics and cosmeceuticals: An overview. *Natural Products Chemistry and Research, 3* (2), 1–8.
53. Jun, X., (2006). Application of high hydrostatic pressure processing of food to extracting lycopene from tomato paste waste. *High Pressure Research, 26*, 33–41.
54. Jun, X., (2009). Caffeine extraction from green tea leaves assisted by high pressure processing. *Journal of Food Engineering, 94*, 105–109.
55. Kamisah, Y., Othman, F., Qodriyah, H. M. S., & Jaarin, K., (2013). *Parkia speciosa* hassk: A potential phytomedicine. *Evidence-Based Complementary and Alternative Medicine, 2013*, 1–9.
56. Pandey, K. B., & Rizvi, S. I., (2009). Plant polyphenols as dietary antioxidants in health and disease. *Oxidative Medicine and Cellular Longevity, 2*, 270–278.
57. Khaw, K. Y., Parat, M. O., Shaw, P. N., & Falconer, J. R., (2017). Solvent supercritical fluid technologies to extract bioactive compounds from natural sources: A review. *Molecules, 22* (7), 1186–1190.
58. Krinsky, N. I., & Yeum, K. J., (2003). Carotenoid-radical interactions. *Biochemical and Biophysical Research Communications, 305* (3), 754–760.
59. Kulshreshtha, G., Burlot, A. S., & Marty, C., (2015). Enzyme-assisted extraction of bioactive material from *Chondrus crispus* and *Codium fragile* and its effect on herpes simplex virus (HSV-1). *Marine Drugs, 13*, 558–580.

60. Kumoro, A. C., & Hasan, M., (2008). Extraction of herbal components-the case for supercritical fluid extraction. *Teknik, 29* (3), 180–183.
61. Landbo, A. K., & Meyer, A. S., (2001). Enzyme-Assisted extraction of antioxidative phenols from black currant juice press residues (*Ribes nigrum*). *Journal of Agricultural and Food Chemistry, 18*, 3169–3177.
62. Le, A. V., Parks, S. E., Nguyen, M. H., & Roach, P. D., (2018). Effect of solvents and extraction methods on recovery of bioactive compounds from defatted gac (*Momordica cochinchinensis Spreng.*) seeds. *Separations, 5*, 39–41.
63. Lebovka, N. I., Shynkaryk, M., & Vorobiev, E., (2007). Moderate Electric field treatment of sugar beet tissues. *Biosystems Engineering, 96*, 47–56.
64. Li, B. B., Smith, B., & Hossain, M. M., (2006). Extraction of phenolics from citrus peels, II: Enzyme-assisted extraction method. *Separation and Purification Technology, 48*, 189–196.
65. Li, S., Fu, Y., Zu, Y., Zu, B., Wang, Y., & Efferth, T., (2009). Determination of paclitaxel and its analogues in the needles of taxus species by using negative pressure cavitation extraction followed by HPLC-MS-MS. *Journal of Separation Science, 32*, 3958–3966.
66. Liu, D., Vorobiev, D., Savoire, R., & Lanoiselle, J. L., (2011). Extraction of polyphenols from grape seeds by unconventional methods and extract concentration through polymeric membrane. *11th International Congress on Engineering and Food, 2011*, 1939–1940.
67. Liu, W., Fu, Y., Zu, Y., Kong, Y., Zhang, L., Zu, B., & Efferth, T., (2009). Negative-pressure cavitation extraction for the determination of flavonoids in pigeon pea leaves by liquid chromatography-tandem mass spectrometry. *Journal of Chromatography A, 1216*, 3841–3850.
68. Liu, Z., Esveld, E., Vincken, J. P., & Bruins, M. E., (2019). Pulsed electric field as an alternative pre-treatment for drying to enhance polyphenol extraction from fresh tea leaves. *Food and Bioprocess Technology, 12*, 183–192.
69. López, M. D., Jordán, M. J., & Pascual-Villalobos, M. J., (2008). Toxic compounds in essential oils of coriander, caraway, and basil active against stored rice pests. *Journal of Stored Products Research, 44*, 273–278.
70. Manach, C., Hubert, J., Llorach, R., & Scalbert, A., (2009). The Complex Links Between Dietary Phytochemicals and Human Health Deciphered by Metabolomics. *Molecular Nutrition and Food Research, 53*, 1303–1315.
71. Martin-Smith, M., & Sneader, W. E., (1969). Biological activity of the terpenoids and their derivatives: Recent advances. *Progress in Drug Research, 13*, 11–100.
72. McGraw, K. J., & Ardia, D. R., (2003). Carotenoids, Immunocompetence, and the information content of sexual colors: An experimental test. *The American Naturalist, 162*, 704–712.
73. Miniati, E., (2007). Assessment of phenolic compounds in biological samples. *Annal Instituto Superiore di Sanita (Annal Higher Health Institute), 43*, 362–368.
74. Murphy, M., (1999). Plant products as antimicrobial agents. *Clinical Microbiology Reviews, 12*, 564–582.
75. Nagarajan, J., Ramanan, R. N., & Raghunandan, M. E., (2017). *Nutraceutical and Functional Food Components* (pp. 259–296). New York: Elsevier Inc.
76. Nagendra-Chari, K. L., & Manasa, D., (2013). Enzyme-assisted extraction of bioactive compounds from ginger (*Zingiber officinale Roscoe*). *Food Chemistry, 139*, 509–514.
77. Ncube, B., & Van-Staden, J., (2015). Tilting plant metabolism for improved metabolite biosynthesis and enhanced human benefit. *Molecules, 20*, 12698–12731.

78. Ng, L. T., Ko, H. H., & Lu, T. M., (2009). Potential Antioxidants and tyrosinase inhibitors from synthetic polyphenolic deoxybenzoins. *Bioorganic and Medicinal Chemistry, 17*, 4360–4366.
79. Noorolahi, Z., Sahari, M. A., Barzegar, M., Doraki, N., & Naghdi-Badi, H., (2013). Evaluation antioxidant and antimicrobial effects of cinnamon essential oil and echinacea extract in kolompe. *Journal of Medicinal Plants, 12*, 14–28.
80. Nyamien, Y., Belleville, M. P., Coulibaly, A., Adima, A., & Biego, G., (2017). Extraction and concentration of bioactive compounds of *Cola nitida* using membrane processes: Analysis of operating parameters and membrane fouling. *European Journal of Nutrition and Food Safety, 7*, 85–101.
81. Opara, E. I., & Chohan, M., (2014). Culinary herbs and spices: Their bioactive properties: The contribution of polyphenols and the challenges in deducing their true health benefits. *International Journal of Molecular Sciences, 15*, 19183–19202.
82. Owen, R. J., & Palombo, E. A., (2007). Anti-listerial activity of ethanolic extracts of medicinal plants, *Eremophila alternifolia* and *Eremophila duttonii*, in food homogenates and milk. *Food Control, 18*, 387–390.
83. Paduch, R., Kandefer-Szerszeń, M., Trytek, M., & Fiedurek, J., (2007). Terpenes: Substances Useful in human healthcare. *Archivum Immunologiae et Therapiae Experimentalis (The Archive Immunologie and Experimental Therapies), 55* (5), 315–320.
84. Paduch, R., Trytek, M., Król, S. K., & Kud, J., (2016). Biological activity of terpene compounds produced by biotechnological methods. *Pharmaceutical Biology, 54* (6), 1096–1107.
85. Pananun, T., & Montalbo-Lomboy, M., (2012). High-power ultrasonication-assisted extraction of soybean isoflavones and effect of toasting. *LWT-Food Science and Technology, 47*, 199–207.
86. Pandey, K. B., & Rizvi, S. I., (2009). Plant Polyphenols as Dietary Antioxidants in Human Health and Disease. *Oxidative Medicine and Cellular Longevity, 2*, 270–278.
87. Panozzo, A., Lemmens, L., & Van, L. A., (2013). Microstructure and bioaccessibility of different carotenoid species as affected by high pressure homogenization: A case study on tomatoes. *Food Chemistry, 141*, 4094–4100.
88. Pereira, R. N., Rodrigues, R. M., & Genisheva, Z., (2016). Effects of ohmic heating on extraction of food-grade phytochemicals from colored potato. *LWT-Food Science and Technology, 74*, 183–503.
89. Poojary, M., Barba, F., Aliakbarian, B., & Donsì, F., (2016). innovative alternative technologies to extract carotenoids from microalgae and seaweeds. *Marine Drugs, 14*, 1–34.
90. Popova, I. E., Hall, C., & Kubátová, A., (2009). Determination of lignans in flaxseed using liquid chromatography with time-of-flight mass spectrometry. *Journal of Chromatography A, 1216*, 217–229.
91. Prado, J. M., Prado, G. H. C., & Meireles, M. A., (2011). Scale-up study of supercritical fluid extraction process for clove and sugarcane residue. *The Journal of Supercritical Fluids, 56*, 231–237.
92. Puri, M., Sharma, D., & Barrow, C. J., (2012). Enzyme-assisted extraction of bioactives from plants. *Trends in Biotechnology, 30*, 37–44.
93. Qilong, R., Huabin, X., & Zongbi, B. A. O., (2013). Recent advances in separation of bioactive natural products. *Chinese Journal of Chemical Engineering, 21*, 937–952.

94. Quagliariello, V., Iaffaioli, R. V., & Falcone, M., (2016). Effect of pulsed electric fields-assisted extraction on anti-inflammatory and cytotoxic activity of brown rice bioactive compounds. *Food Research International, 87*, 115–124.
95. Radojcic, I., Glivetic, T., Delonga, K., & Vorkapic, J., (2008). Glucosinolates and their potential role in plant. *Periodicum Biologorum, 110*, 297–309.
96. Ramawat, K. G., Dass, S., & Mathur, M., (2009). *Herbal Drugs: Ethnomedicine to Modern Medicine* (pp. 153–171). New York, NY: Springer.
97. Ribeiro, A., Estanqueiro, M., Oliveira, M., & Sousa, L. J., (2015). Main benefits and applicability of plant extracts in skin care products. *Cosmetics, 2*, 48–65.
98. Roohinejad, S., Koubaa, M., & Barba, F. J., (2016). Negative pressure cavitation extraction: A novel method for extraction of food bioactive compounds from plant materials. *Trends in Food Science and Technology, 52*, 98–108.
99. Roselló-Soto, E., Barba, F. J., & Parniakov, O., (2015). High voltage electrical discharges, pulsed electric field, and ultrasound assisted extraction of protein and phenolic compounds from olive kernels. *Food and Bioprocess Technology, 8*, 885–894.
100. Rostagno, M. A., Villares, A., & Guillamón, E., (2009). Sample preparation for the analysis of isoflavones from soybeans and soy foods. *Journal of Chromatography A, 1216*, 2–29.
101. Sasidharan, S., Chen, Y., Saravanan, D., Sundram, K. M., & Yoga, L. L., (2011). Extraction, isolation, and characterization of bioactive compounds from plant extracts. *African Journal of Traditional, Complementary, and Alternative Medicine, 8*, 1–10.
102. Sastry, S. K., Shynkaryk, M., & Somavat, R., (2011). Ohmic and Moderate electric field processing: Developments and new applications. *Paper Presentation at the 11th International Congress on Engineering and* Food (p. 3). Athens, Greece.
103. Scalbert, A., & Williamson, G., (2000). Chocolate: Modern Science investigates an ancient medicine dietary intake and bioavailability of polyphenols. *Journal of Medicinal Food, 1*, 2073–2085.
104. Selvamuthukumaran, M., & Shi, J., (2018). Recent advances in extraction of antioxidants from plant by-products processing industries. *Food Quality and Safety, 1*, 61–81.
105. Sensoy, I., & Sastry, S. K., (2004). Extraction using moderate electric fields. *Journal of Food Science, 69*, 7–13.
106. Serrano, A., Ros, G., & Nieto, G., (2018). Bioactive Compounds and extracts: Traditional herbs and their potential anti-inflammatory health effects. *Medicines, 5* (3), 1–9.
107. Singh, A. K., Srivastava, A., Kumar, V., & Singh, K., (2018). Phytochemicals, medicinal, and food applications of shatavari (*Asparagus racemosus*): An updated review. *The Natural Products Journal, 8*, 32–44.
108. Spigno, G., & De Faveri, D. M., (2009). Microwave-assisted extraction of tea phenols: A phenomenological study. *Journal of Food Engineering, 93*, 210–217.
109. Kumar, S., Vivek, V., Sujata, S., & Ashish, B., (2012). Herbal cosmetics: Used for skin and hair. *Inventi Rapid, 4*, 1–7.
110. Taucher, J., Baer, S., & Schwerna, P., (2016). Cell disruption and pressurized liquid extraction of carotenoids from microalgae. *Journal of Thermodynamics and Catalysis, 7*, 158–164.
111. Tian, H., Li, W. Y., Xiao, D., Li, Z. M., & Wang, J. W., (2015). Negative-pressure cavitation extraction of secoisolariciresinol diglycoside from flaxseed cakes. *Molecules, 20*, 11076–11089.

112. Todisco, S., Tallarico, P., & Gupta, B. B., (2002). Mass transfer and polyphenols retention in the clarification of black tea with ceramic membranes. *Innovative Food Science & Emerging Technologies, 3*, 255–262.
113. Toepfl, S., Mathys, A., Heinz, V., & Knorr, D., (2006). Potential of high hydrostatic pressure and pulsed electric fields for energy efficient and environmentally friendly food processing. *Food Reviews International, 22*, 405–423.
114. Tonthubthimthong, P., Chuaprasert, S., Douglas, P., & Luewisutthichat, W., (2001). Supercritical CO_2 extraction of nimbin from neem seeds: An experimental study. *Journal of Food Engineering, 47*, 289–293.
115. Valdés, A., Mellinas, A. C., & Ramos, M., (2015). Use of herbs, spices, and their bioactive compounds in active food packaging. *RSC Advances, 5*, 40324–40335.
116. Vinatoru, M., Toma, M., Radu, O., Filip, P. I., Lazurca, D., & Mason, T. J., (1997). The use of ultrasound for the extraction of bioactive principles from plant materials. *Ultrasonics Sonochemistry, 4*, 135–139.
117. Wang, S., (2017). Anti-skin-aging activity of a standardized extract from panax ginseng leaves *in vitro* and in human volunteer. *Cosmetics, 4*, 18–23.
118. Wang, Y., & Wang, X., (2015). Binding, stability and antioxidant activity of quercetin with soy protein isolate particles. *Food Chemistry, 188*, 24–29.
119. Weinert, M., & Fernando, G. W., (1989). Antiferromagnetism of CuO_2 layers. *Physical Review B, 39* (1), 835–838.
120. Xiao, J., Capanoglu, E., Jassbi, A. R., & Miron, A., (2016). Advance on the flavonoid C-glycosides and health benefits. *Critical Reviews in Food Science and Nutrition. 56*, 29–45.
121. Zeng, H., Wang, Y., Kong, J., Nie, C., & Yuan, Y., (2010). Ionic liquid-based microwave-assisted extraction of rutin from Chinese medicinal plants. *Talanta, 83*, 582–590.
122. Zhang, H. F., Yang, X. H., & Wang, Y., (2011). Microwave-assisted extraction of secondary metabolites from plants: Current status and future directions. *Trends in Food Science and Technology, 22*, 672–688.
123. Zhang, Z., Pang, X., Xuewu, D., Ji, Z., & Jiang, Y., (2005). Role of peroxidase in anthocyanin degradation in litchi fruit pericarp. *Food Chemistry, 90*, 47–52.
124. Zhou, H. Y., & Liu, C. Z., (2006). Microwave-assisted extraction of solanesol from tobacco leaves. *Journal of Chromatography A, 1129*, 135–139.

CHAPTER 10

BIOACTIVE COMPOUNDS FROM *IN-VITRO* CULTURE OF *SWERTIA CHIRAYITA* (ROXB. EX FLEM.) KARSTEN: IDENTIFICATION AND QUANTIFICATION

RITUPARNA KUNDU CHAUDHURI and DIPANKAR CHAKRABORTI

ABSTRACT

Swertia chirayita (or *S. chirata*) is traditionally used for its bioactive molecules in health care. Efficient regeneration methods of *S. chirata* under *in vitro* condition provided novel tools of plant production enriched with secondary metabolites. HPTLC and LC/ESI-MS/MS were used to detect and quantify the secoiridoid glycosides and mangiferin from *in vitro* grown *S. chirata* plant samples. Enhanced synthesis of mangiferin (2.93 ± 0.6%) during vegetative growth of tissue culture raised plants with reference to the donor plant was confirmed through HPTLC. The LC/ESI^+-MS/MS studies have confirmed that *in vitro* grown plants were equally capable to synthesize mangiferin, amarogentin, amaroswerin, sweroside, and swertiamarin.

10.1 INTRODUCTION

Plant-based drugs from secondary metabolites constitute more than 25% of approved new drugs during last 30 years. Also, 50% of the commercially successful medicinal components were developed based on knowledge from plant secondary metabolites and their structures [9, 26].

Swertia chirayita (Roxb. ex Flem.) Karsten or *Swertia chirata* Buch.-Ham. Ex Wall. (Family: *Gentianaceae*; Hindi: *Chirayata, Charaita*; English: felworts) is potential ethnomedicinal herb reported in Indian Pharmaceutical codex, British, and American Pharmacopoeias. Conventional and native medication systems (such as: Ayurveda, Unani, and Siddha) have also recognized the safe and potent use of this herb as a crucial medicinal plant. This plant is used as a purifier of blood in lever diseases. Also, it has anti-inflammatory, anti-platelet, anti-cancerous, antifungal, and anti-malarial activities [1]. The stimulation of central nervous system is another application of *S. chirata* extract [8].

Among all species of this plant in India, *S. chirata* (Appendix-I) holds most significant position for its demand in national and international market because of high yield of bioactive compounds. More than 400 tons of dried plants are being exported from India to the foreign market annually and the demand is increasing @ 10% every year. Indiscriminate harvesting and overexploitation of this plant from the wild by traders has enlisted the plant as critically endangered in their natural habitat, while medicinal plant board of India has tagged it as a priority plant.

Keeping in view its demand in foreign and national market and critically endangered status in India, sustainable utilization of this plant is a prerequisite with various medicinally important active principles. Various methods of *in vitro* cultures have been demonstrated by different authors in recent times [17, 23]. However, micro-propagation utilizing different explants (such as: nodal meristem, leaf, and immature seed culture) [2–4] of this plant for conservation of the germplasm has already been established.

Plants belonging to "genus: *Swertia*" have the presence of following bioactive compounds:

- 1,3,5,8-tetrahydroxyxanthone;
- 1,5,8-trihydroxy-3-methoxyxanthone;
- 1,7-dihydroxy-3,8-dimethoxyxanthone;
- 1,8-dihydroxy-3,5-dimethoxyxanthone;
- 1-hydroxy-2,3,4,7-tetramethoxyxanthone;
- 1-hydroxy-2,3,5,7-tetramethoxyxanthone;
- 1-hydroxy-3,5,8-trimethoxyxanthone;
- 1-hydroxyl-2,3,4,6-tetramethoxyxanthone;
- Amarogenitine;
- Balanophonin;
- Mangeferin;
- Maslinic acid;
- Oleanolic acid;
- Sawertiamarine; and
- Sumaresinolic acid.

Earlier phytochemical investigations on the plant resulted in the isolation of mangiferin and other secoiridoid glycosides (such as: amarogentin, amaroswerin, swerosideand swertiamarin. The amarogentin and amaroswerin have the similar structure, except the substitution of R- by -H in amarogentin and-OH in amaroswerin Similarly, R is substituted by -H in sweroside and R is substituted by -OH in swertiamarin, respectively (Figure 10.1).

Mangiferin (a xanthone-C-glycoside) has taxonomically exclusive characteristics from biogenesis and distribution point of view. Xanthones are group of compounds that exhibit biological and pharmacological activities, such as: antitumor, antiviral, antioxidant, antidiabetic, immuno modulatory, cytotoxic, anti-inflammatory, antimicrobial, and antifungal [5, 7, 14]. The secoiridoid glycosides (Figure 10.1) on the other hand occur in limited families of angiosperms and have been used as essential chemotaxonomic marker [24]. Those glycosides possess bioactivities, such as: antibacterial, anti-cholinergic, chemo preventive and anti-hepatitis.

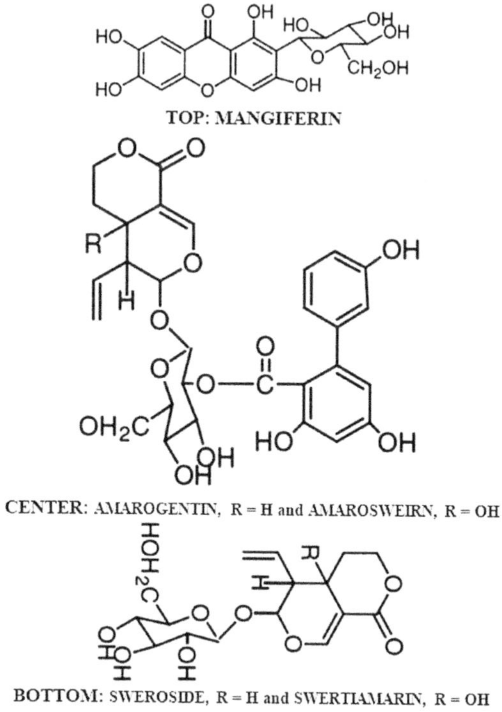

TOP: MANGIFERIN

CENTER: AMAROGENTIN, R = H and AMAROSWEIRN, R = OH

BOTTOM: SWEROSIDE, R = H and SWERTIAMARIN, R = OH

FIGURE 10.1 Biochemical structures: Top: Mangiferin; center: Amarogentin, R = H and amarosweirn, R = OH; bottom: Sweroside, R = H and swertiamarin, R = OH.

Source: Modified and adapted from Ref. [25].

Very few attempts have been made in the past to detect bioactive compounds from the *in vitro* grown *S. chirata* cultures. Synthesis of amarogentin from root culture of *Agrobacterium rhizogenes* transformed *S. chirata* has been reported by Keil et al. [15]; while Ishimaru et al. [12] reported amarogentin, amaroswerin, and four xanthones in *S. japonica* root culture.

In this chapter, the *Swertia chirayita (*or *S. chirata*) plants raised through various explants were analyzed to identify and quantify its secondary metabolites; and micro propagation of this plant from axillary meristem was evaluated for level of secondary metabolites.

10.2 MATERIALS AND METHODS

10.2.1 QUANTITATIVE ANALYSIS OF MANGIFERIN THROUGH HPTLC FROM IN VIVO AND IN VITRO GROWN PLANTS

In this research study, two different experiments were carried out with *in vivo* and *in vitro* grown plants. In first experiment, mangiferin was quantitated from field grown and *in vitro* grown plant material (shoots from node and callus regenerated plants) [2, 4] through high performance thin layer chromatography (HPTLC) technique. The second experimental set up was conducted on *in vitro* grown plants to detect and quantify mangiferin, amarogentin, amaroswerin, sweroside, and swertiamarin through liquid chromatography/electro spray ionization tandem mass spectroscopy (LC/ESI-MS/MS) technique. In the second experiment, the generated fragmentation patterns from these compounds were analyzed and compared with similar compounds from field grown plants in an earlier report [24]. Therefore, presence of these bioactive compounds was confirmed from *in vitro* culture.

S. chirata (a temperature sensitive plant is grown in different elevation and agro-climatic conditions in the hills) were collected from Sikkim, and were grown aseptically in the controlled environment. The phytochemical analyses were performed from different tissues for selective bioactive compounds using advanced analytical instruments, such as: HPTLC and LC/ ESI^+-MS/MS.

About 1 g of dry weight of samples was subjected to extraction, in three batches: (i) from 5 months old field grown mother plant (collected during the month of June, from Gangtok, India 27°19′N, 88°36′E), (ii) one-month-old *in vitro* grown node, and (iii) callus regenerated plants developed from the donor collected from the same field and at same time. The regeneration protocol was followed based on previous reports [2, 4].

One gram of leaf tissue (each obtained from field grown donor plant, node culture and callus culture regenerated plants) was extracted with 30 mL of 80% methanol at 60°C water bath for 30 min in a round bottom flask. Each sample was replicated three times. Extracts were filtered with What man filter paper No. 1. Evaporation of methanol was allowed from the extracted sample solution using a rotary evaporator (Eyela, N-1100, China) at 45°C under reduced pressure. The solid mass thus obtained was dissolved in 10 ml of methanol.

10.2.1.1 CONDITIONS FOR HIGH PERFORMANCE THIN LAYER CHROMATOGRAPHY

Thin-layer chromatography (TLC) aluminum sheet silica gel 60-F-254 20×10 cm or 10×10 cm with 0.25 cm thickness (Merck KgaA; 1.05554.0007) was used. Standards in a range of known amounts along with samples were layered in 6 mm wide bands and were placed at 10 mm from the base of the TLC plate. Banding was performed using an automated TLC applicator (Linomat 5, from CAMAG Muttenz, Switzerland) having a flow of nitrogen from a syringe with a delivery speed of 100 µL/s. The mobile phase for chromatography was prepared in a ratio with "Ethyl acetate: Formic acid: Acetic Acid: water (100:11:11:27)." The temperature was maintained at 22±2°C and relative humidity at 50%. A CAMAG twin trough glass chamber that was already saturated with mobile phase vapor for 15 minutes was used for developing the plates.

Densitometric scanning of the dried plates (at 65°C for 15 min duration) was performed in absorbance and reflectance mode at 254 nm with slit width of 5 mm× 0.45 mm at a scanning speed used of 20 mm/s. Data resolution (100 µm/step) was performed with CAMAG model-3 TLC scanner. Further acquired data were processed with WINCATS 4 software. The 0.1 mg mL^{-1} of standard mangiferin (Sigma, Aldrich) stock solution was prepared. Also different concentrations of marker solutions (0.5, 1, 1.5, 2, 2.5 µg) were prepared from the stock solution and were applied in duplicates in different lanes with Linomat-5 applicator to construct standard curve and to determine the linearity. Calibration equation was developed using regression analysis; and peak were determined against each corresponding concentration (Figure 10.2).

The extract was spotted thrice on the HPTLC plates. The R_f values were generated and overlaying spectra of the bands from samples and that of the standard confirmed the presence of mangiferin (Figure 10.3). The software WINCATS-4 was used to analyze purity of the peaks.

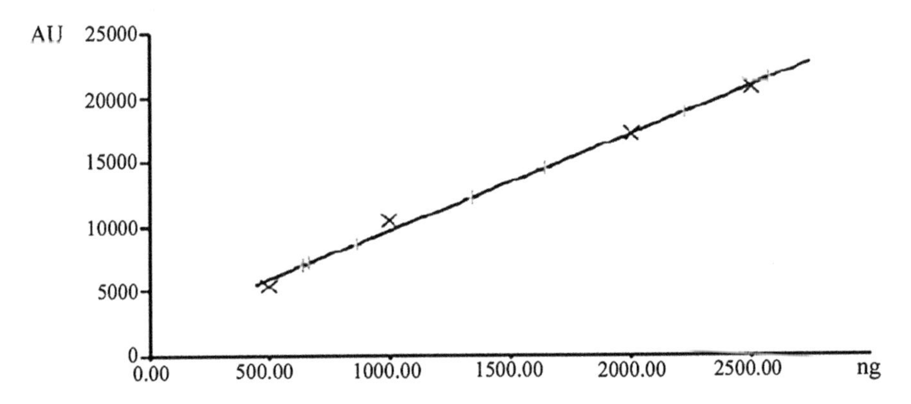

FIGURE 10.2 Calibration plot for 0.5 to 2.5 μg (or 500 to 2500 ng) of mangiferin per spot.

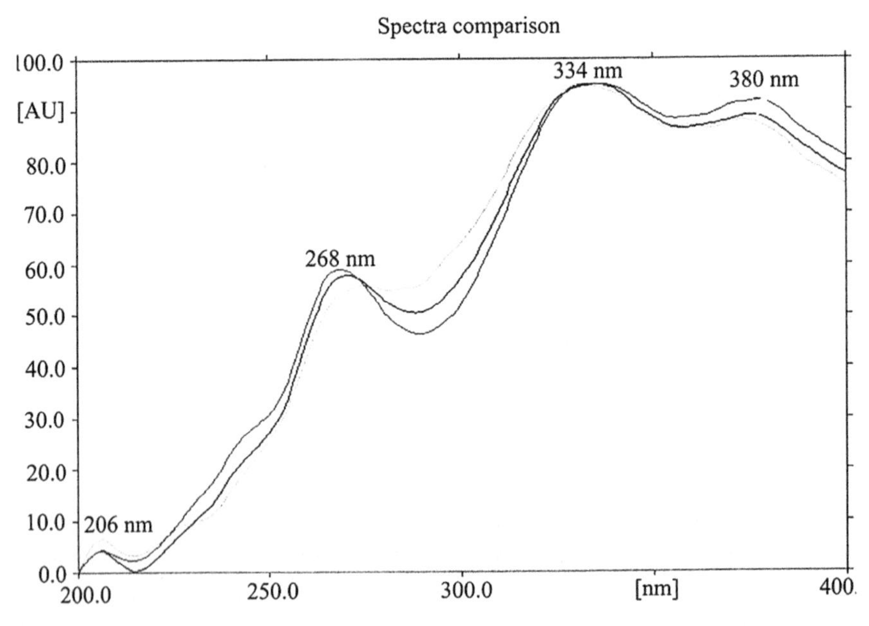

FIGURE 10.3 Spectral overlay of mangiferin in S. chirata samples for standard (lighter lines) and in vitro grown (darker lines).

10.2.2 QUALITITATIVE ANALYSIS OF MANGIFERIN, AMAROGENTIN, AMAROSWERIN, SWEROSIDE, AND SWERTIAMARIN THROUGH LC/ESI$^+$-MS/MS FROM CULTURE GROWN PLANTS

Donor plants (collected from Sikkim-Gangtok: 27°19′N and 88°36′E) were micro-propagated through axillary meristem culture in the month of June; and leaves of *In vitro* grown shoots (one month old) from basal medium were taken and dried at 60°C and were grounded to powder and one g of this powder was taken for extraction of bioactive compounds. Extraction procedure was the same as mentioned above for HPTLC analysis.

Acetonitrile (HPLC grade), Glacial acetic acid (Merck, France), ultra-pure water (18.2 MΩ cm^{-1}) from MilliQ Plus (Millipore, USA) were used for HPTLC. The 90% of ultra-pure water and 10% (v/v) of acetonitrile were used to prepare a stock solution of methyl extract (1000 mg mL^{-1}). Further dilution of stock solution was performed using acetonitrile to prepare the working solution.

10.2.2.1 CONDITIONS FOR LC-MS AND LC-MS/MS ANALYSES

Waters (Alliance-2695) HPLC system joined to a Quattro LC triple quadrupole mass spectrometer (Waters, Milford, MA, USA) were used to perform LC-MS and LC-MS/MS analyses. This HPLC system was connected with pneumatic electro spray source ionization (ESI). Data collection and its processing were carried out by Mass Lynx NT 4.1 system.

A varying sampling cone voltage (starting from 2 kV to 60 kV) was applied for ESI mass spectra. In order to optimize the parent ion and carry out Collision-induced dissociation (CID) experiment, specific value of the cone voltage was set. The electron spray (ES) source potentials were as follows: 2.75 kV in positive mode and 3.0 kV in negative mode, extractor at 2 V, with source block at 120°C and desolvation gas at 350°C.

Nebulization and desolvation (250 and 450 L h^{-1}, respectively) was performed with Nitrogen. CID was achieved using Argon (Ar) when 3.5×10^{-3} Torr pressure was applied with 5–45 eV collision energy (CE). Liquid chromatography conditions were standardized by performing an isocratic elution with 3.5 min run time. In an isocratic mode of delivery of the mobile phase, one mL min^{-1} flow rate was allowed with the help of a Perkin Elmer (Series 200) pump. A reverse-phase column (Uptisphere C18 5 µm, 150 × 2 mm I. D., Interchrom, France) attached to precolumn (Interchrom) under two different mobile phase conditions was used for separation and detection of the compounds under interest.

- **First Condition (Condition A):** It describes a gradient of 0.5% acetic acid mixture of acetonitrile and water:
  - solvent A: acetonitrile/water (95/5, v/v);
  - solvent B: acetonitrile/water (5/95, v/v);
  - 0–5 min 10% A in B;
  - 5–30 min 10–70% of A;
  - 30–35 min 70% of A (linear gradient);
  - 35–50 min 100% of A;
  - Final equilibration for 20 min at 10% of A.

  Acetic acid was used for separation and detection of mangiferin, amarogentin, and sweroside (Figure 10.1) in order to obtain more intensive $[M+H]^+$ ions for these compounds.
- **The Second Mobile Phase Condition:** It was applied without acetic acid (condition B), and all other conditions were same as in condition A for the separation and better detection of amaroswerin and swertiamarin (Figure 10.1) compounds as their sodiated ions $[M+Na]^+$. The flow rate was fixed at 0.2 mL min^{-1} with a constant column temperature at 24°Cand UV detection at 280 nm. The compounds were introduced using a Waters Alliance-2695 separation module (Milford, MA, USA) attached to Waters-2487 dual UV detector.

10.3 RESULTS AND DISCUSSION

10.3.1 ANALYSIS OF MANGIFERIN THROUGH HPTLC

The mangiferin is a C-glucosyl xanthone (1,3,6,7-tetrahydroxyxanthone-C2-beta-d-glucoside)that exhibits antidiabetic, anti-HIV, anticancer, immunomodulatory, and antioxidant activities [1]. In an earlier review article by Miura [19], presence of amarogentin and xanthones was not detected from callus culture in *S. japonica.* In another related species, *S. pseudochinensis* swertiamarin was only detected and compared with the donor plant. It was observed that swertiamarin content from leaves of five months old regenerated plants under cultural conditions was enhanced nearly eight folds in comparison to that of a two-year-old wild plant.

In the present study, an appreciable amount of mangiferin was detected through HPTLC from a culture grown *S. chirata* plant-lets, while compared with the standard (Table 10.1; Figures 10.4–10.7).

TABLE 10.1 Mangiferin Content in Donor and *In Vitro* Grown *S. Chirata* Raised from Node Explant and Immature Seed Callus

Source	Percentage (%) of Mangiferin
Donor	0.62 ± 0.3*
In vitro grown Node regenerated plants	2.93 ± 0.6*
In vitro grown Callus regenerated plants	2.24 ± 0.6*

*Values represent mean ± standard deviation of three experimental replications each with 5 biological replications. Data were scored after one month of inoculation for *in vitro* grown plants and five months old field-grown plants.

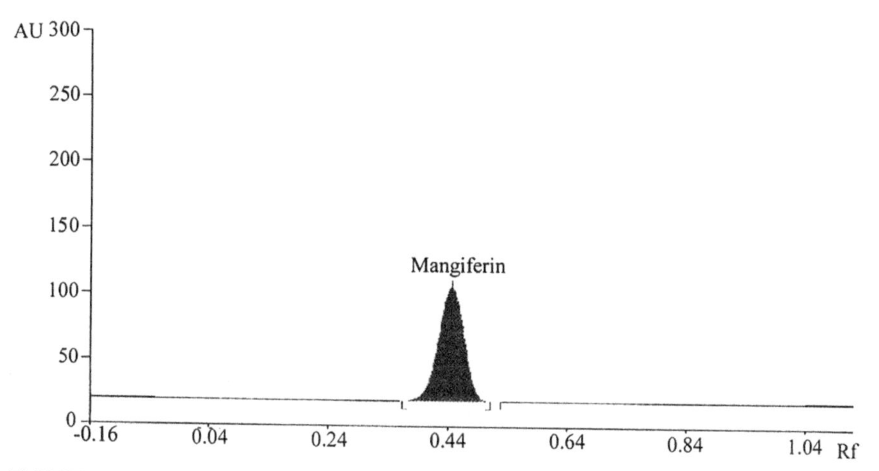

FIGURE 10.4 HPTLC Chromatogram for mangiferin content in standard treatment.

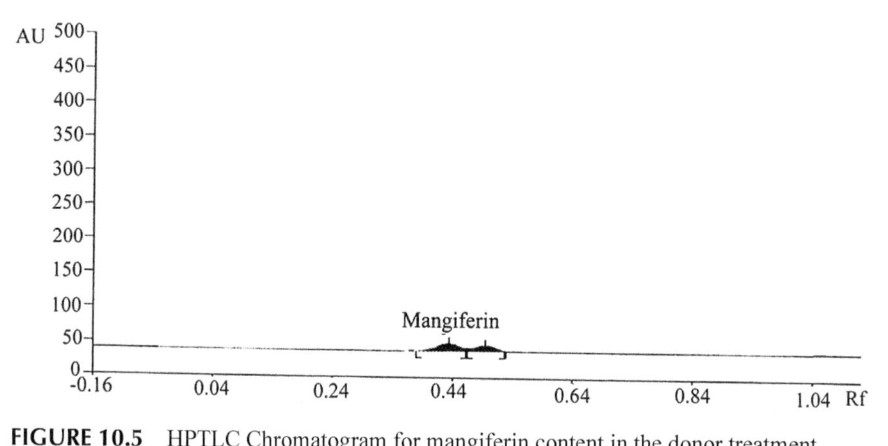

FIGURE 10.5 HPTLC Chromatogram for mangiferin content in the donor treatment.

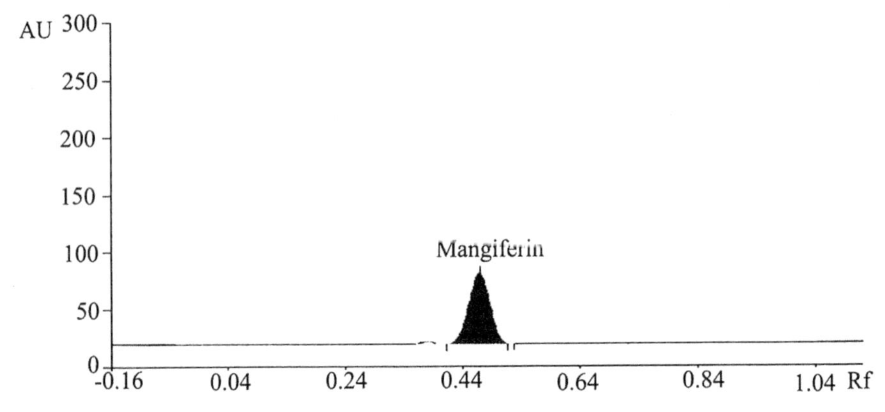

FIGURE 10.6 HPTLC Chromatogram for mangiferin content in node regenerated plants.

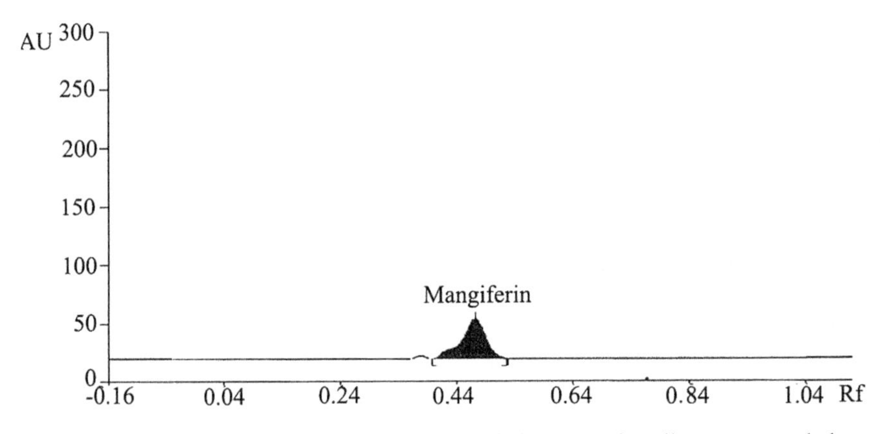

FIGURE 10.7 HPTLC Chromatogram for mangiferin content in callus regenerated plants.

Quantitative analysis revealed that mangiferin accumulation in culture grown plants was higher than that found in the donor plant (Figure 10.5). Nearly five-fold increase in mangiferin concentration was detected among the *in vitro* plants raised from the nodal explants (Figure 10.6) compared to the donor plant (Figure 10.5). The 0.62% of mangiferin from the donor plant and 3% of mangiferin from node regenerants (Figure 10.6) were obtained.

Callus regenerated plants yielded nearly 2% of mangiferin (Figure 10.7). For the analytical results in Figures 10.4 to 10.7, the mother plant showed lower synthesis of these important bioactive molecules. Synthesis of natural products in field grown plants is dependent on age of the plant and agro-climatic conditions.

The donor plants were collected prior to its full maturity and this may be one possible reason for synthesis of lower quantity of mangiferin. In a similar experiment of chemical analyses from five batches of fully mature field grown plants of *S. chirata* harvested during the month of September-October, Suryawanshi et al. [25] reported 1.7 ± 1.1% of mangiferin and 0.84 ± 0.3% of amarogentin in the wild plants. These authors used the LC/ESI-MS/MS system and reported [21] variation in mangiferin content during vegetative (1.23 to 2.67%) and flowering (2.7–4.3%) stages among various populations of *S. chirata*. These variations within the plants may be due to difference in location and time of collection.

It was observed that *in vitro* micro-propagated and callus-regenerated plants synthesized mangiferin in much higher amounts under aseptic cultural conditions. Nearly 3% and 2% of mangiferin were obtained from the node and callus regenerated plants, respectively; and these values were higher than 0.62% of mangiferin for the mother plant. Suryavanshi et al. [25] and Pandey *et al.* [21] obtained 1.23 to 2.67% of mangiferin from the field grown plants.

The mangiferin content was 2.7% at the flowering stage in *S. chirata* collected in Sikkim, while micro-propagated plants of *S. chirata* had the similar concentration under *in vitro* conditions indicating a scale-up in the concentration of bioactive compound. In this chapter, the ingredient, and cultural conditions that support the aseptic cultures were probably able to trigger the synthesis of bioactive molecules, during *in vitro* propagation of the plants through various explants.

The germ plasm used in this chapter may also be an important factor for enhancement of bioactive compounds. A review of *Swertia* literature did not reveal such findings from *in vitro* grown shoots. The presently described results indicate that the quality improvement of this critically endangered medicinal plant through plant biotechnological methods is possible. Also *in vitro* culture of this plant can be exploited effectively for extraction of mangiferin without depletion of natural resources.

10.3.2 ANALYSIS FOR LC/ESI-MS AND LC/ESI$^+$-MS/MS

The chemical structures of biologically active components present in methanol extract of *S. chirata* leaf tissues have been described earlier (Figure 10.1) in this chapter. *In vitro* grown micro-propagated plants were extracted at a concentration of 10 μg mL^{-1} and were scanned to obtain the LC/ESI-MS spectra. To confirm the presence of phytomolecules (mangiferin, amarogentin, amaroswerin, sweroside, and swertiamarin) in tissue culture grown plants, authors of this chapter compared their results with the published report by Suryavanshi et al. [24], who analyzed these bioactive compounds (mangiferin, amarogentin, amaroswerin, sweroside, and swertiamarin) from field grown plant of *S. chirata* through LC-ESI$^+$-MS-MS fragmentation spectra of specific ions ($[M+H]^+$ and $[M+Na]^+$) of these molecules. The chromatography condition was described earlier, where mobile phase comprising of acetonitrile, MilliQ water, 0.5% glacial acetic acid (condition A) and the similar composition without glacial acetic acid (condition B) were used. Both positive and negative mode analyses for *in vitro* grown micro-propagated plants through LC/ESI-MS were performed and spectrometric data are presented between *m/z* 200 to 700 (Table 10.2).

An Extracted Ion Chromatogram (EIC) of specific ions $[M+H]^+$ or $[M+Na]^+$ showed three peaks at 8.45, 8.55 and 22.98 with condition A (Figure 10.8) and two peaks at 13.24 and 24.38 with condition B (Figure 10.9). The obtained MS spectra were used to assign the molecular weights to mangiferin, amarogentin, amaroswerin, sweroside, and swertiamarin.

After comparing the results with the published data [13], it was concluded that *in vitro* plants are capable of synthesizing all five types of bioactive compounds (mangiferin, amarogentin, amaroswerin, sweroside, and swertiamarin) that were also synthesized by the mother plants. Ions having similar retention time might be obtained from matrix as adducts or fragments of these components. Presence of these five bioactive components was confirmed as follows (Figures 10.10 and 10.11):

1. (MW 422, RT 8.45 min.): mangiferin;
2. (MW586, RT 22.98 min.): amarogentin;
3. (MW 602, RT 24.38 min.): amaroswerin;
4. (MW 358, RT 8.55 min.): sweroside; and
5. (MW 374, RT 13.24 min.): swertiamarin.

Positive MS spectra for mangiferin, amarogentin, and sweroside are shown in Figure 10.10. These three components showed highly abundant $[M+H]^+$ ions at m/z 423, 587 and 359 peaks, which were similar to the earlier published reports.

The positive MS spectra for amaroswerin and swertiamarin are shown in Figure 10.11. These two components produced $[M+Na]^+$ at m/z 625 and 397, respectively. Also, these two compounds have the potential to form alkali metal ions, primarily with sodium ions. This explains why the LC was used under condition B (without acid) for these two compounds. mangiferin, amarogentin, sweroside, amaroswerin, and swertiamarin.

TABLE 10.2 LC/ESI-MS Results for Positive and Negative Modes (Cone Voltage of 40 to 60 V)

Bioactive Compound			ESI Positive Mode					ESI Negative Mode					
—	MW	t_R (min.)	$[M+H]^+$	$[M+Na]^+$	$[M+NH_4]^+$	$[M+K]^+$	$[MNa^+ + ACN]$	$[M-H]^-$	$[M+Na-2H]^-$	$[M+Cl]^-$	$[M+HCOO]^-$	$[M+OAc]^-$	$[glu]^-$
Mangiferin	422	8.45	423	—	—	—	—	421	—	—	—	—	—
Amarogentin	586	22.98	587	609	604	625	650	585	—	621	—	—	—
Amaroswerin	602	24.38	603 (unstable)	625	620	641	666	601	—	637	—	—	—
Sweroside	358	8.55	359	—	—	—	—	—	379	377	401	—	—
Swertiamarin	374	13.24	375	397	—	413	—	373	—	409	—	433	179

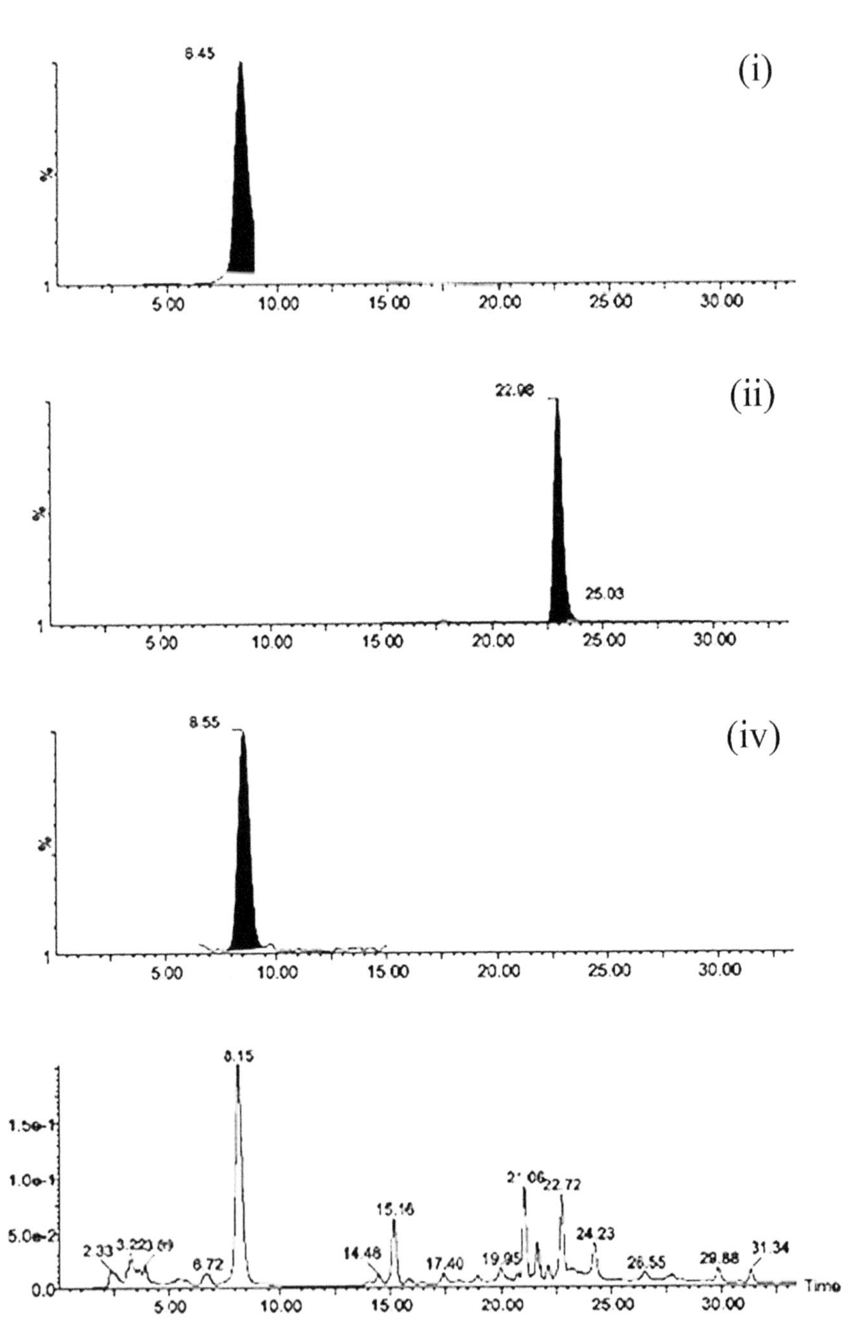

FIGURE 10.8 UV Chromatogram obtained with acetic acid gradient (280 nm, bottom) and extracted specific ion traces (ESI+ mode, MH+) of mangiferin (i), amarogentin (ii) and sweroside (iv) compounds.

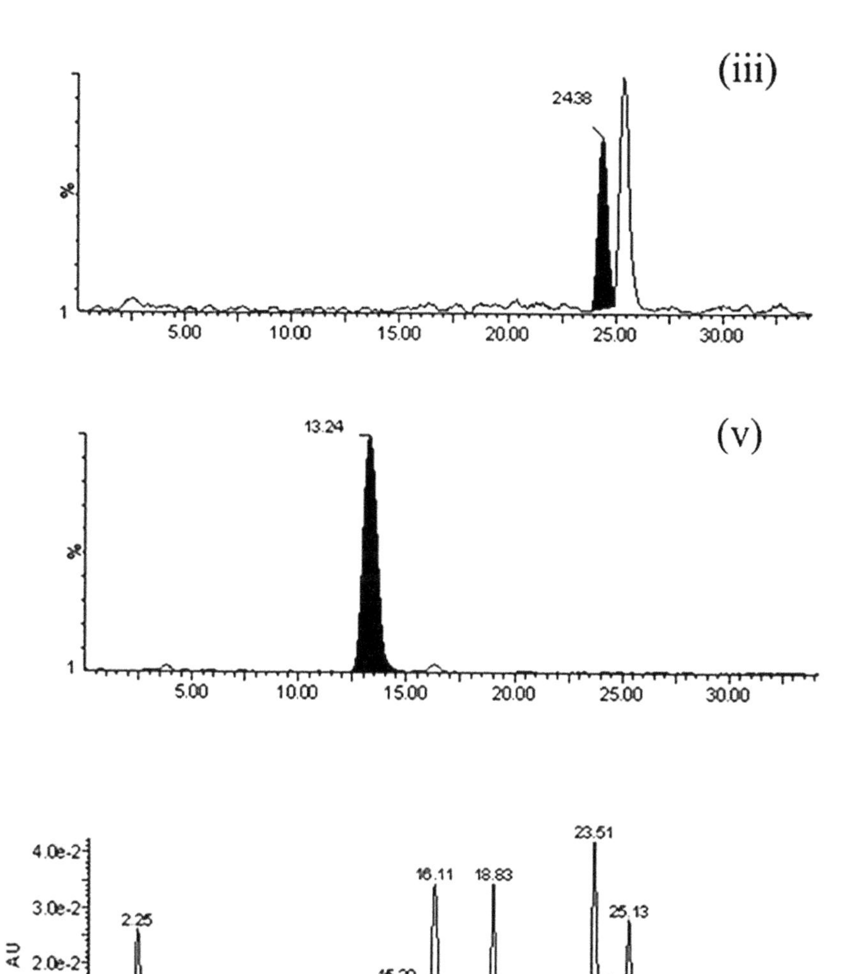

FIGURE 10.9 UV Chromatogram obtained without acetic acid gradient (280 nm, bottom) and extracted specific ion traces (ESI+ mode, MNa+) of amaroswerin (iii) and swertiamarin (v).

The development of protonated molecules was replaced by metallic adduct formation for amaroswerin, because the $[M+H]^+$ species were not found. It was present with comparatively higher relative abundance in swertiamarin. Such results could be attributed towards acidic nature of iridoid glycosides.

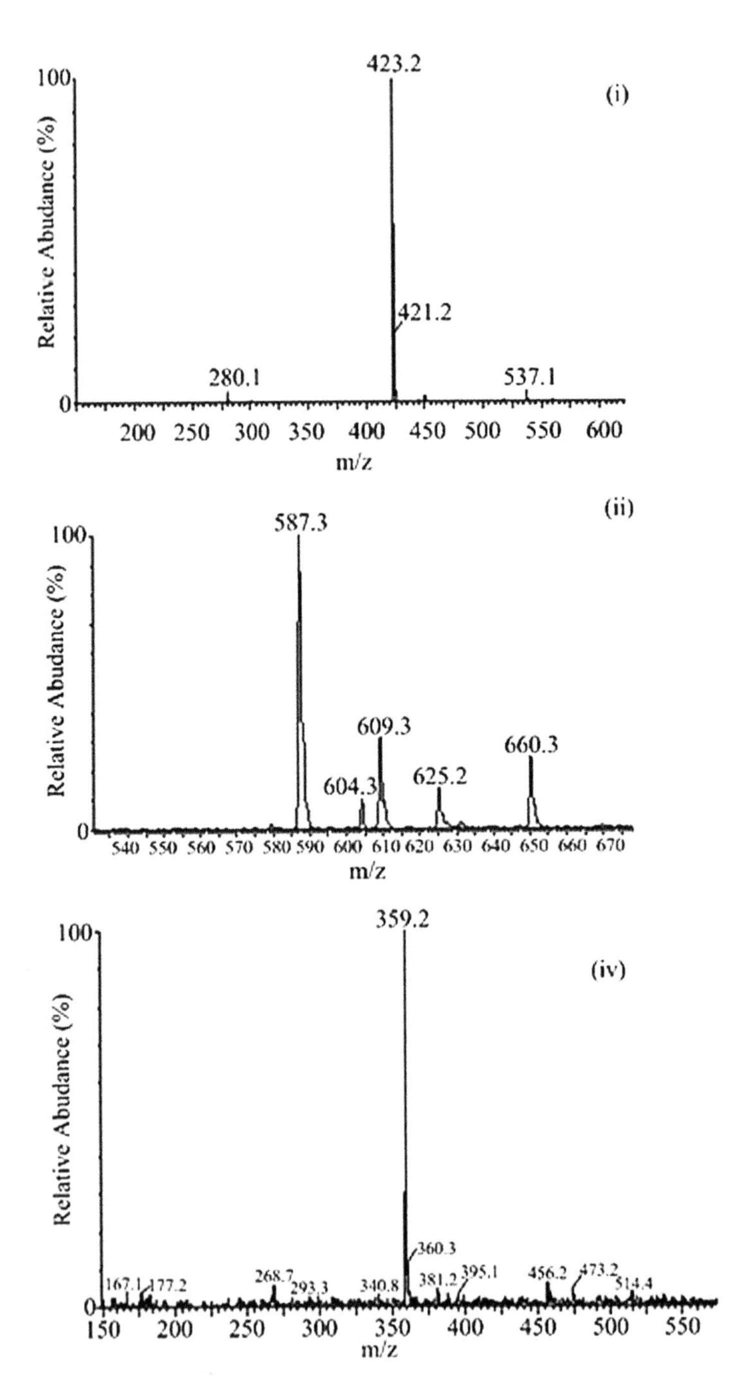

FIGURE 10.10 The detection of ions in LC-ESI-MS spectra at cone voltage of 40V mangiferin (i) amarogentin (ii) and sweroside (iv) for compounds.

Negative LC/ESI-MS data exhibited the presence of peak, which corresponded to deprotonated molecule [M-H]⁻ in all spectra. This was a major peak for all components except sweroside and swertiamarin. The chlorine adduct was also observed for all the components except for mangiferin. In the sweroside, it was present with high abundance. Formate for sweroside and acetate for swertiamarin adducts were relatively abundant. It was also observed that the fragment ion of a single sugar unit retained the charge at 179 m/z for swertiamarin.

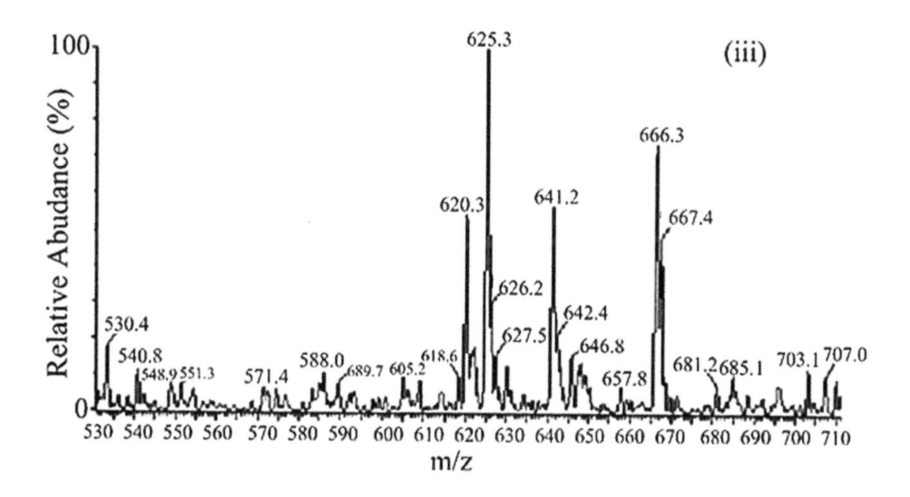

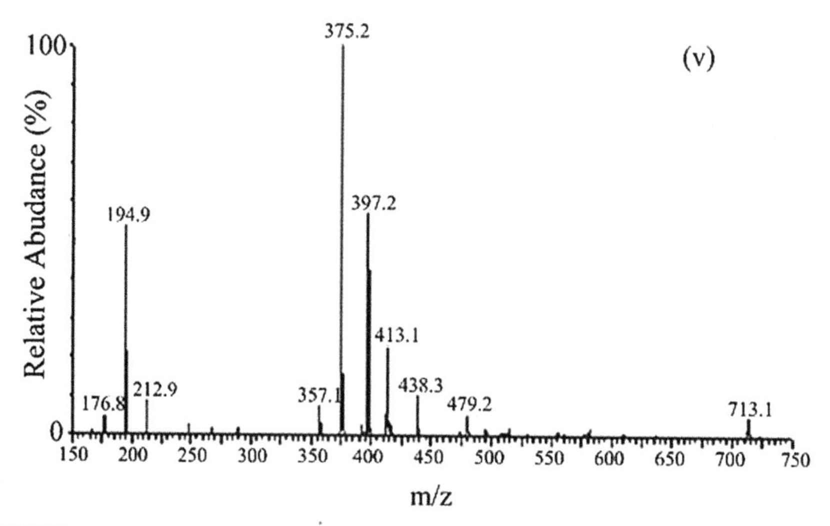

FIGURE 10.11 The detection of ions in LC-ESI-MS spectra at cone voltage 40V amaroswerin (iii) and swertiamarin (v) for compounds.

Thus in-source fragmented compounds provided sufficient details of structural configuration required for their identifications. This information along with results of mass signals is presented in Table 10.2. Further confirmation of detected compounds can be obtained through tandem mass spectrometry that is an essential tool, where first quadrupole is used as mass filter and second quadrupole serves as a collision cell.

The ESI-MS/MS was performed with the Quattro-LC instrument by selective CID in the daughter mode for mangiferin, amarogentin, and sweroside, amaroswerin, and swertiamarin compounds that were identified in this chapter with chromatographic parameters. The dissociation within 5 to 45 eV was studied for each compound and the closest pattern generated at one specific CE condition has been presented here. Major product ions found in positive ESI-MS/MS product ion spectrum of protonated/ sodiated ion for mangiferin, amarogentin, and sweroside, amaroswerin, and swertiamarin are presented in Table 10.3. The daughter spectra (CID) are shown in Figures 10.12 and 10.13.

TABLE 10.3 LC/ESI$^+$-MS/MS Characterization Profile of *S. chirata* Components at Various Ionization Modes (CE (Collision Energy) in eV)

Compounds	MW	Precursor Ions	Product Ions m/z with Relative Abundance (%) in Brackets	CE (eV)
Mangiferin	422	423	405 (70), 387 (80), 369 (60), 357 (100), 327 (90), 303 (80), 273 (75)	15
Amarogentin	586	587	391 (100), 373 (50), 359 (100), 311 (75), 247 (40), 229 (100), 213 (6), 197 (100), 127 (18)	12
Amaroswerin	602	625	397 (100), 255 (19), 251 (20), 217 (9), 185 (4), 165 (15)	30
Sweroside	358	359	341 (15), 311 (75), 197 (100), 179 (25), 127 (20)	15
Swertiamarin	374	397	255 (50), 235 (15), 217 (95), 203 (12), 185 (18), 165 (100)	25

The CID spectra in this chapter were also similar to the ones in earlier published report [26]. Some differences in relative abundance were found due to different experimental conditions and different instruments.

Detection of mangiferin, amarogentin, amaroswerin, sweroside, and swertiamarin from *in vitro* grown *S. chirata* cultures have not been studied earlier through LC coupled to the MS method. This research study confirms presence of mangiferin, amarogentin, amaroswerin, sweroside, and swertiamarin in the tissue culture grown plants under *in vitro* conditions through LC/ESI-MS and MS/MS techniques that agrees with the previously published report for *in vivo* plants. Similar study to identify tentative metabolites through LC technique coupled with MS and ESI has been described earlier for two different *Medicago species, Arabidopsis thaliana, and* four *Polygonum species* [10, 11, 18].

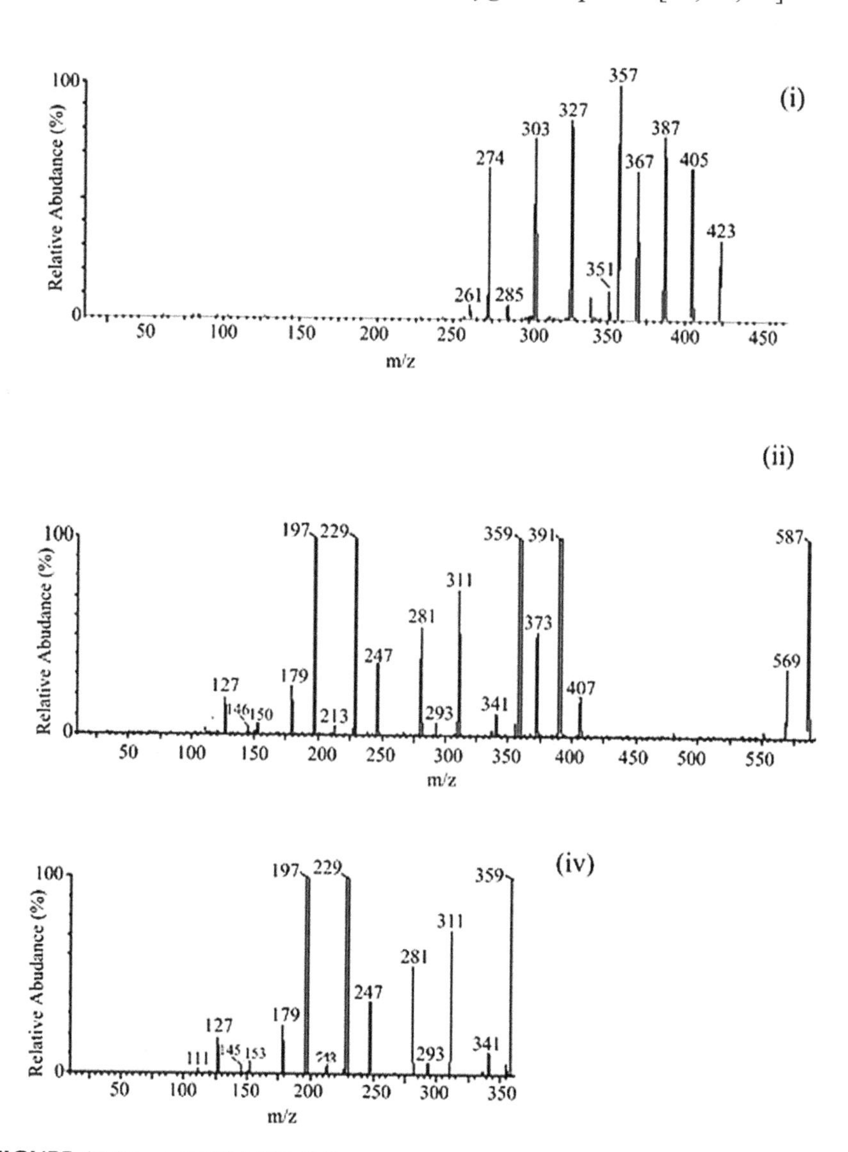

FIGURE 10.12 LC-ESI-MS/MS fragmentation patterns generated from protonated parent ions of Mangiferin (423), amarogentin (587), and sweroside (359) in extract of S. chirata plant.

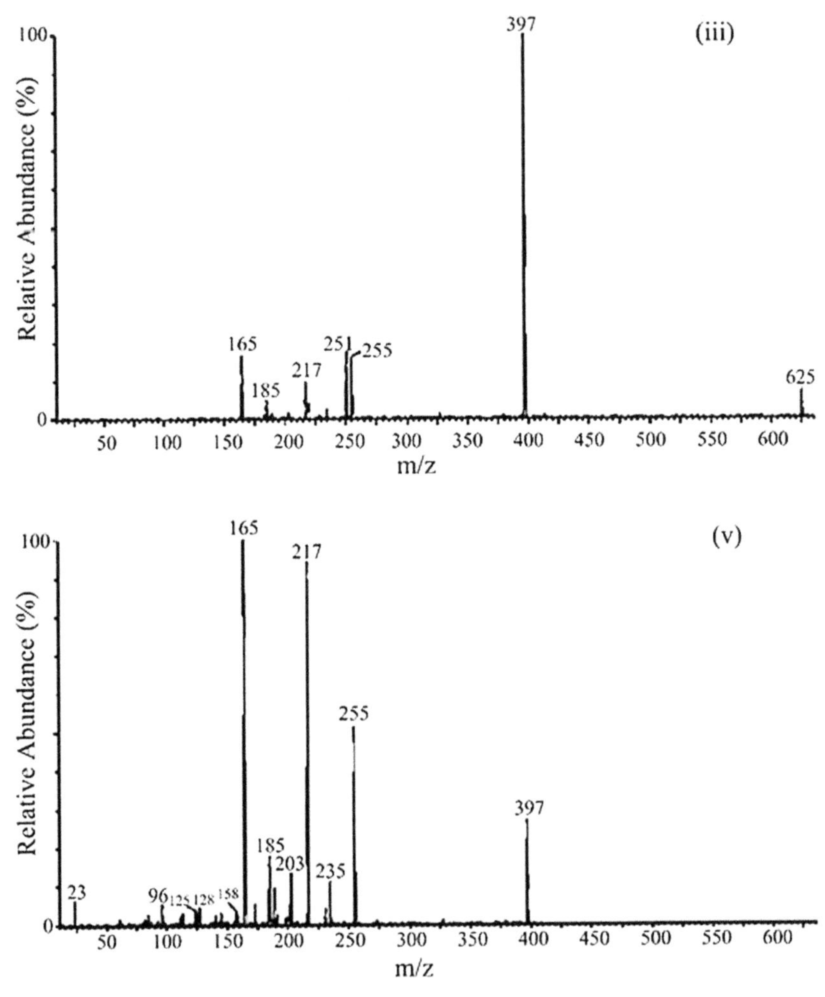

FIGURE 10.13 LC-ESI-MS/MS fragmentation patterns generated from sodiated parent ions of amaroswerin (625) and swertiamarin (397) in extract of S. chirata plant.

The study of plant secondary metabolites is a demanding task and plant biotechnology coupled with advanced analytical instruments can provide opportunities for controlling the enhanced production of natural bioactive components in higher amounts compared to for the intact plants. Enhancement of natural product synthesis in *in-vitro* cultures have been reported in *Hypericum perforatum, Pueraria montana, Crataegus laevigata, C. monogyna, Glycyrrhiza glabra, Morinda citrifolia, Catharanthus roseus, Dioscorea deltoidea, Coleus blumei,* and *Taxus* by various investigators [6, 13, 16, 22].

In a recent study with *S. chirata*, interception of ethylene and activation of antioxidant activity has been reported by supplementation of silver nanoparticles under *in vitro* conditions [23]. Zenk suggested importance of selection of elite germplasm either for traditional cultivation or for development of improved production system for plant bioactive chemicals using genetic and cultural modifications [27]. Bioreactor technology may be beneficial for scaling up biomass and henceforth procurement of natural products [20].

10.4 SUMMARY

Enhanced synthesis of mangiferin was confirmed in tissue culture plants through HPTLC. The LC/ESI$^+$-MS/MS studies confirmed that *in vitro* plants were equally capable to synthesize mangiferin, amarogentin, amaroswerin, sweroside, and swertiamarin from *S. chirata.* Chemical investigation of plant secondary metabolites and plant biotechnology coupled with advanced analytical instruments can provide novel and accurate information for the identification and quantification of bioactive compounds. The in-depth screening of quality germplasm from natural populations through chemical analyses along with their multiplication and conservation for metabolic profiling is suggested future line of research.

ACKNOWLEDGMENTS

The authors acknowledge Professors David Tepfer, Lucien Kerhoas and Walter Kollmann for carrying out the LCMSMS at INRA, Versailles-France; and Prof. Timir Baran Jha at Kolkata, India for his advice.

KEYWORDS

- **callus culture**
- **collision energy**
- **electron spray**
- **extracted ion chromatogram**
- **micropropagation**
- **seed culture**

REFERENCES

1. Brahmachari, G., Mandal, S., & Gangopadhyay, A., (2004). *Swertia* (Gentianaceae): Chemical and pharmacological aspects. *Chemistry and Biodiversity, 1*, 1627–1651.
2. Chaudhuri, R. K., Pal, A., & Jha, T. B., (2008). Conservation of *Swertia chirata* through direct shoot multiplication from leaf explants. *Plant Biotechnological Reports, 2*, 213–218.
3. Chaudhuri, R. K., Pal, A., & Jha, T. B., (2007). Production of Genetically uniform plants from nodal explants of *Swertia chirata* buch. ham. ex wall.: Critically endangered medicinal herb. *In Vitro Cellular and Developmental Biology, Plant, 43*, 467–472.
4. Chaudhuri, R. K., Pal, A., & Jha, T. B., (2009). Regeneration and characterization of *Swertia chirata* buch. ham. ex wall. Plants from immature seed cultures. *Scientia Horticulturae, 120*, 107–114.
5. Chen, J. W., Zhu, Z. Q., Hu, T. X., & Zhu, D. Y., (2002). Structure-Activity relationship of natural flavonoids in hydroxyl radical-scavenging effects. *Acta Pharmacologica Sinica (APS), 23*, 667–669.
6. Estrada, K. R., Vidal-Limon, H., Hidalgo, D., & Moyano, E., (2016). Elicitation, an effective strategy for the biotechnological production of bioactive high-added value compounds in plant cell factories. *Molecules, 21*, 182–189; online; doi: 10.3390/molecules21020182.
7. Gales, L., Sousa, M. E., & De Pinto, M. M., (2002). Naturally Occurring 1,2,8-trimethoxyxanthone and biphenyl ether intermediates leading to 1,2-Dimethoxyxanthone. *Acta Crystallographica, C57*, 1319–1322.
8. Ghosal, S., Sharma, P. V., & Jaiswal, D. K., (1978). Chemical constituents of gentianaceae: Xxiii, tetra oxygenated and pentaoxygenated xanthone-o-glucosides of *Swertia angustifolia* buch ham. *Journal of Pharmaceutical Sciences, 67*, 55–60.
9. Gomez-Galera, S., Pelacho, A. M., & Gene, A., (2007). The genetic manipulation of medicinal and aromatic plants. *Plant Cell Reports, 26*, 1689–1715.
10. Huang, W. Y., Cai, Y. Z., Xing, J., Corke, H., & Sun, M., (2008). Comparative Analysis of bioactivities of four *polygonum* species. *Planta Medica, 74*, 43–49.
11. Huhman, D. V., & Sumner, L. W., (2002). Metabolic profiling of saponins in *Medicago sativa* and *Medicago truncatula* using HPLC coupled to an electrospray Ion-trap mass spectrometer. *Phytochemistry, 59*, 347–360.
12. Ishimaru, K., Sudo, H., Satake, M., & Shimamura, K., (1990). Phenyl glucosides from a hairy root culture of *Swertia japonica*. *Phytochemistry, 29*, 3823–3825.
13. Jha, T. B., & Ghosh, B., (2016). *Plant Tissue Culture Basic and Applied* (p. 183). Kolkata, India: Platinum Publishers.
14. Jiang, D. J., Tan, G. S., Ye, F., Du, Y. H., Xu, K. P., & Li, Y. J., (2003). Protective effects of xanthones against myocardial ischemia-reperfusion injury in rats. *Acta Pharmacologica Sinica, 24,* 175–178.
15. Keil, M., Hartle, B., Guillaume, A., & Psiorz, M., (2000). Production of amarogentin in root cultures of *Swertia chirata*. *Planta Medica, 66,* 452–457.
16. Kirakosyan, A., (2007). Plant biotechnology for the production of natural products. In: Cseke, L. J., Kirakosyan, A., Kaufman, P. B., Warber, S., Duke, J. A., & Brielmann, H. L., (eds.), *Natural Products from Plants* (pp. 221–262). Boca Raton, FL: CRC Press.

17. Kumar, V., & Staden, J. V., (2015). Review of *Swertia chirayita* (Gentianaceae) as a traditional medicinal plant. *Frontiers in Pharmacology, 6,* online: https://www.frontiersin.org/articles/10.3389/fphar.2015.00308/full (accessed on 6 August 2020).
18. Lahaye, E. V. R., Degenkolb, T., & Zerjeski, M., (2004). Profiling of *Arabidopsis* secondary metabolites by capillary liquid chromatography coupled to electrospray ionization quadrupole time-of-flight mass spectrometry. *Plant Physiology, 134,* 548–559.
19. Miura, H., (1991). Biotechnology in agriculture and forestry. In: Bajaj, Y. P. S., (ed.), *Medicinal*, and *Aromatic Plants III* (Vol. 15; pp. 451–463). Berlin Heidelberg: Springer.
20. Paek, K. Y., Hosakatte, M. N., & Zhong, J. J., (2014). *Production of Biomass and Bioactive Compounds Using Bioreactor Technology* (p. 709) Dordrecht, Heidelberg: Springer. doi: 10.1007/978-94-017-9223-3.
21. Pandey, D. K., Basu, S., & Jha, T. B., (2012). Screening of different east Himalayan species and populations of *Swertia* L. based on exomorphology and mangiferin content. *Asian Pacific Journal of Tropical Biomedicine*, S1450–S1456.
22. Sabater-Jara, A. B., Onrubia, M., & Moyano, E., (2014). Synergistic effect of cyclodextrins and methyl jasmonate on taxane production in *Taxus x Media* cell cultures. *Plant Biotechnology Journal, 12*, 1075–1080.
23. Saha, N., & Duttagupta, S., (2018). Promotion of shoot regeneration of *Swertia chirata* by biosynthesized silver nanoparticles and their involvement in ethylene interceptions and activation of antioxidant activity. *Plant Cell Tissue Organ Culture, 134* (2), 289–300.
24. Suryawanshi, S., Asthana, R. K., & Gupta, R. C., (2007). simultaneous estimation of mangiferin and four secoiridoid glycosides in rat plasma using liquid chromatography tandem mass spectrometry and its application to pharmacokinetic study of herbal preparation. *Journal of Chromatography. B, Analytical Technologies in the Biomedical and Life Science, 858*, 211–219.
25. Suryawanshi, S., Mehrotra, N., Asthana, R. K., & Gupta, R. C., (2006). Liquid chromatography/tandem mass spectrometric study and analysis of xanthone and secoiridoid glycoside composition of *Swertia chirata*, a potent antidiabetic. *Communication Mass Spectrometry, 20*, 3761–3768.
26. Terryn, N., Van, M. M., Inze, D., & Goosens, A., (2006). Functional genomics approaches to study and engineer secondary metabolites in plant cell cultures. In: Bogers, L. J., Craker, L. E., & Jange, D., (eds.), *Medicinal*, and *Aromatic Plants* (pp. 291–300). Netherlands: Springer.
27. Zenk, M. H., (2005). Chasing the enzymes of secondary metabolism: Plant cultures as a pot of gold. *Phytochemistry, 30*, 3861–3863.

APPENDIX I

Typical *Swertia chirayita* plant; this image is modified from [https://ayur-wiki. org/Ayurwiki/File:Whf_blue_08. jpg], licensed under the Creative Commons Attribution-Share Alike 3.0 Unported license.

INDEX

B

C

D

I

Q

R

S

V

W

X

Z